Antonio León Sánchez

THE DISCRETE REALITY OF PHYSICAL SPACE

Collected published and unpublished works 2020-2025

Revised version. September 2025

Revised version: September 2025
First edition: June 2024
ISBN 9798327194533
All right reserved
Salamanca and Santiago del Collado (Ávila), Spain.

Table of Contents

		Page
1	**Important warning**	**1**
2	**Conventions and symbols**	**3**
	2.1 Conventions	3
	2.2 Symbols	6
3	**Finite versus infinite**	**9**
	3.1 Introduction	9
	3.2 Zeno, Aristotle and Cantor	9
	3.3 The actual and the potential infinity	10
	3.4 The infinity of the Axiom of Infinity	16
	3.5 A short proof of inconsistency	18
	3.6 The axiom of infinity is inconsistent	21
	3.7 Conclusion	25
4	**Discrete Magnitudes and Functions**	**27**
	4.1 Introduction	27
	4.2 Democritus' argument	27
	4.3 The ultraviolet catastrophe	28
	4.4 Discrete magnitudes and functions	29
	4.5 Some discrete conclusions	34
	4.6 Discrete arithmetic	35
5	**Infinite regress**	**39**
	5.1 Introduction	39
	5.2 The Aristotelian infinite regress	39
	5.3 The universe is consistent	41
	5.4 Münchhausen Theorem	45
	5.5 Theorem of the First Element	46
	5.6 Infinite regress of proofs	46
	5.7 Infinite regress of definitions	48
	5.8 Infinite regress of causes	49
6	**The Formal Scenario**	**51**
	6.1 Introduction	51

6.2 Formal elements for a new theory of space and time 51

7 Space in ancient Greece — 57
7.1 Introduction 57
7.2 The first cosmologists 58
7.3 Parmenides and Zeno of Elea 61
7.4 The Pythagoreans 63
7.5 The atomists 66
7.6 Space according to Plato 68
7.7 The Aristotelian space 70
7.8 Space in post-Aristotelian Classical Greece 74

8 Euclidean space — 77
8.1 Introduction 77
8.2 Euclid 77
8.3 Euclid's Elements 78
8.4 The enigma of the parallel straight lines 80
8.5 The definition of straight line 82
8.6 Physical space is Euclidean 84

9 Space light and Gold — 89
9.1 Introduction 89
9.2 The Neoplatonics 89
9.3 Arab and Judeo-Christian ideas about space 92
   Science in the Roman Empire 92
   Indian science in the 5th-13th centuries 94
   Arab science in the 7th-13th centuries 95
   Latin translations of the Greek authors 96
   Atomic theory of space in Kalam 97
   Judeo-Christian ideas about space 98

10 Newton absolute space — 103
10.1 Introduction 103
10.2 The formal language of Newton's Principia 103
10.3 Space, time and motion in the Principia 105
10.4 Critique of Newton's absolute space 114

11 Questioning Leibniz's Principle of Sufficient Reason — 119
11.1 Two Leibniz's Principles 119
11.2 The formal setting of the discussion 120
11.3 The Principle of Sufficient Reason 121

12 Newton's bucket and absolute rotations — 125
12.1 A real Newton's experiment 125
12.2 Criticism of Newton's bucket experiment 127

12.3 A thought experiment: Newton's rotating globes 130
12.4 Mass and Mach's Principle 132

13 Space in the XVIII and XIX centuries — 135
13.1 Introduction 135
13.2 The initial success of Newtonian absolute space 136
13.3 The nature of space according to Kant 137
13.4 Mach's Principle 139
13.5 The birth of non-Euclidean geometries 141
13.6 Michelson-Morley experiment 143

14 Infinity, language, and non-Euclideangeometries — 149
14.1 Introduction 149
14.2 The inevitable incompleteness of human knowledge 150
14.3 Straight lines and parallelism 151
14.4 Language abuses in non-Euclidean geometries 152

15 Theories of inertial relativity — 157
15.1 Physics and mathematics 157
15.2 The Newton-Maxwell relativistic conflict 158
15.3 Lorentz theory of inertial relativity 160
15.4 Einstein's theory of special relativity 162
15.5 Relativistic consequences on space and time 165
15.6 Experimental confirmations of special relativity 167
15.7 Space in the 20th century 168

16 Discrete versus continuous — 171
16.1 Introduction 171
16.2 The problem of the continuous 171
16.3 The spacetime continuum 173
16.4 Finite but non-computable natural numbers 176
16.5 Numbers with infinitely many decimals 179
16.6 Discrete and continuous magnitudes 181
16.7 Inconsistency of the actual infinite divisions 182
16.8 Finite lengths and distances 185
16.9 Pythagoras Discrete Theorem 187

17 Physical versus geometrical space — 191
17.1 The mathematical language of physics 191
17.2 Geometrical points and physical points 193
17.3 Physical space 196
17.4 A relativistic conflict on the reality of space 198
17.5 Gravity from the CALM perspective 201
17.6 Expanding geometrical space and physical space 204
17.7 Fields and CALMs 205

18 Cellular Automata Like Models — 207
- 18.1 Introduction — 207
- 18.2 Indivisible units of space and time — 207
- 18.3 The problem of change — 211
- 18.4 Canonical changes — 213
- 18.5 Discrete versus continuous — 215
- 18.6 A discrete model: cellular automata — 216

19 Universal preinertia — 221
- 19.1 Introduction — 221
- 19.2 Definition of Preinertia — 221
- 19.3 Photons are preinertial — 224
- 19.4 Preinertia and absolute motion — 231
- 19.5 Two key questions — 234
- 19.6 Preinertia and the nature of light — 235
- 19.7 The speed of light and absolute motion — 237

20 Zeno Dichotomies — 239
- 20.1 Introduction — 239
- 20.2 Introductory definitions — 239
- 20.3 Zeno Dichotomy II — 242
- 20.4 Zeno Dichotomy I — 244
- 20.5 Conclusion — 246

21 Achilles, the tortoise and the speed of light — 247
- 21.1 Introduction — 247
- 21.2 *AT*-points — 249
- 21.3 Zeno Dichotomy — 249
- 21.4 Zeno Contradiction — 250
- 21.5 The Axiom of Infinity and Zeno Contradiction — 250
- 21.6 A discrete solution to Zeno Contradiction — 251

22 Infinity, physics and language — 255
- 22.1 Introduction — 255
- 22.2 Infinity and ordinary language — 256
- 22.3 As firm as a rock — 257
- 22.4 Points and instants of the spacetime continuum — 258
  - The size and dense order of points and instants — 258
  - The Dimension Problem — 260
  - The zero point energy — 261
  - 22.4.1 ω-ASYMMETRY — 262
  - On the universal constants — 263
- 22.5 The problem of change — 264
- 22.6 Anything but discrete — 264
- 22.7 The Aristotelian infinite regress — 265

22.8 Conclusion 267

23 Gravitational Waves as Empirical Proofsof Space Reality   269
23.1 Introduction: Gravitational waves 269
23.2 Space deformations 271
23.3 Physical space is a real physical object 273
23.4 Physical space is discrete 274
23.5 Time is a discrete magnitude 277
23.6 Absolute motion 279

24 The substance of physical space   285
24.1 Introduction 285
24.2 Physical space is real and discrete 286
24.3 On the substantiality of physical space 288
24.4 Cellular Automata Like Models 292
24.5 Additional reasons for the paradigm shift 294

25 On space deformations   297
25.1 Introduction 297
25.2 Points have neither extension nor shape 299
25.3 Experiments and theories 299
25.4 Expanding and contracting the space continuum 300
25.5 The relativistic contraction of space 302
25.6 The expansion of intergalactic space 305
25.7 The gravitational deformation of space 307

26 On time deformations   309
26.1 Introduction 309
26.2 Instants have neither duration nor contiguity 310
26.3 The model R^+ of time and the problem of change 311
26.4 The Ives-Stiwell experiment 311
26.5 Relativistic dilation of time 314
26.6 Relativistic local simultaneity 315
26.7 Consequences of the relativistic time deformations 318
26.8 Time in CALM 319

27 The Shame of Physics   321
27.1 Introduction 321
27.2 Preinertia: the vectorial inheritance of motion 323
27.3 An elementary preamble on rotations 326
27.4 A preinertial argument on the nature of motion 327
27.5 Consequences on the theory of special relativity 330

28 A Special Relativity Inconsistency   333
28.1 The memory of a historic debate 333

28.2 Rotations are always absolute motions 334
28.3 Real and apparent velocity changes 335
28.4 Inconsistency of the non-Causal Relativism 337
28.5 On the empirical confirmation of special relativity 339

29 Two fallacies in modern physics — 341
29.1 On spooky actions and double slits 341
29.2 Two false assertions in modern physics 342
29.3 Discussion 343

30 Discrete conclusions — 345
30.1 Introduction 345
30.2 Some classical questions to start with 345
30.3 Real or fictitious? 349
30.4 Continuous or discrete? 351
30.5 Special relativity is not compatible with discreteness 353
30.6 A discreet model to start with 354

31 The pending revolution in physics — 357
31.1 It is impossible to exaggerate the importance of 357
31.2 A revolution in three words 358

A Hilbert machine — 363
A.1 Hilbert Hotel 363
A.2 Hilbert machine 364
A.3 Hilbert machine contradiction 366
A.4 Discussion 367

B A disturbing supertask — 369
B.1 Introduction 369
B.2 The last disk 370
B.3 Discussion 371

C Proving Unproved Euclidean Propositions — 375
C.1 Introduction 376
C.2 Conventions and general fundamentals 376
C.3 Foundational basis of Euclidean geometry 378
  Fundamentals on lines 378
  Fundamentals on straight lines 386
  Fundamentals on planes 391
  Fundamentals on distances 398
  Fundamentals on circles 398
  Fundamentals on angles 401
  Fundamentals on polygons 405

List of Figures 409

Bibliographic References 411

Alphabetical index 433

1. Important warning

This book brings together a series of published and unpublished works written by the author between 2020 and 2023. Although they are not presented in the chronological order in which they were written, they are arranged to form a functional and coherent text on the real nature of physical space, it is what is called in Spanish a facticio[1]. For this reason, the reader will find several repetitions throughout the book. I could have eliminated these repetitions, but in exchange for modifying the original texts and their corresponding autonomy, which seems to me a worse solution. Besides, we learn by repeating.

As its main thesis, the book argues about the reality of physical space and its finite and discrete nature (although this is not the only thing it argues about). A reality that is already very difficult to deny, despite the opinion of certain relevant physicists who continue to deny that space is a real physical object. Indeed, the empirical detection of gravitational waves implies the empirical detection of the vibrations of space itself. And what does not exist simply does not have empirically detectable properties.

Although this is not a book on the history of science, it does include some chapters on the history of the concept of space, from the preSocratics to the relativistic spacetime continuum, which is an infinitist concept legitimated by the Axiom of Infinity. For this reason, chapter discussing the consistency of this axiom are also included. Although contemporary physicists usually do not pay attention to it, the Axiom of Infinity is the key to modern physics. Here, and in other works of the author, it is shown to be an inconsistent axiom, which will have very significant consequences in a good part of physical theories.

The inconsistency of the actual infinity serves, among other things, to demonstrate the discrete nature of space (and time). For this rea-

---

[1] Said of a book or volume: A collection of various books or printed matter (DRAE). I have not been able to find a satisfactory English translation.

son, the first three chapters of the book demonstrate the elements that constitute the formal setting for all the discussions that follow in the rest of the book. The fourth chapter sets out an ordered list of their respective statements. The discrete nature of space allows us to deduce some basic properties of the constituent substance of real physical space. Of course, the space substance must be different from, but not indifferent to, ordinary matter. The spatial substance might even be the generator of ordinary matter. In this sense, models similar to cellular automata are proposed here to initiate the discussion on the new discrete and finitist paradigm of the physical world.

Finally, the concept of complete totality is used throughout the book and, although the union of the two words, "complete" and "totality", is sufficiently explicit, it can also be formally defined: A complete totality is *a set defined by comprehension in which every element that should be in the set, is in the set.*

2. Conventions and symbols

2.1 Conventions

P1 To facilitate explanations and discussions, some paragraphs of this book will be consecutively numbered (as this one). They will be referred to by the number that appear at the beginning of each paragraph, preceded by the letter P. For instance, P1 refers to this paragraph. As with the proofs, these numbered paragraphs end with the symbol □. For the same reason, all equations will also be consecutively numbered within each chapter, although in this case the numbers will be put in brackets on the right side of each equation:

$$f(i) = a_i \text{ (example of equation)} \qquad (1)$$

Equations will be referred to by their corresponding numbers in brackets: the above equation would be referred to by (1). As usual, numbers in straight parentheses will indicate bibliographical references. In bibliographic references, the abbreviation p. will be used to indicate page or pages. □

Theorems, definitions, corollaries, etc. will be successively numbered. In most cases they will be named by proper names. The symbol "□" will be used to indicate the end of the demonstration of a statement when the demonstration follows the statement. To facilitate reading and minimize errors (related to punctuation) the initial letter of all substantives in the proper names of theorems, corollaries, definitions, principles, axioms and conclusions will be written in capital letters.

When the same explanation serves to two different alternatives, only one of the alternatives will be explained, adding in parentheses the word, or words, that would have to be changed in the given explanation to be the explanation of the other alternative. For example: If the first (last) item in the list is an even (odd) number, the list begins (ends) with an even (odd) number.

All symbols used in the book are listed at the end of this chapter. The ellipsis, symbolically represented by three dots ..., will often be used to denote the rest of the elements of a set or sequence that obviously follow the indicated elements. The logical expression "if, and only if" will be written "iff" when convenient. The expression "actual infinity" refers to one of the types of infinity, the other being the potential infinity. Both are introduced and explained in Chapter 3.

It will be inevitable the use of a few number of primitive concepts, i.e. concepts that cannot be defined in terms of other more basic concepts. That is the case, for instance, of point, line or set. The word "collection" will be used in a general sense to refer to sequences, sets, lists, tables, etc.

Definition 1 (of Complete Totality) *A complete totality is a set defined by comprehension in which every element that meets the definition of membership is in the set, so that to a complete totality of a certain type of elements, it is not possible to add new elements of that type because it already contains all of them.*

Most of the collections, mainly sequences and sets, will be ω-ordered (as the sequence 1, 2, 3, ... of the natural numbers in their natural order of precedence). In a few cases they will be ω^*-ordered (as in the case of the increasing sequence of negative integers ... -3, -3, -1). The sets used in the demonstrations, for example the real interval $(0, 1)$, or the set $\mathbb{Q}^+$ of the positive rational numbers, will always be the simplest possible in each occasion.

As usual, to put into a correspondence a set A with another set B means to pair off each element of the A set with an element of the set B. All correspondences will be injective, and in most cases surjective (bijections or one-to-one correspondences). Unless otherwise indicated, the sets $\mathbb{N}$ (natural numbers), $\mathbb{Z}$ (integer numbers), $\mathbb{Q}$ (rational numbers), $\mathbb{A}$ (algebraic numbers) and $\mathbb{R}$ (real numbers), and any of their subsets, will always be considered in their natural order of precedence, that is, ordered by their increasing magnitudes or values. In the case of $\mathbb{N}$, the natural order of precedence is the ω-order (a case of well-order defined in Chapter 20). In all the other cases, excluding $\mathbb{Z}$, the order of precedence is a dense order (see P2) that is not a well order.

In most cases, we will use the word "denumerable" to refer to the infinity of the set $\mathbb{N}$ of the natural numbers and to the infinity of any other set or sequence that can be put into a one to one correspondence with $\mathbb{N}$. The words "enumerable" or "numerable" can also be used with the same meaning. Although the word "countable" is also used to refer to finite or denumerable infinite sets, it will not be used here in

2.1 Conventions

order to avoid confusions. Finally, the terms "non-countable" or "non-denumerable" will be used to denote the infinities greater than the denumerable infinity.

Although formally unacceptable, Euclid defined two capital concepts in geometry: the concept of line [150, Definition 2, p. 153] and the concept of straight line [150, Definition 4, p. 153], being the second a particular case of the first; and being both of them currently assumed as primitive, undefinable, concepts. Languages maybe evolving from their most popular use that, unfortunately is not always the most correct one [130]. That could be the reason why in English, *line* and *straight line* came to mean the same thing, and now there is no English word to denote the original Euclidean concept of line, a universal concept that applies to all types of lines. For this reason, in the English edition of this book, the word "line*" will be used to refer to the general geometric object that Euclid called line. Thus, and still being a primitive concept, a line* (línea in Spanish) can be understood as any uni-dimensional continuum of points. Although it is possible to give a formally productive definition of straight line [196, 200], it will not be necessary to do so in this book, so that they can continue to be understood as a particular type of lines whose lengths are the shortest of all possible lines joining any two given points. No matter how redundant, straight lines will always be referred to by "straight lines". As usual, real and rational lines* and straight lines will be used to denote lines* and straight lines whose points represent respectively densely ordered sets (see P2) of real numbers and of rational numbers.

P2 In all discussions and arguments, time, distances and lengths will be assumed to be Euclidean and represented by real numbers and intervals of real numbers. As usual, a finite interval (a, b) is said finite if its extension $b - a$ is finite, even if the interval is infinitely dense, which means that between any two elements (points, instants, numbers) of the interval, the interval contains infinitely many different elements. This is the case of all intervals of rational and real numbers in their corresponding natural order of precedence. An element inside an interval will be an element of the interval different from its endpoints. □

Although supertasks will be introduced in Chapter 3, they will start to be used from the first chapters. A supertask consists of performing an infinite number of actions or tasks (for example counting numbers, or removing balls from a box containing balls) in a finite interval of time, which, unless otherwise indicated, will be the real interval (t_a, t_b). The successive actions a_1, a_2, a_3, ... of the infinite sequence of actions $\langle a_i \rangle$ will be supposed to be carried out in the successive instants t_1, t_2, t_3, ... of an ω-ordered, strictly increasing and convergent sequence of instants $\langle t_i \rangle$ within the interval (t_a, t_b), being t_b the limit of the sequence

$\langle t_i \rangle$. Every action a_i of $\langle a_i \rangle$ will be assumed to be performed in the precise instant t_i of $\langle t_i \rangle$, and all of them will be instantaneous.

Needless to say, all arguments in this book are of a conceptual nature, even when they make use of material artifacts as machines, boxes, balls and the like, all of which have to be understood as theoretical devices to illustrate the arguments and to facilitate discussions.

2.2 Symbols

The followings symbols and notations could be used in what follows:

MT: Modus Tollens

*: Thomson' lamp on.

o: Thomson's lamp off.

c: Thomson's lamp clicked.

$\mathbb{N}$: set of the natural numbers in their natural order of precedence.

$\mathbb{Z}$: set of the integer numbers in their natural order of precedence.

$\mathbb{Q}$: set of the rational numbers in their natural order of precedence.

$\mathbb{Q}^+$: set of the positive rational numbers in their natural order of precedence.

$\mathbb{A}$: set of the algebraic numbers in their natural order of precedence.

$\mathbb{R}$: set of the real numbers in their natural order of precedence, and real straight line.

$\mathbb{R}^+$: set of the positive real numbers in their natural order of precedence.

$\mathbb{R}^3$: Euclidean tridimensional space.

$\mathbb{R}^n$: Euclidean n-dimensional space.

$|A|$: cardinal of the set A.

...: ellipsis.

$\in$: belongs.

$\notin$: does not belong.

$\subset$: subset.

$\supset$: superset.

$\not\subset$: not subset.

$\cup$: union of sets.

$\cap$: intersection of sets.

2.2 Symbols

$P(A)$: power set of the set A (set of all subsets of A).

$\aleph_0$: aleph-null, the smallest transfinite cardinal.

$2^{\aleph_0}$: power of the continuum.

ω: omega, the smallest transfinite ordinal.

$2\omega, 3\omega, \omega_1 \ldots$: ordinals greater than ω.

$2^{2^{\aleph_0}}, \aleph_1, \aleph_2 \ldots$: cardinals greater than $\aleph_0$.

∞: infinity, the improper real number.

(a, b): open interval or segment.

$[a, b]$: closed interval or segment.

$(a, b]$: right closed interval or segment.

$[a, b)$: left closed interva or segmentl.

I_o: 0-interval, interval whose left endpoint is 0.

$\langle q_n \rangle, \langle q_i \rangle \ldots$: ω-ordered sequence $q_1, q_2, q_3, \ldots$

$\sum_{i=1}^{n} x_i$: sum of n terms: $x_1 + x_2 + \cdots + x_n$.

$\sum_{i=1}^{\infty} x_i$: sum of infinite terms: $x_1 + x_2 + x_3 + \ldots$

$\lim_{n \to \infty} a_n$: limit of the sequence $\langle a_n \rangle$.

$\lim_n a_n$: limit of the sequence $\langle a_n \rangle$.

$\langle D_n(x) \rangle$: ω-ordered sequence of definitions of x.

$D_i(x)$: ith definition of x.

$\langle D_i(x) \rangle_{i=1,2,\ldots n}$: first n definitions of x.

${}^k S_i$: ith element of a collection at the kth definition of the collection.

$|x|$: absolute value of x.

$\min(a, b)$: least of the two values in brackets.

$\forall$: for all.

$\exists$: exists.

$\Rightarrow$: logic inference.

$\Leftrightarrow$: logic double inference.

iff: if, and only if.

$\neg$: logic negation.

∨: logic or.

∧: logic and.

∴ therefore.

□: end of a proof.

3. Finite versus infinite

3.1 Introduction

This chapter proves a fundamental result for the rest of the book:

The Hypothesis of the Actual Infinity is inconsistent.

A result that changes everything in mathematics and physics. I think that is why there is so much resistance to accepting it. The first demonstrations of this inconsistency have been available for more than twenty-five years, but so far their echo has been practically nil. Obviously, the infinitist mathematics of modern physics (which has never been tested) will be seriously affected by the inconsistency of the Hypothesis of the Actual Infinity (fortunately, experimental physics can only be finitist and discrete). Some of the mathematical and physical consequences of the inconsistency of the actual infinity will be discussed in this and the following chapters of the book. One such physical consequence will be proved in this chapter: In a consistent reality, there can only be a finite number of universes (if there are several), each with a finite number of physical objects.

3.2 Zeno, Aristotle and Cantor

Fortunately there is an abundant and excellent literature on the history of infinity (for instance: [376, 229, 318, 35, 300, 69, 223, 248, 251, 190, 191, 2, 252, 249, 66, 363, 22, 298]). The details of that story will not be necessary here, although three of its most relevant protagonists could be remembered as historical references:

a) Zeno of Elea (490-430 BC), a pre-Socratic philosopher that made use for the first time of the mathematical infinity when defending Parmenides' thesis on the impossibility of change. We know Zeno work (near forty arguments, including his famous paradoxes against the possibility of change [3, 71]) through his doxographers: Plato, Aristotle, Diogenes Laertius or Simplicius. The infinity in Zeno arguments is the actual infinity, although Zeno is obviously not doing infinitist mathematics, but logical reasoning in which infinite collections of points and instants appear. Zeno arguments only work properly if these collections are considered as complete infinite totalities (Zeno dichotomies are discussed in Chapter 20).

b) Aristotle (384-322 BC), one of the most influential thinkers of western culture. He introduced, in a broad sense, the notion of *one to one correspondence* just when he was trying to solve some of Zeno paradoxes [18, Books III-VII]. He also introduced the basic distinction between the potential and the actual infinity. A distinction that will be analyzed in the next section.

c) Georg Cantor (1845-1918), mathematician co-founder, together with R. Dedekind and G. Frege, of set theory at the end of the XIX century. His work on transfinite numbers [56] (cardinals and ordinals) lays the foundations of modern infinitist mathematics. He inaugurated the so called paradise of the actual infinity, where, according to D. Hilbert, infinitists will inhabit forever [157, p. 170]:

> Wherever there is the slightest prospect of fruitful concepts and conclusions, we will carefully track them, cultivate them, support them and make them usable. No one shall be able to drive us out of the paradise that Cantor has created for us.

From Zeno to Aristotle, the infinity discussed was usually the actual infinity, although this notion was far from being clearly established before Aristotle. From Aristotle to Cantor, there were defenders of both types of infinity (actual and potential), although with a certain hegemony of the potential infinity, especially since the 13th century, after Aristotle had been *christianized* by the medieval scholastics. In those pre-infinitist times, the same arguments could be used in support of one or the other infinity (for example, the arguments based on the correspondence between the points of a circle and the points of one of its diameters). But there is still no theory of mathematical infinity. The first mathematical theory of infinity appears at the end of the XIX century, Bolzano, Dedekind and especially Cantor being its most relevant founders. From Cantor until today, the hegemony of the actual infinity has been almost absolute and, moreover, free from serious criticism.

3.3 The actual and the potential infinity
(This section includes published texts by the author [212, p. 32-35])

In common parlance, the word infinite is used to refer to the quality of being immense, gigantic, unlimited, etc. C. F. Gauss (*Princeps Mathematicorum* [375, p. 1188]) said that infinity is a way of speaking (C.

3.3 The actual and the potential infinity

F. Gauss, Letter to astronomer H.C. Schumacher, 12 July 1831 [127, Vol. II, p. 268]):

> I protest against the use of infinite magnitude as something completed, which is never permissible in mathematics. Infinity is merely a *façon de parler* [a way of speaking], the true meaning being a limit which certain ratios approach indefinitely close, while others are permitted to increase without restriction.

The consideration of an infinite magnitude (or an infinite sequence, for instance of numbers) as something completed is what we call *actual infinity* since Aristotle, who introduced the distinction between the potential infinity and the actual infinity [16, 17, Books III, VIII]. It is remarkable the fact that in the above quotation, Gauss implicitly includes the distinction between both infinities (see below in this section). I have the impression that most physicists think, like Gauss, in terms of the potential infinity, without worrying about the fact that they are building physics with the mathematics of the actual infinity.

As will be seen below, it is possible to give a precise definition of the concept infinity, albeit based on a primitive concept: the concept of set. However, the concept of set could be defined in operational, non-Platonic terms [212, p. 360]:

Definition 2 (of Set) *A set is the theoretical object that results from a mental grouping of different arbitrary objects previously defined.*

This definition has the advantage of avoiding the entanglements caused by self-reference, simply by requiring that the elements to be grouped be previously defined, which seems quite reasonable if we intend to know what we are grouping. It is convenient, on the other hand, to consider that some sets exist as complete totalities, i.e. as sets satisfying the following:

Definition 3 (of Complete Totality) *A complete totality is a set defined by comprehension in which every element that satisfies the corresponding membership definition of the set is in the set.*

In consequence, to a complete totality of a certain type of elements, it is not possible to add new elements of that type because it already contains *all of them*.

But returning to the concept of infinity, and apart from Gauss's opinion, the word "infinite" has also a precise meaning based on the primitive concept of set:

Definition 4 (of Infinite Set) *A set is said infinite if it can be put into a one to one correspondence with one of its proper subsets.*

which is the well known Dedekind's definition of infinite set [77, p. 115], an important element of the foundations of modern infinitist mathematics, which began its development at the end of the 19th century. As is well known, the controversial history of the (philosophical and) mathematical infinity has its roots in the pre-Socratic times, although here we are not interested in the details of that history (there is an abundant and excellent literature on the history of infinity, for instance: [376, 229, 318, 35, 300, 69, 223, 248, 251, 190, 191, 2, 252, 249, 66, 363, 22, 298]).

Now I will try to explain the distinction between the two infinities, the actual and the potential. The set of the natural numbers and *supertask theory* are two suitable instruments to evidence such a distinction. The set of the natural numbers needs no presentation. With respect to supertask theory it must be recalled that it is an infinitist theory based, as set theory, on the Axiom of Infinity (introduced in Section 3). It originated as a consequence of a seminal discussion about the possibility, or impossibility, of performing an infinite number of actions (tasks) in a finite time interval. [345, 38, 344, 364, 27].

Although the main objective of supertask theory was, and continue to be, the discussion on the actual infinity, its physical implications (including special relativity) have also been discussed in the last years [274, 284, 288, 312, 141, 143, 142, 284, 285, 286, 86, 287, 263, 7, 8, 289, 369, 160, 84, 85, 263, 83, 323]. In short:

> A supertask consists in performing an infinite sequence of actions $\langle a_i \rangle$ within a finite time interval $[t_a, t_b)$, each action a_i being performed at the precise instant t_i of a strictly increasing and convergent sequence of instants $\langle t_i \rangle$ within $[t_a, t_b)$, being t_b the mathematical limit of $\langle t_i \rangle$.

where the elements of $\langle a_i \rangle$ and $\langle t_i \rangle$ are ordered in the same way as the set of the natural numbers in their natural order of precedence: ω-order: 1, 2, 3,.... Notice in this ordering the set exist as a complete totality (Definition 3) and each element n has an immediate successor $n+1$ (Peano's Axiom of the Successor, [272, p. 1]), where immediate successor is defined according to:

Definition 5 (of Immediate Successor) *All elements of an ordered set A succeeding (preceding) a given element n of A are successors (predecessors) of n in the considered order of A. An element n of an ordered set A is said the immediate successor (predecessor) of another element m of A if n succeeds (precedes) m in the considered ordering of A and no other element of A exists between m and n in that ordering.*

3.3 The actual and the potential infinity

We are now in the appropriate position to analyze the difference between the actual and the potential infinity. Indeed, consider the list L_n of the natural numbers in their natural order of precedence:

$$L_n = 1, 2, 3, \ldots \tag{1}$$

The list L_n can be considered in two different ways:

a) As a complete totality, i.e. as a list in which every element that could be in the list, is in the list (actual infinity).

b) As an unlimited and uncompletable totality (potential infinity).

According to the Hypothesis of the Actual Infinity, the list L_n of the natural numbers in their natural order of precedence 1, 2, 3,... exists as a *complete totality*, i.e as a totality that contains, all at once, all natural numbers. The ellipsis (...) in $1, 2, 3, \ldots$ stands for *all* natural numbers. For all. The word "actual" in *actual infinity* means, therefore, that all elements of an infinite collection as L_n, exist all at once (in the *act*), as a complete totality. In consequence, the list L_n of the natural numbers in their natural order of precedence is considered as a complete totality despite the fact that no last number completes the list. To assume the Hypothesis of the Actual Infinity means, therefore, to assume that it is possible to complete the incompletable, as Aristotle would surely say [17, p. 291]. Or that the incompletable can exist as complete.

To emphasize this sense of completeness, let us consider the task of counting the successive elements of L_n, i.e. the successive natural numbers 1, 2, 3,... in their natural order of precedence. In agreement with the Hypothesis of the Actual Infinity we could count *all* natural numbers in a finite time, for example in an hour, or in a millisecond. The task of counting all natural numbers in a finite time interval, even in less than a second, is an example of supertask:

- *Count each of the successive natural numbers 1, 2, 3... at each of the successive instants t_1, t_2, t_3... of a strictly increasing sequence of instants $\langle t_i \rangle$ within the finite real interval (t_a, t_b), being t_a and t_b any two instants such that $t_a < t_b$, and t_b the mathematical limit of the sequence $\langle t_i \rangle$. For instance, the classical sequence defined by:*

$$t_n = t_a + (t_b - t_a) \frac{2^n - 1}{2^n} \tag{2}$$

As we will now prove, at t_b all natural numbers would have been counted. All. In effect, let each natural number n of the list L_n be counted at the precise instant t_n of $\langle t_i \rangle$. Being t_b the limit of $\langle t_i \rangle$, t_b is the first instant after all instants of $\langle t_i \rangle$, and all those instants do exist as a complete totality according to the Hypothesis of the Actual Infinity. So, the one to one correspondence f between L_n and $\langle t_i \rangle$ defined by:

$$f(n) = t_n, \ \forall n \in L_n \tag{3}$$

proves that at t_b all natural numbers of the list L_n has been counted. All. The reader can easily imagine why ellipsis and correspondences between sets are the key instruments for demonstrations in infinitist mathematics. Note, on the other hand, that the fact of pairing the elements of two infinite sequences (in our case the one of natural numbers and the other of instants) does not prove both sequences exist as complete totalities. They could also be potentially infinite with the same number of elements, a possibility usually ignored in modern infinitist mathematics.

The alternative to the Hypothesis of the Actual Infinity is the Hypothesis of the Potential Infinity, which rejects the existence of *complete* infinite totalities, and then the possibility to count all natural numbers. From this perspective, the natural numbers result from the *endless* process of counting: it is always possible to count a number greater than any given number (Peano's Axiom of the Successor [272, p. 1]). But it is impossible to complete the process of counting all of them, simply because there is not a last natural number to complete the process. So, the complete list of all natural numbers makes no sense, simply because it is incompletable.

The word "potential" in *potential infinity* means, therefore, that the elements of an infinite collection do not exist all at once, but potentially, as possible. The potential infinity is *the unlimited*, as the list L_n of the natural numbers in their natural order of precedence, but only finite collections can be considered as complete totalities, as large as wished but always finite. Similarly, only finite natural numbers can be considered, as large as wished but always finite. For the potential infinite there is not a last natural number (it is always possible to consider a number greater than any previously considered number), but neither is there the complete collection of *all* natural numbers. Contrarily to the actual infinity, the potential infinity assumes the incompletable cannot be completed, cannot exist as complete, precisely because it is not completable.

In short, the actual infinite hypothesis states that the infinite collections are complete totalities, even if no last element completes the collection, as in the case of the ordered list of the natural numbers. On the contrary, the hypothesis of the potential infinite proposes that the infinite collections do not exist as complete totalities, the only complete totalities are the finite totalities, though they can be unlimited in the number of their possible elements. All of which can be summarized in the following definition:

3.3 The actual and the potential infinity

Definition 6 (of Actual and Potential Infinity) *An ordered collection of elements is infinite if there is no last (first) element that completes (initiates) it. The collection is actually infinite if it is considered a complete totality, and potentially infinite if it is not considered a complete totality.*

Where collection is by set, succession, sequence, list, etc. To be formally precise, the words *set, succession, sequence*, etc. should be replaced by the more general word *collection*. However, for the sake of brevity, it will not be necessary to do so. Therefore, in what follows all of them will be interchangeable with each other, unless otherwise specified.

The potential infinity (the 'improper' or 'non-genuine' infinity as Cantor called it [57, p. 70]) has never deserved the attention of contemporary mathematics. The infinity in Dedekind's Definition 4 of infinite set is the actual infinity (see next section). The infinitely many elements of an infinite set exist all at once, as a complete totality. Dedekind's Definition 4 is, therefore, based on the violation of the old Euclidean Axiom of the Whole and the Part (the whole is greater than the proper part) [106]. Set theory has been built on that violation.

The hegemony of the actual infinity in contemporary mathematics is absolute. As absolute as the submission of physics to infinitist mathematics. Some authors proceed as if the existence of complete infinite totalities had been formally demonstrated. Obviously, if that were the case we would not need the Axiom of Infinity to legitimize the existence of such infinite totalities. The Hypothesis of the Actual Infinity is just a hypothesis, not a proven fact. And physics should not be subject to infinitist mathematics. In fact, and in agreement with P. Dirac, it should not be subject to any kind of mathematics at all [81, p. VIII]:

> Mathematics is only a tool and one should learn to hold physical ideas in one's mind without reference to the mathematical form.

The three most important "proofs" of the existence of actual infinite totalities (by Bolzano, Dedekind and Cantor) are illustrative of what we could call *naive infinitism*. They also explain why modern infinitist mathematics had finally to establish the existence of actual infinite sets by an arbitrary law, i.e. by means of an arbitrary axiom (the Axiom of Infinity, which is introduced in the next section).

- Bolzano's proof goes as follow (taken from [249, p 112]):

> One truth is the proposition that Plato was Greek. Call this p_1. But then there is another truth p_2, namely the proposition that p_1 is true [But then there is another truth p_3, namely the proposition that p_2 is true]. And so *ad infinitum*. Thus the set of truths is infinite.

But the existence of an endless process (p_1 is true, then p_2 is true, then p_3 is true, then ...) does by no means prove the existence of a final result as a complete totality. At best it proves the existence of an endless (potentially infinite) process. But it does not prove the existence of an actual infinite totality.

- Dedekind's proof is similar (taken from [249, p 113]):

  Given some arbitrary thought s_1, there is a separate thought s_2, namely that s_1 can be object of thought [there is a separate thought s_3, namely that s_2 can be object of thought]. And so ad infinitum. Thus the set of thoughts is infinite.

The above comment on Bolzano proof also applies here. Dedekind gave another proof a little more detailed, albeit with the same formal defect, based on his definition of infinite set [77, p. 115].

- And finally, Cantor's proof: ([147, p 25], [249, p. 117]):

  Each potential infinite presupposes an actual infinity.

or ([55, p. 404] English translation [305, p. 3]):

  ... in truth the potential infinity has only a borrowed reality, insofar as a potentially infinite concept always points towards a logically prior actually infinite concept whose existence it depends on.

But this is an opinion, not a formal proof. It is now clear why the existence of an actual infinite set had to be finally established by law; that is, by means of an axiom.

Let us, finally, state a conventional use of the expressions actual infinite and potential infinity in this and the subsequent chapters of this book. From now on, and for sake of simplicity, the actual infinity will be referred to simply as infinity or actual infinity, while the potential infinity will always be referred to as potential infinity. Or put another way, the word "infinity" will always mean actual infinity, unless it is preceded by the word "potential", in which case it will obviously mean potential infinity. For the sames reasons of simplicity, the world "universe" will always denote the observable universe.

3.4 The infinity of the Axiom of Infinity

Nothing we have been able to observe and measure so far has been infinite. Nor has it been possible to divide anything into an infinite number of parts. On the other hand, and after more than twenty-seven centuries of arguments and discussions, it was not possible to prove (or disprove) the existence of the actual infinities. Infinitism had no choice but to accept that existence in axiomatic terms by means of

3.5 A short proof of inconsistency

the Axiom of Infinity. An axiom that simply states the existence of an infinite set:

Axiom 1 (of Infinity (ordinary language)):

There exists an infinite set.

Or in abstract, symbolic, terms:

Axiom 2 (of Infinity (abstract form)):

$$\exists A : \emptyset \in A \land \forall a \in A \, (a \cup \{a\} \in A) \tag{4}$$

that reads: there exists a set A such that $\emptyset$ (the empty set) belongs to A and for every element a in A, the element $a \cup \{a\}$ also belongs to A. Although it is not explicitly declared the type of infinity involved in the set A, it can be easily proved that it is the actual infinity:

Theorem 1 (of the Actual Infinity) *The infinity in the Axiom of Infinity can only be the actual infinity.*

Proof: Since potentially infinite sets do not exist as complete totalities (Definitions 3 and 7), only two subsets with the same number of elements of the same potentially infinite set could be put into a one to one correspondence, and then Dedekind Definition 4 is not satisfied, because we would have a one to one correspondence between two proper subsets of a potentially infinite set, in the place of a one to one correspondence between a set and one of its proper subsets. In consequence, the infinity involved in the Axiom 1 of Infinity can only be the actual infinity. □

Obviously, an axiom is just an axiom, i.e. a statement that can be accepted or rejected. Some relevant authors as L.E.J. Brouwer, C. Hermite, S. Kleene, J. König, L. Kronecker, H. Poincaré, A. Robinson, L. Wittgenstein, or H. Weyl, among others, rejected the Axiom of Infinity, more or less explicitly. H. Poincaré went so far as to say that (quoted in [249, p. 121], [76, p. 1]):

> infinity is a perverse pathological illness that would one day be cured.

But the vast majority of contemporary mathematicians and physicists do not question the Axiom of Infinity. Indeed, in our days the criticism of the actual infinity is practically non-existent. And infinitism has become a current of thought absolutely hegemonic and quite intolerant of dissent, as if the existence of the actual infinite had been proven. And no, it has not been proven; it has been assumed. And one has the right and the duty to question that assumption, without being insulted and ostracized for it (as is currently the case).

3.5 A short proof of inconsistency

Over the last 30 years, and from different perspectives (set theory, supertask theory, transfinite cardinals, transfinite ordinals, transfinite arithmetics, geometry) I have developed more than forty formal proofs of the inconsistency of the Hypothesis of the Actual Infinity [212]. This section includes one of them, chosen for its brevity and simplicity: the next Theorem 5. First, however, it is necessary to consider the following formal elements:

Definition 7 (of the Types of Sets) *A set is finite if it has a definite and finite number of elements. A set of elements of a certain type is potentially infinite if it always contains a finite number of elements of that type and any finite numbers of new elements of that type can always be added to it, without the set ceasing to be finite and without it being necessary to change its name.*

Definition 8 (of the Types of Infinities) *The actual infinity is the infinity of the infinite sets. The potential infinity is the infinity of the potentially infinite sets.*

Definition 9 (of Inconsistent Set) *A set is inconsistent if a contradiction can be deduced from the number of its elements, or from the number of elements of at least one of its proper subsets.*

Corollary 1 (of Inconsistent Sets) *A set with the same number of elements as an inconsistent set, is also inconsistent.*

Proof: It is an immediate consequence of Definition 9. □

Definition 10 (of Denumerable Set) *A set is denumerable if its cardinal is the smallest infinite cardinal $\aleph_0$ of the infinite set of all natural numbers. An infinite set is non-denumerable if its cardinal is greater than the smallest infinite cardinal $\aleph_0$.*

Definition 11 (of ω-Ordered Sets) *A set is ω-ordered if being denumerable, it has a first element, each element has an immediate successor and an immediate predecessor, except the first one which has no predecessor.*

Theorem 2 (of Denumerable Sets) *It is always possible to define a one-to-one correspondence between any two denumerable sets.*

Proof: Let A and B be any two denumerable sets. They have the same number of elements: $\aleph_0$ elements (Definition 10). So their respective elements can be put into one-to-one correspondence, i.e. each of the different elements of A can be paired with a different and exclusive element of B, so that all elements of A and B result exclusively paired. □

3.5 A short proof of inconsistency

Theorem 3 (of non-Denumerable Sets) *Every non-denumerable set has denumerable proper subsets.*

Proof: Let X be any non-denumerable set. Since its cardinal is greater than $\aleph_0$ (Definition 10), X contains proper subsets with only $\aleph_0$ elements, all of which are denumerable proper subsets of X (Definition 10). □

Theorem 4 (of Indexation) *The elements of a denumerable set can be reordered with the same order as the elements of any other denumerable set.*

Proof: Let $A = \{a, b, c, \dots\}$ and $B = \{\alpha, \beta, \dots\}$ be any two denumerable sets. There exists at least one bijection f between the elements of A and B (Theorem 2). Consequently, f pairs each element k of A with a unique and exclusive element, say δ, of B, which can be used to exclusively index that element k of A, so that element k can be rewritten as a_δ. Consequently, the elements of the set A can be reordered and rewritten to define the set $A' = \{a_\alpha, a_\beta, a_\gamma, \dots\}$ which has exactly the same elements as A, and ordered in the same way as the elements of B. □

The infinity of infinite sets is the actual infinity, not the potential infinity (Theorem 6 of the Axiom of Infinity). This implies the existence of certain infinite sets that are also complete totalities (Definition 3). For example the set $\mathbb{N}$ of ALL natural numbers in their natural order of precedence. It is not possible, then, to add new natural numbers to the set $\mathbb{N}$ of natural numbers because it already contains them all. And the same is true of many other numerical or non-numerical sets. For many authors, the existence of these ordered and complete totalities without a last element that completes them (or without a first element that initiates them) is a proven conclusion independent of the Axiom of Infinity. It is not. It is an existence assumed and legitimized by the Axiom of Infinity. Their existence is, therefore, as debatable as the Axiom of Infinity itself. So it is as legitimate to argue about that axiom as it is to argue about the existence of those complete totalities. This fully justifies the following:

Theorem 5 (of Denumerable Infinity) *All denumerable sets are inconsistent.*

Proof: Let A be any denumerable set. The set A allows us to define the set A' with the same elements as A but reordered as the set $\mathbb{N}$ of natural numbers in their natural order of precedence: $A' = \{a_1, a_2, a_3, \}$ (Theorem 4). The open interval of rational numbers $(0, 1)$ is densely ordered in the natural order of precedence (represented by the symbol <) defined by the natural values of the rational numbers. It is also a denumerable set, so there exists a bijection f between A' and $(0, 1)$

(Theorem 2). Consequently, $(0,1)$ can be reordered and rewritten as the set $\mathbb{Q}_{01} = \{q_{a_1}, q_{a_2}, q_{a_3}, \dots\}$, where $q_{a_i} = f(a_i), \forall a_i \in A'$, and the successive elements $q_{a_1}, q_{a_2}, q_{a_3}, \dots$ of $\mathbb{Q}_{01}$ are ordered by the successive natural numbers in their natural order of precedence, and not by their respective values as rational numbers. Let x now be a rational variable defined initially as q_{a_1}. And let the value of x be <-compared (i.e., compared according to the values of the rational numbers) with the successive elements of the set $\mathbb{Q}_{01}$, with x being redefined as the compared element q_{a_i} if, and only if, $q_{a_i} < x$.

For short, let us call comparison* this <-comparison and redefinition of x if, and only if, the value of the compared element is smaller than the current value of x. It is immediate to prove that for each natural number v it is possible to perform the first v comparisons* of x with the first v successive elements of $\mathbb{Q}_{01}$. Indeed, if it were not possible, there would be at least one natural number $n \leq v$ such that x could not be compared* with q_{a_n}, which is impossible because q_{a_n} is a rational number of $\mathbb{Q}_{01}$ that can be compared* with the current value of x, which is also a rational number. Once all possible comparisons* of x with the successive elements $q_{a_1}, q_{a_2}, q_{a_3}, \dots$ of $\mathbb{Q}_{01}$ have been made, the current value of x, whatever it may be, could only be the smallest rational number of that set. Indeed, if once performed all possible comparisons* of x with the successive elements of $\mathbb{Q}_{01}$ the current value of x were not the smallest rational number of $\mathbb{Q}_{01}$, there would be at least one element q_{a_n} in $\mathbb{Q}_{01}$ such that $q_{a_n} < x$. But that is impossible because n is a natural number; the first n comparisons* have been carried out; and therefore x was compared* with q_{a_n} and redefined as q_{a_n}; and in all subsequent comparisons*, x could only be redefined with values smaller than q_{a_n}. Therefore, it is impossible for $q_{a_n} < x$. But, on the other hand, it is also immediate to prove that once all possible comparisons* of x with the successive elements of $\mathbb{Q}_{01}$ have been made, the current value of x is not the smallest rational number of that set: every element of the infinite set $\{x/2, x/3, x/4 \dots\}$ is an element of $\mathbb{Q}_{01}$ smaller than x. This contradiction proves that the set A', defined exclusively with the elements of A, is inconsistent. Therefore A' and A are inconsistent (Definition 9). And A being any denumerable set, it must be concluded that all denumerable sets are inconsistent. □

Although the consistency of a mathematical proof of infinite steps is universally accepted without the need to perform all of its infinite steps, the theory of supertasks considers the possibility of performing them in finite time. In the case of the above successive comparisons* of x with each successive q_{ai} would be performed at each successive instant t_i of a strictly increasing and convergent sequence $\langle t_i \rangle$ of instants within the finite time interval (t_a, t_b), whose limit is t_b. The instant t_b is the first instant after all instants of $\langle t_i \rangle$, and therefore the first instant after having performed all possible comparisons* of x with the successive elements of Q_{01}. At the instant t_b the rational variable x will still be a rational variable with a certain value, whatever it is; and not, for example, an elephant (in which case anything could be proved). The problem is that the value of x at the instant t_b is and is not the least rational of Q_{01}.

Corollary 2 (of Inconsistent ω-Order) *ω-ordered sets are inconsistent.*

Proof: Since ω-ordered sets are also denumerable sets (Definition 11), they are inconsistent (Theorem 5). □

From the previous theorems and corollaries, we can immediately deduce, among many others, the following results:

3.6 The axiom of infinity is inconsistent

The above Theorem 5 proves the inconsistency of any denumerable set. It is then immediate to prove the following results:

Theorem 6 (of the Axiom of Infinity) *The Axiom of Infinity is inconsistent.*

Proof: Let us write the set A defined in Axiom 2:

$$\exists A : (\emptyset \in A \wedge \forall a \in A\, (a \cup \{a\} \in A)) \tag{5}$$

as:

$$A = \{a, s_1(a), s_2(a), s_3(a), \ldots\} \tag{6}$$

where:

$$s_1(a) = a \cup \{a\} \tag{7}$$
$$s_2(a) = s_1(a) \cup \{s_1(a)\} \tag{8}$$
$$s_3(a) = s_2(a) \cup \{s_2(a)\} \tag{9}$$
$$s_4(a) = s_3(a) \cup \{s_3(a)\} \tag{10}$$
$$s_5(a) = s_4(a) \cup \{s_4(a)\} \tag{11}$$
$$\ldots$$

Consider now the set $\mathbb{N}$ of the natural numbers, which is denumerable, and the set A defined by (6), which is the set whose existence claims the Axiom of Infinity. The one to one correspondence f between the denumerable set $\mathbb{N}$ and A defined according to:

$$f(n) = s_n(a),\ \forall n \in \mathbb{N} \tag{12}$$

proves that A is also an inconsistent set (Theorem 5 and Corollary 1). □

And from Theorems 1 and 6 it immediately follows the next three corollaries:

Corollary 3 (of the Inconsistent Infinity) *The actual infinity is inconsistent.*

Proof: It is an immediate consequence of Theorems 1 and 6. □

Corollary 4 (of the Actual Infinite Sets) *All actual infinite sets are inconsistent.*

Proof: It is an immediate consequence of Theorems 1 and 6. □

Corollary 5 (of Infinite Divisibility) *The actual infinite divisibility of any formal or physical object is inconsistent.*

Proof: From the actual infinite divisibility of any formal or physical object can only result an inconsistent infinite set of parts (Corollary 4). So that actual infinite divisibility is inconsistent. □

Corollary 6 (of Consistent Collections) *A set can be either a finite complete totality or a potentially infinite and uncompletable totality. Otherwise it is inconsistent.*

Proof: It is an immediate consequence of Definition 6 and Corollary 4. □

Let us now recall the following definition:

Definition 12 (of Densely Ordered Sets) *If no element of a strictly ordered set has an immediate predecessor nor an immediate successor, the set is said to be densely ordered or to define a continuum.*

We can now prove the following:

Theorem 7 (of Inconsistent Dense Order) *Densely ordered sets are inconsistent.*

Proof: Let X be a densely ordered set. Suppose X is finite. It will have a finite number of elements, say n. Let x_1 and x_2 be two elements of X such that x_2 is a successor of x_1. Since x_2 cannot be the immediate successor of x_1, there will exist between x_1 and x_2 at least one other successor x_3 of x_1. Since x_3 cannot be the immediate successor of x_1, there will exist between x_1 and x_3 at least one other successor x_4 of x_1. By repeating this argument $n-2$ times we will arrive at a successor x_{n-2} of x_1 that would have to be its immediate successor, which is impossible. Therefore, X cannot be finite. And being infinite it is inconsistent (Corollary 4). □

3.6 The axiom of infinity is inconsistent

Corollary 7 (of the Inconsistent $\mathbb{Q}$ and $\mathbb{R}$) *When considered as complete infinite totalities, the set $\mathbb{Q}$ of the rational numbers and the set $\mathbb{R}$ of the real numbers are both inconsistent.*

Proof: It is an immediate consequence of Corollary 4, and also of Theorem 7, because they are densely ordered sets. □

Theorem 8 (of the Inconsistent Continuum) *The spacetime continuum is inconsistent.*

Proof: The spacetime continuum is the Cartesian product (cross product) of sets $\mathbb{R}^4 = \mathbb{R} \times \mathbb{R} \times \mathbb{R} \times \mathbb{R}$, each of whose factors is the set $\mathbb{R}$ of real numbers. Consequently it is an inconsistent set (Corollaries 7 and 1). □

The above results on the inconsistency of the infinite sets, including the inconsistency of the continuum and of densely ordered sets, will change everything. So deconstructing the arguments that follow here and in the subsequent articles of this series of articles, will involve proving the falsity of Theorem 5 (and the falsity of each of the more than 40 independent proofs included in [212]).

Let us now consider the following:

Definition 13 (of Discrete Sets) *A set is discrete if it has a first element, a last element and each of its elements (except the first one) has an immediate predecessor and (except the last one) an immediate successor.*

And then, let finally prove the following

Theorem 9 (of Discrete Sets) *All discrete sets are finite.*

Proof: Let A be any discrete set:

$$A = \{a, s_1(a), s_2(a), s_3(a) \ldots s_v(a)\} \tag{13}$$

where $s_1(a)$ is the immediate successor of a; $s_2(a)$ the immediate successor of $s_1(a)$; $s_3(a)$ the immediate successor of $s_2(a)$; and so on. If an element $s_n(a)$ has a finite number n of predecessors, then its immediate successor $s_{n+1}(a)$ has also a finite number $n+1$ of predecessors: all n predecessors of $s_n(a)$ plus $s_n(a)$. And since the element $s_1(a)$ has a finite number of predecessors, just 1 predecessor, the element a, we can inductively conclude that all elements of A, including its last element, have a finite number of predecessors. Therefore, A has a finite number of elements. □

Theorem 10 (of the Strictly Ordered Sets) *Every strictly ordered set is discrete.*

Proof: Let a be any element of any strictly ordered set A, and suppose A has not a last element. Since a is not the last element of A, there exist successors of a in A. Let us consider one such successors and denote it by a_1. For the same reasons as in the case of a, we can consider and denote by a_2 any successor of a_1 in A. For the same reasons as in the case of a_1, we can consider and denote by a_3 any successor of a_2 in A. For the same reasons as in the case of a_2, we can consider and denote by a_4 any successor of a_3 in A. We would thus have a sequence of successors of a: $a_1, a_2, a_3, a_4 \ldots$ in which there is not a last element. The bijection f between A and the ω-ordered set $\mathbb{N}$ defined by $f(a_i) = i$ proves that A, like $\mathbb{N}$, would be infinite, and therefore inconsistent (Corollary 4). Consequently, A has a last element. Exactly the same argument now referring to the predecessors of a, proves also that A has a first element. Let a now be any element of A other than the last element of A. Suppose that a has not an immediate successor. Let a_1 be any successor of a. Since a_1 is not the immediate successor of a there will exist another successor a_2 of a between a and a_1. Since a_2 is not the immediate successor of a there will exist another successor a_3 of a between a and a_2. The same argument above shows that the sequence of successors $a_1, a_2, a_3 \ldots$ of a is inconsistent. Therefore a has an immediate successor. The same argument now referring to any element b different from the first element of A proves that b has an immediate predecessor. Consequently, A is discrete (Definition 13). □

Theorem 11 (of Discrete Sets) *Every set is either discrete or discretely orderable.*

Proof: Let A be any set. If it is strictly ordered, it is a discrete set (Theorem 10). If it is unordered and consistent, it will have a finite number n of elements. By a bijection f, each of its elements can be paired with a different natural number of the set $\mathbb{N}_n$ of the first n natural numbers in their natural order of precedence. The set A^* defined by f^{-1}:

$$A^* = \{f^{-1}(1), f^{-1}(2) \ldots f^{-1}(n)\} \tag{14}$$

is an ordered version of A, and therefore a discrete version of A (Theorem 10). □

As noted above, more than forty other different and independent arguments included in [212] reach the same conclusion about the inconsistency of the Hypothesis of the Actual Infinity subsumed in the Axiom of Infinity. This infinity is what Aristotle would surely call infinite by addition. In Chapter: 16, it will be proved the inconsistency of the other Aristotelian infinitude: the infinite by division, which was the type of infinite involved in the formalized version of Zeno's Dichotomies I and II [52, 53, 358, 359, 312, 165, 363, 71, 236, 139, 140, 374, 141, 143, 142, 238, 237, 226, 227, 269, 6, 288, 312, 165, 323].

Physical models and theories work reasonably well (even very well) until the infinities appear. But physicists do not usually concern themselves with the formal consistency of the infinitist mathematics that they use in all their models and theories. Nor do they concern themselves with another problem essential to a consistent explanation of the physical world: the problem of the infinite regress (of proofs, definitions, and causes). As will be seen throughout this series of articles, it is possible to modify the infinitist models and theories used in physics by finitist and discrete versions in such a way that they remain compatible with all the accumulated empirical knowledge about the physical world. And, at the same time, they are much simpler, more physical and less extravagant than their infinitist counterparts.

3.7 Conclusion

If any one of the more than forty proofs of the inconsistent nature of the actual infinity given in [212] is right, then the Hypothesis of the Actual Infinity is inconsistent. One of those arguments has been reproduced here so that the reader can directly evaluate the possibility that, in fact, the Hypothesis of the Actual Infinity, and then the Axiom of Infinity were inconsistent. If so, we might draw our first two cosmological conclusions:

Corollary 8 (of the eternal universe) *A consistent universe cannot be eternal.*

Proof: In an eternal universe time would be infinite, with an actual infinite number of, for instance, seconds. Therefore, an eternal universe would contain inconsistent infinite sets of time units (Corollary 4), and then it would be inconsistent. □

Corollary 9 (of the Finite Number of Universes) *In a consistent reality only a finite number of universes could exist.*

Proof: It is an immediate consequence of Corollary 4. □

Corollary 10 (of the Finite Number of Cycles) *In a coherent reality there can only be a finite number of cycles of creation-destruction of universes.*

Proof: It is an immediate consequence of Corollary 4. □

Corollary 11 (of the Finite Universe) *A consistent universe cannot contains an actual infinite number of physical objects.*

Proof: It is an immediate consequence of Corollary 4. □

According to the Standard Model there exists a finite number of different elementary particles (six quarks, six leptons and five bosons), each with a different finite mass. Therefore, the following is also true:

Corollary 12 (of the Finite Mass-Energy) *The mass and the energy of the observable universe cannot be actually infinite.*

Proof: It is an immediate consequence of the Standard Model, Theorem 11 and the mass-energy relation. □

Theorem 12 (of the Finite Universe) *A consistent universe cannot contains an actual infinite number of physical objects.*

Proof: It is an immediate consequence of Corollary 4. □

According to the Standard Model there exists a finite number of different elementary particles (six quarks, six leptons and five bosons), each with a different finite mass. Therefore, the following is also true:

Corollary 13 (of the Finite Mass-Energy) *The mass and the energy of the observable universe cannot be actually infinite.*

Proof: It is an immediate consequence of the Standard Model, Theorem 12 and the mass-energy relation. □

4. Discrete Magnitudes and Functions

4.1 Introduction

One of the most ubiquitous and frequent concepts in the primary and secondary literature of the physical sciences is undoubtedly the concept of the spacetime continuum, considered as a 4-dimensional continuum of inextensive points and instants of zero duration. The theories of relativity, for example, are theories of the spacetime continuum. But even quantum mechanics, whose quantum surname alludes to the discrete, has been developed with the same infinitist mathematics of the spacetime continuum. The consideration of discrete spaces is still very much in the minority in contemporary physics, and that of discrete time is practically non-existent.

In this chapter, and after recalling the first (pre-Socratic) argument in favor of the discrete and the first modern consideration of energy as a discrete magnitude, we denounce a truly scandalous situation in the physics of our days that, as far as I know, no one has pointed out: the existence of a large number of mathematical functions whose outputs should be discrete values that are impossible because they involve continuous variables, such as space and time. And it is not a question of accuracy or approximation, but of representation of a discrete reality by means of an indiscrete language. If physicists forced themselves to express in a discrete language the discreteness of matter, electromagnetic energy and electricity, they would surely end up discovering the inconsistency of their infinitist mathematical language.

4.2 Democritus' argument

Many of the problems that arise today in physics were already raised in pre-Socratic Greece. Among them are those related to the finite or infinite divisibility of things. One such argument is that of Democritus concerning the divisibility of matter. Collected and recalled by Aristotle [18, A2, 316a], and slightly modified, Democritus' argument is as follows:

> Let us suppose that matter can be divided to infinity, and let us imagine that we do indeed successively divide a piece of matter as long as it is possible to do so. Could there remain extensive particles of that matter? The answer is no, because otherwise we would not have chopped up those remaining large pieces of matter. Therefore, we must continue chopping up those remaining large pieces of matter. But then we would get inextensive particles of that matter. And with inextensive particles of matter, it is impossible to reconstruct an extensive piece of matter. Therefore, we cannot think that matter is made of points without extension. Therefore, there must be minimally extended particles of matter that cannot be divided: the atoms.

Until the first years of the twentieth century some renowned scientists continued to deny the existence of atoms, such as the physicist E. Mach and the mathematician G. Cantor. From the latter are the following words [57, p. 78]:

> I cannot regards them [the atoms] as existent either in concept or in reality no matter how many useful thingshave up to a certain limit been accomplished by means of this fiction.

But, as is well known, the atomic theory ended up being universally accepted around that same time. Today no one doubts that, for example, the smallest possible amount of iron is exactly one atom of iron. This does not imply that this iron atom can be broken into the subatomic particles that form it, but these particles are no longer iron particles.

4.3 The ultraviolet catastrophe

In 1900 M. Planck published a paper [276] that can be considered as the first major step towards a new science in which the discrete takes center stage: quantum mechanics. Some 27 years later it was already the most successful science in the mathematical description of physical phenomena at the scale of the very small, specially at the atomic and subatomic scale. Another thing is the physical interpretation of its mathematical formalism, still to be solved and still with several alternative interpretations. In my opinion, it is important to highlight a fact that is almost never emphasized: the mathematical formalism of quantum mechanics is the formalism of the infinitist mathematics of the actual infinity. Or in other words: the science of the discontinuous built with the mathematics of the continuous.

Going back to the origins, to Planck's pioneering work of 1900, it is worth remembering that a very common method of solving physical problems by means of infinitist mathematics (differential and integral calculus, for instance) consists in trying first a discrete solution in order to make discreteness tends to zero and find there (in the continuum scenario) the correct solution. This was the method M. Planck was using to solve the so called ultraviolet catastrophe, an apparently unsolvable problems in those days, at the beginning of the XX century, just in 1900. Surprisingly enough, the correct solution appeared much more before discreteness vanishes in the infinitist scenario of the continuum. What we now call Planck constant gave the correct solution at the particular value of 6.626068×10^{-34} m^2 Kg s^{-1}. The key to the solution was to consider, as Planck did, rather as an artifice of calculation, that the electromagnetic energy was discrete, with indivisible minima (quanta), so that the electromagnetic energy E could be expressed by the simple equation:

$$E = h\nu \qquad (1)$$

where h is Plack's constant and ν is the frequency of the electromagnetic radiation. Although Planck's discrete solution to the ultraviolet catastrophe was initially taken as provisional, it immediately led to the birth of quantum mechanics, the most successful science ever developed by man (as is often said about this discipline).

But as we have just indicated, quantum mechanics, the science of discreteness par excellence, the science where the indivisible and extensible minima play a fundamental role, is also made of infinitist mathematics: the mathematics of the continuum, where spacetime plays a capital role and where the indivisible and extensible minima of space and time are meaningless. This incompatibility is surely the cause of another apparently unsolvable problem: the incompatibility between quantum mechanics and the general theory of relativity. In S. Majid words [228, p 73]:

> The continuum assumption on space and time seems then to be the root of our problems in quantum gravity.

But physicists never question the formal consistency of the actual infinity, as if that consistence were a proved fact. Evidently that is not the case, otherwise the Axiom of Infinity would be unnecessary. The Hypothesis of the Actual Infinity, the belief that the infinite sets exist as complete totalities (Definition 3), is just a hypothesis. Brouwer, Poincaré or Wittgenstein, among others, rejected it.

4.4 Discrete magnitudes and functions

Although the concept of discrete magnitude needs no presentation, it is useful to give a formal definition of it in order to be able to use it in the discussions that follow in this and other chapters of the book:

Definition 14 (of Discrete Magnitudes) *A magnitude is discrete if any of its values is an integer multiple of an indivisible and invariant minimum.*

That in the year 2024 physics has not yet discovered preinertia, the most universal property of all physical objects, including photons, is described in Chapter 19 as a shame for contemporary physics. The reasons justifying that label is given there. But preinertia is not the only shameful issue for contemporary physics. Here we will see another one, and in this case the reader will be able to check it in any physics text available to him, or even on the Internet.

We already know that in physics there are discrete magnitudes with indivisible minima, for example the electromagnetic energy or the electric charge. Let M_d be one of these discrete magnitudes, and suppose that it is defined by a continuous function f of three variables x_1, x_2 and x_3, one of which, for example x_3, is a continuous variable such as time:

$$M_d = f(x_1, x_2, x_3) \qquad (2)$$

Since f includes a continuous variable, the variable x_3, its output cannot be discrete but continuous, when it should be discrete because M_d is a discrete magnitude. For the output of f to always be a discrete output, f cannot contain continuous variables. Yet physics is full of continuous functions with continuous variables that should give discontinuous, discrete, outputs: integer multiples of indivisible minima. To give a very simple example:

$$q = 4\pi\epsilon_o Ur/Q \qquad (3)$$

where q and Q are electric charges, π an irrational numerical constant, ϵ_o a physical constant (dielectric constant), U the electric potential energy, and r a continuous variable (distance). The output of equation 3 should be, but cannot be, discrete because of the assumed continuous nature of r (continuum spacetime).

There is in modern physics a multitude of cases similar to the above: functions whose output corresponds to a discrete magnitude involving continuous variables, such as space and time, which make the necessary discrete outputs impossible. And in none of these cases do physicists consider the incompatibility of the corresponding discrete outputs of the functions with the continuous nature of some of their variables. The problem is not a quantitative one, related to the accuracy of the results. It is a qualitative problem of representation; continuous functions are not the appropriate expression of physical

4.4 Discrete magnitudes and functions

phenomena that produce discrete magnitudes. To consider that they are, or that they are valid approximations, distorts the physical nature of the relationships between the physical magnitudes involved in physical phenomena. Or put in other words, it allows us to go and stay in the wrong direction in understanding the physical world.

The embarrassment noted above is the fact that for more than a century thousands of physicists have written tens of thousands (or hundreds of thousands) of functions of the type (2) and (3) just denounced here, without in any case having considered the inconsistencies between the continuous inputs of their variables and the necessary discrete outputs of their results. As if these inconsistencies were not of the slightest importance. But it is immediately clear that they are: confronting these inconsistencies would force us to look for new, more realistic ways of expressing the relationships between physical magnitudes, relationships that would surely reveal new aspects of the represented physical reality. It is this new search that is blocked by the mathematical routine, which comes to have the force of an intolerant religious creed. An inevitable consequence of the Pythagorean-Platonic extremism anchored in human science for more than twenty centuries.

Modern physics assumes the discrete nature of a certain number of essential physical magnitudes, such as electromagnetic energy, electric charge, or angular momentum of electrons, of atomic nuclei and atoms, each of them with one indivisible minimum unit (which generically could be called quantum). At the same time, and together with these discrete magnitudes, others are used which, like space and time, are considered continuous and densely ordered: between each two of their values there is another different value, which makes the existence of minimum indivisible units impossible in these continuous magnitudes. A significant problem arises here which, as far as I know, no one has considered.

Although the inevitable errors in these impossible mathematical relationships are negligible, there is a fundamental theoretical error when it is considered that these mathematical relationships between physical variables, such as the one given by the equation (3), describe in formal terms the reality of the physical world. In short, it is impossible to define in formal terms a discrete magnitude by means of indiscrete (continuous) variables. It is then immediate to prove the following results:

Theorem 13 (of the Discrete Physical Laws) *If the mathematical output of a physical law is a discrete magnitude, its definition cannot contain continuous variables.*

Proof: If the mathematical definition of a physical law contains any continuous variable, its output (result) will always be continuous, not discrete as required for a discrete magnitude. □

As is well known, ordinary matter is made up of atoms (118 in total, 92 of which are found in known nature), all of which are made up of the same three subatomic particles: electrons, protons and neutrons:

a) Electron (elementary particle):

  1.- Mass: $9.109383701 \times 10^{-31}$ kg.

  2.- Electric charge: $e^- = -1.602176634 \times 10^{-19}$ C.

b) Proton (2 quarks u y 1 quark d):

  1.- Mass: $1.672621923 \times 10^{-27}$ kg.

  2.- Electric charge: $p^+ = +1.602176634 \times 10^{-19}$ C.

c) Neutron (2 quarks d y 2 quark u):

  1.- Mass: $1,674927498 \times 10^{-27}$ kg.

  2.- Electric charge: 0 C.

where C is the unit of electric charge of the International System of Units (the charge that for one second passes through the cross section of a conductor when the electric current is one ampere). So, each subatomic particle has a certain mass and a certain electric charge, both invariant and well defined. All electric charges are integer multiples of the same elementary charge: of the electron if it is negative, or of the proton if it is positive. Therefore the electric charge is a discrete quantity. (Definition 14).

The same is true for atoms: they all have an *electronic mass* (the sum of the masses of all its electrons), a *protonic mass* (the sum of the masses of all its protons) and a *neutronic mass* (the sum of the masses of all its neutrons), which in all three cases are discrete magnitudes (Definition 14). The same is true for molecules and macroscopic objects. There are in addition atomic, molecular and macroscopic masses, which are also discrete magnitudes (after correction for the effects of electromagnetic and nuclear strong and weak interactions). Since there is a mathematical relationship between the magnetic force ($\vec{F}$), the magnetic field ($\vec{B}$), the velocity ($\vec{v}$) and the electric charge q:

$$\vec{F} = q(\vec{v} \times \vec{B}) \qquad (4)$$

all these four magnitudes must be discrete (Theorem 13). In short, electric charge, electromagnetic energy, electronic mass, protonic mass, neutronic mass, atomic, molecular and macroscopic masss, magnetic force, magnetic field and velocity, angular momentum of electrons, of atomic nuclei and of atoms, are discrete magnitudes. As it could not be otherwise, there is an enormous empirical evidence for the existence of discrete magnitudes, as in the cases just given. And taking

4.4 Discrete magnitudes and functions

into account all physical magnitudes that are mathematically directly or indirectly related to them should be as well discrete, it is inevitable to consider the following important formal consequence:

Theorem 14 (of Discrete Magnitudes) *All physical magnitudes are discrete, except those that are not mathematically related to discrete magnitudes, in which case they could be continuous.*

Proof: There are discrete magnitudes, such as electric charge or electromagnetic energy. Therefore, all physical magnitudes that are mathematically related to them must also be discrete (Theorem 13 of the Discrete Physical Laws). And the same is true for all physical quantities that are mathematically related to the latter. And the same is true for all physical quantities that are mathematically related to the latter. And so on. □

Comment: taking into account the discrete magnitudes already pointed out and those which must be discrete according to Theorem 14, one could state that all physical magnitudes are discrete. A conclusion that agrees with the finite results demonstrated in Chapter 3. But, as we will see below, other physico-mathematical problems remain to be solved.

Protons are formed by two *up quarks* and one *down quark*, while neutrons are formed by one *up quark* and two *down quarks*. Although in nature quarks always appear forming protons and neutrons (hadrons), we can speculate about their electric charges for the sole purpose of highlighting the problems that immediately follow from infinitist arithmetic. If we represent the electric charge of an up quark by q_u and that of a down quark by q_d, we can write:

$$\left.\begin{array}{l}\text{neutron } n^o : \quad 2q_d + 1q_u = 0 \\ \text{proton } p^+ : \quad 2q_u + 1q_d = e^+\end{array}\right\} \quad (5)$$

System of equations whose immediate solution leads to the corresponding electric charges of up quark and down quark:

$$q_d = -\frac{1}{3} e^+ \quad (6)$$

$$q_u = \frac{2}{3} e^+ \quad (7)$$

But it happens that 1/3 and 2/3 are rational numbers with an infinite number of ω-ordered decimal places:

$$\frac{1}{3} = 0.33333\ldots \qquad \frac{2}{3} = 0.66666\ldots \tag{8}$$

and if one considers as a complete totality the infinite numbers of each of them, which is what it corresponds to do according to the Axiom of Infinity, then the consistency of those numbers is bound up with the consistency of that axiom. And we have already seen in this book that this axiom is inconsistent (Theorem 6 of the Inconsistent Axiom of Infinity, page 21). One would then have to decide and justify the number of digits to take in the above definitions of q_u and q_d. This and many other problems related to the existence of different quanta of different physical magnitudes will have to be posed and solved in the framework of a new finitist and discrete arithmetic. The last section of this chapter deals, albeit very briefly, with this possible new arithmetic.

4.5 Some discrete conclusions

The above Theoem 14of the Discrete Magnitudes is anything but irrelevant. Among many others, the following fundamental results are immediately drawn from it:

Theorem 15 (of Discrete Space and Time) *Space and time magnitudes are discrete, each with an indivisible and invariant minimum.*

Proof: Since space and time magnitudes, as distances or durations, are involved in the definition of discrete physical magnitudes, according to Theorem 14 they must be discrete entities, each with an indivisible and invariant minimum (Definition 14), otherwise the defined discrete magnitudes could not be discrete. □

In this book, and from now on, the minimum, indivisible and invariant unit of space will be called qusit (quantum space unit) and the minimum, indivisible and invariant unit of time will be called qutit (quantum time unit).

Theorem 16 (of Discrete the Threshold) *The laws of physics do not apply in spaces smaller than the minimum unit of space nor in times smaller than the minimum unit of time, both being of non-zero extension (duration).*

Proof: If the laws of physics could be applied to intervals of space smaller than the minimum unit of space, then that minimum unit it would be neither invariant nor indivisible, which is impossible (Theorem 15). The same argument holds for time. □

Although Theorem 16 of the Discrete Threshold has not been explicitly stated in contemporary physics, its statement has broad theoretical and empirical support. As will be seen throughout the pages of this book, it is a fundamental result for the construction of discrete models of the universe.

Corollary 14 (of the Physical Laws) *The laws of physics apply to all regions of space and time, provided they are not less than their respective indivisible minimum units.*

Proof: It is an immediate consequence of Theorem 16. □

Theorem 17 (of Adjacency) *No space exists between any two successive space minimum units, and no time elapses between two successive time minimum units.*

Proof: Let AB and CD be two successive space minimum units (simplified to a one dimensional version) and assume they are not adjacent, i.e assume that $0 < BC$. BC must be less than the space minimum unit, otherwise AB and CD would not be two successive space minimum units. Consequently, AD would not be an integer multiple of the space minimum units, which is impossible according to Definition 14 and Theorem 15. □

Other important consequences of Theorem 14 of the Discrete Magnitudes will be deduced in the following chapter.

4.6 Discrete arithmetic

Although related to the modern spacetime continuum, the problem of the continuous has a Pythagorean origin [231]. In my opinion, its importance in the history of science has not been sufficiently appreciated. The firsts Pythagorean believe in the existence of indivisible geometrical points with an extension δ greater than zero, consequently they believed that all lengths would have to be commensurable: the ratio between any two of these lengths, say L_1 and L_2, would be a ratio between two natural numbers [231, pp. 11-16]:

$$L_1 = n_1\delta; \quad L_2 = n_2\delta \tag{9}$$

$$\frac{L_1}{L_2} = \frac{n_1\delta}{n_2\delta} = \frac{n_1}{n_2} \tag{10}$$

Somewhat later, the Pythagorean discovered the existence of non commensurable lengths: the length of the diagonal L_d of a square with the length of its side. For example, if the length of the side is 9δ, we would have:

$$L_d = \sqrt{9^2\delta^2 + 9^2\delta^2} \tag{11}$$

$$= 9\delta\sqrt{2} \tag{12}$$

$$\frac{L_d}{L_s} = \frac{9\delta\sqrt{2}}{9\delta} = \sqrt{2} \tag{13}$$

Unfortunately, they did not consider the possibility of a discrete arithmetic, for instance:

$$L_d = \lfloor \sqrt{9^2\delta^2 + 9^2\delta^2} \rfloor \tag{14}$$

$$L_d = \delta \lfloor \sqrt{9^2 + 9^2} \rfloor \tag{15}$$

$$= 9\delta \lfloor \sqrt{2} \rfloor \tag{16}$$

$$= 9\delta \tag{17}$$

$$\frac{L_d}{L_s} = \frac{9\delta}{9\delta} = 1 \tag{18}$$

where $\lfloor$mathematical expression$\rfloor$ stands for the integer part of the mathematical expression. As we will see in Chapter 16 equations (14)-(18) represent the discrete version of Pythagoras theorem.

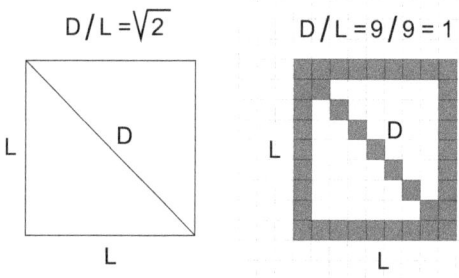

Figure 4.1 – Left: In continuous geometry the diagonal D and the side L of a square are not commensurable. Right: In discrete geometry the diagonal D and the side L of a square are commensurable.

On the other hand, although in the same direction of the discrete, the existence of different discrete magnitudes with their corresponding indivisible minima (quanta) raises new arithmetical problems that for the moment, and for the shameful reason given above, are completely ignored. But we will have to consider them and find the way to express those relations, and in the search for those relations it would not be strange that we would find new aspects or details of physical reality. If, for example, three discrete magnitudes M_1, M_2 and M_3, whose respective quanta are q_1, q_2 and q_3, are related in the form:

$$M_1 = M_2 M_3 \tag{19}$$

It should be accomplished:

$$n_1 q_1 = n_2 q_2\, n_3 q_3; \quad n_1, n_2, n_3 \in \mathbb{N} \tag{20}$$

4.6 Discrete arithmetic

$$q_1 = \frac{n_2 n_3}{n_1} q_2 q_3 \qquad (21)$$

$$q_2 = \frac{n_1}{n_2 n_3} \frac{q_1}{q_3} \qquad (22)$$

Consequently, the rational numbers $n_2 n_3 / n_1$ and $n_1 / n_2 n_3$ could only have a finite number of decimal places. This is just one example of the kind of problems that would have to be solved in a discrete arithmetic that correctly expresses the relationships between the different quanta of the different discrete physical magnitudes. In fact, and according to the inconsistency of the actual infinity, all physical magnitudes should be discrete if they are formally consistent. This type of discrete arithmetic is yet to be developed in formal and universal terms. The fact that its necessity has not even been raised is surely related to our perception of the physical world as essentially continuous, not discontinuous.

Indeed, it seems reasonable to assume that we model reality as a continuous system because we perceive it as a continuous system. The problem is that this perceived continuity is illusory. In fact, our brain takes a time greater than zero ($\approx$13 ms [282]) to process each visual image (the base of the well known α, β, γ and δ movements, and of ϕ-phenomenon [102]), so that *a continuum* of visual images is physiologically impossible. The same illusory perception happens with motion when observed in a film. And in the same way a film is a discontinuous sequence of photograms, natural motion could also be a discontinuous sequence of changes in position, which is perceived as continuous by our brains and our physical instruments.

5. Infinite regress

5.1 Introduction

This chapter introduces an inductive principle (the Principle of Directional Evolution) from which several formal results about the consistent nature of the observable universe are proved, including the Theorem of the Consistent Universe, the Theorem of Formal Dependence and the Theorem of the First Element, all of which are fundamental to our own analysis of physical space to be developed from the next Chapter onward. The most important result demonstrated in this chapter is undoubtedly the existence of FIRST CAUSES that cannot be explained in terms of other causes deduced from our knowledge of the observable universe. Or in other words, the existence of first causes in all physical phenomena; causes that cannot be explained in terms of physical phenomena; causes that cannot be explained by human understanding. A logical result that goes far beyond the content of this book.

5.2 The Aristotelian infinite regress

The Aristotelian infinite regress of arguments says that since statements do not prove themselves, in order to prove a statement you will always need at least one other statement. And the same goes for that last statement(s). An inevitable indefinite regression of statements appears, an endless regress of statements that makes it impossible to complete the demonstration of the veracity of any initial statement. This is the reason why all sciences need basic statements whose truth is admitted without proof; these basic statements are the axioms, postulates and principles or fundamental laws. In some cases these primitive statements are obvious, in others they are arbitrary and not obvious at all, and in others they have more or less inductive confirmation.

The infinite regress of statements (arguments) extends to definitions and causes for the same reasons. Although rarely considered, the infinite regress of arguments, whether extended to definitions and causes or not, is a serious limitation of human knowledge, much more serious than Gödel's famous incompleteness theorems. But at the same time, and also considering the inconsistency of the actual infinity and infinite divisibility, it suggests the way we should go to explain the physical world, by clearly stating what can and cannot be explained in formal terms. In particular, it is proved here that every history either has an initial and unexplained instant or is inconsistent.

Although an infinite regress is any infinite sequence of elements that are recursively related (each element in the sequence is related in some, and always the same, way to its immediate predecessor in the sequence), we are interested here only in the particular cases of demonstrations, definitions, and causes. The case of demonstrations was treated by Aristotle [13, I.3], and it will be the case with which we begin our discussion, also making use of the famous Münchhausen Trilemma. The discussion will then be extended to definitions and causes.

Making use of the inconsistency of the actual infinity, the need for primitive concepts, axioms, fundamental laws, and inductive principles will be demonstrated. The Theorem of the First Element will also be proved. The common thread of all these demonstrations will be the Theorem 20 of Formal Dependence, directly deduced from the Principle 1 of the Directional Evolution. That theorem establishes that statements do not prove themselves; concepts do not define themselves; and objects and causes are not the cause of themselves. Obviously, the alternative would be an inevitable collection of nonsense incompatible with science.

As indicated above, the infinite regress of arguments was discovered more than twenty-two centuries ago. And for the reasons discussed below, it is a serious limitation of human knowledge. However, most contemporary scientists ignore it. For example, no science has established the list of its primitive concepts. The curious thing is that at the same time that modern science ignores these logically unavoidable limitations, it pays exaggerated attention to other, much less general limitations, almost always related to the contradictory self-reference [201].

In the particular case of physicists (experimental and theoretical), it is rare to see them concerned with these formal limitations, as if these limitations did not limit anything. This is not the best attitude, because until the problems posed by such limitations are resolved, physics cannot be built on adequate foundations. Experimental data drive and force the adjustments of physical theories, but (at least so far) do not determine all of their formal foundations.

Even more dramatic is the attitude of physics toward its most fundamental problem: the problem of change (physics is essentially the science of change). Raised by Parmenides and Zeno of Elea, no one has been able to explain how a simple change of position of a material object occurs. Physics has completely forgotten that it has this problem. Under these conditions, disagreements, even incompatibilities, between some physical theories are not strange.

5.3 The universe is consistent

The observable universe contains billions and billions of objects of the same type: galaxies, stars, planets, minerals, chemical elements... evenly distributed and of very different ages. This is only possible if the same things have always happened in the universe, and in such a way that the same consequences have always occurred under the same conditions. Something that has been suspected for at least a couple of centuries: the naturalists of the 19th century already assumed the Principle of Actualism-Uniformism, which suggests the same conclusion:

The laws of nature are the same in all places and times.

Even this general principle can be deduced in formal terms from another even more general principle. For, in effect, the observable universe has been producing the same type of objects throughout its history (over 13.8 billion years): there are planets, stars, galaxies etc. OF ALL AGES. And all these objects have been EVOLVING IN THE SAME WAY. In addition, there is a law of thermodynamics (though with numerous versions), the *Second Law* of thermodynamics, that also points in the same direction regarding the evolution of the heat-energy interconnections. All this, and taking into account that for most of its history the universe had no rational observers amply justifies proposing the following:

Principle 1 (of Directional Evolution) *The observable universe always evolves independently of its rational observers and in the same direction of increasing its global entropy.*

where entropy can be replaced with isotropy [202]. This primordial directionality has made possible other local directional evolutions that seem to go in the opposite direction: the creation and evolution of open systems that exchange matter and energy with their surroundings. In these systems there is a remarkable decrease in isotropy, but in return there is an even more remarkable increase in the isotropy of their surroundings, so that the final balance is a directional evolution of the universe in the sense of the above Principle of Directional Evolution. This is the case for crystalline minerals and for all self-organizing systems we call living beings [192, 193]. It could be said that the isotropic evolution of the universe produces strongly anisotropic residues, among which you, kind reader, find yourself (Figure 5.1).

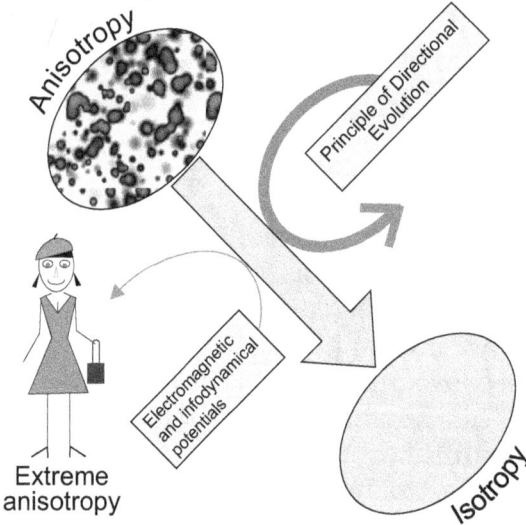

Figure 5.1 – Anisotropy as a residual product of the isotropic evolution of the universe.

As will be seen below, from the Principle of Directional Evolution we immediately deduce some fundamental results that detail the way in which the universe evolves. To begin with, the following definition is proposed:

Definition 15 (of the Consistent Set of Laws) *A set of physical laws is consistent if under the same conditions it always leads to the same results.*

It is now immediate to prove the following:

Theorem 18 (of the Consistent Universe) *The universe evolves under the control of a unique set of invariant and consistent physical laws.*

Proof: If the physical laws governing the evolution of the universe were not an invariable set of consistent laws, changes would occur with equal frequency in all directions, and no progress would be possible in any of them. Thus, directional evolution would not be possible, which violates the Principle 1 of Directional Evolution. Thus, the universe evolves under the control of a unique set of invariant and consistent physical laws. □

Note: The Theorem of the Consistent Universe could have been chosen as an inductive principle, and from it the Principle of Directional Evolution could be deduced as a theorem. There is overwhelming empirical evidence for both, and in fact their statements are mutually reinforcing inductively and formally. However, the alternative of the Consistent Universe Principle would have to be extended to the past (actually it would be the geological Principle of Actualism-Uniformitarianism) and for that we no longer have the same empirical evidence. On the contrary, we have been able to confirm on many occasions the existence in the observable (and observed) universe of the same objects with different ages, as well as very complex objects, such as the possible reader of this text, whose formation requires millions of years of directional evolution.

Corollary 15 (of Physical Laws) *The laws of physics apply to all regions of space and time.*

Proof: It is an immediate consequence of Theorem 18. □

Theorem 19 (of Identicality) *All particles of the same type have the same properties and behave the same way under the same conditions.*

Proof: It is an immediate consequence of Definition 15, Theorem 18 and Corollary 15. □

Comment: Evidently, this theorem goes against the Principle of the Identity of Indiscernibles, of G.W. Leibniz. Real identical objects exist and are distinguishable from each other because they do not occupy the same places in space, which requires (contrary to Leibniz's view) a real physical space.

Theorem 20 (of Formal Dependence) *No concept defines itself; no statement proves itself; no physical object is the cause of itself; and no cause is the cause of itself.*

Proof: If propositions proved themselves, then any proposition could prove itself, and consistent sets of laws would be impossible, which goes against Theorem 18. If concepts defined themselves, their meanings would be inaccessible to human knowledge, and they could not be used to establish the natural laws that we can only establish with those concepts, which also goes against Theorem 18. If physical objects, processes or causes were the cause of themselves, then anything could exist and anything could happen, so that the directional evolution of the universe would be impossible,which goes against the Principle of Directional Evolution. □

Then it is clear that in our observable universe, under the same conditions, the same results will occur. Another thing is that these results may seem strange or paradoxical to us. In that case, the corresponding laws will always produce the same strangeness or perplexity, as is the case with some aspects of quantum mechanics. Moreover, if the universe worked in a way similar to cellular automata, perhaps there could be two types of laws: the basic laws of the automaton (established before the automaton is started) and the laws that emerge from its evolution, laws that drive the relationships between the also emerging objects of the automaton. (see Chapter 18)

So far, we have not known any exception to the Principle of Directional Evolution of the Universe. Henceforth, and for the sake of simplicity, the directional evolution of the observable universe will simply be referred to as the evolution of the universe. This is the reason why the universe can be described by formal and computational languages. Whether the current infinitist mathematical language of physics is the most appropriate is another matter. And certainly it is not because it is based on an inconsistent axiom (the Axiom of Infinity). Unnecessary as it may seem, the consistent evolution of the universe is of interest beyond evolution itself. For example, it makes a Principle of Discrete Relativity unnecessary:

Theorem 21 (of the Reference Frames) *The laws of physics are the same in all discrete reference frames.*

Proof: Let RF_1 and RF_2 be any two discrete reference frames in relative motion with respect to each other. Their relative velocity will be a consequence of their different absolute velocities with respect to the absolute reference frame of the discrete space and time. Therefore, the observations made with respect to RF_1 and RF_2 can be modified in both cases by considering the absolute velocity of each system with respect to the absolute reference frame of the discrete space and time. By doing so, the same laws will be obtained in both systems, otherwise the Theorem 18 of the Consistent Universe Theorem would not be verified, and therefore neither would the Principle of Directional Evolution. □

Comment The problem is that for now, and due to preinertia (see Chapter 19), it is not possible to observe the absolute velocity of a discrete reference frame. Unless it is possible to do so by referring its motion to the isotropic reference frame of the Cosmic Microwave Background.

In reality, and as incredible as it may seem, the Theorem of the Consistent Universe is absolutely necessary. To understand this necessity, it is enough to remember a very famous phrase of A. Einstein [99, p. 315]:

> The eternal incomprehensible of the world is its comprehensibility

Popularized as:

> The fact that the universe is understandable is a miracle.

But, for the reasons given above, a universe that produces scientists (through a process that here on Earth has lasted several billion years) can only be a universe that evolves in a directional way and, consequently, under the control of a single set of invariant and formally consistent laws. That is, a universe understandable in terms of constant and consistent regularities. The miracle would be, then, that a universe with scientists were not understandable. Another thing is that the mathematical language used to explain the universe is appropriate. In this book I am showing that it is not. This inappropriateness would explain why it is so difficult to understand what should not be so difficult to understand.

5.4 Münchhausen Theorem

Also known as Agrippan Trilemma, the well-known Münchhausen Trilemma is an argument that tries to demonstrate the impossibility of proving the truth of any statement without making use of an arbitrary initial statement. According to this argument, a truth can only be proved by means of:

1. an infinite regress of proofs.
2. a first arbitrary statement.
3. a circular sequence of proofs.

the three of which are formally unsatisfactory. As we will see now, things do not improve with the Theorem 20 of Formal Dependence and the Theorem of the First Element 23. Arguably they get worse because they become a theorem whose consequences are the same as the Münchhausen Trilemma. Here, we will deal only with infinite regress of proofs, definitions, and causes, in relation to which it is immediate to demonstrate the following:

Theorem 22 (of Incompletable Regress) *Every recursive sequence S of proofs, definitions or causes in which there is a last element to be proved (defined, caused) and each element has an immediate predecessor that proves (defines or causes) it, is incompletable.*

Proof: If every element of the sequence has an immediate predecessor, then there is not a first element of the sequence, because this first element would have no immediate predecessor. Therefore, the sequence, if consistent, can only be potentially infinite and then incompletable (Corollary 6). □

In addition, the third option of the Münchhausen Trilemma would be inconsistent because there would be at least one element that proves (defines or is the cause) of itself: For example, if A is the cause of B, which is the cause of C which is the cause of A, the C is the cause of C, which goes against the Theorem of Formal Dependence 20. Therefore, and taking into account the above Theorem of the Incompletable Regress, only the second alternative of the Münchhausen Trilemma could be considered as part of the logical system for the explanation of

5.5 Theorem of the First Element

As mentioned above, an infinite regress is a sequence of recursively related elements such that each element has the same type of relationship to its immediate predecessor. In our case, the elements of the sequence will always be formal elements: arguments, definitions, and causes. And their relations will always be the same formal relation: arguments in the infinite regress of arguments; definitions in the infinite regress of definitions; and causes in the infinite regress of causes.

We will now prove a general result that applies immediately to infinite regresses:

Theorem 23 (of the First Element) *A consistent sequence in which there is a last element and each element has an immediate predecessor is a complete totality only if it has a first arbitrary element without predecessors.*

Proof: Let $S = \ldots S_{3*}, S_{2*}, S_{1*}$ be any sequence with a last element S_{1*} and in which each element S_{n*} has an immediate predecessor $S_{(n+1)*}$, where n^* read last but $n-1$. If S is consistent it can only be finite or potentially infinite (Corollary 1). Therefore, if S is a complete totality it can only have a finite number n of elements (Definition 6). In these conditions, and taking into account that each element S_i of S has exactly one predecessor more than its immediate predecessor $S_{(i+1)*}$, the element S_{1*} has $n-1$, predecessors; the element S_{2*} has $n-2$ predecessors; the element S_{3*} has $n-3$ predecessors; etc. Consequently, the smallest number of predecessors that an element of S can have is $n-(n-1)=1$. That element will be $S_{(n-1)*}$ whose predecessor can only be a first element S_{n*} of the sequence that has no predecessor. So, S has a first element S_{n*} with zero predecessors. □

5.6 Infinite regress of proofs

An immediate, and well known, consequence of Theorems 22 and 23 is that all formal sciences must be founded on a set of axioms, i.e. a set of statements whose veracity is assumed without proof (see next section for the role of definitions and primitive concepts in the foundations of formal sciences). Ideally, the number of axioms should be small and as self-evident as possible (otherwise, they would give rise to an excessively arbitrary and abstract science).

As is well known, classical Euclidean geometry was founded on five geometrical axioms, the fifth of which is the controversial axiom of

5.7 Infinite regress of definitions

parallels, whose statement is anything but self-evident. By contrast, Playfair, Hilbert and the author of this book founded their Euclidean geometries respectively on 30, 20 and 10 axioms (see Appendix C). As expected, the initial set of axioms will have important consequences on the resulting of science [196, 200, 279, 158]. For example, Euclidean and non-Euclidean geometries.

For its part, set theory is based on about ten axioms (depending on the version). One of them is the Axiom of Infinity. Assuming this axiom has enormous consequences, not only in mathematics but also, and above all, in physics, a science that pays little attention to the foundations of its mathematical language, and considers that this language is not an instrument for analyzing observations but a model to which observations must be adapted.

Since the beginning of the 20th century, physics has been built up with a type of mathematics (the infinitist mathematics of the space-time continuum) that assumes the existence of the complete list of the natural numbers in their natural order of precedence (which is a way of stating the Axiom of Infinity), even though there is no last natural number completing the list. They assume, as Aristotle would say, that the incompletable exists as complete.

The formal proofs of the inconsistency of theAxiom of Infinity has been available for more than twenty years, but it is taking too much effort to fight against the infinitist stream that assumes that axiom. An stream of thought absolutely dominant and hostile to dissidence. But in the end, well-constructed proofs will end up imposing their conclusions. The reader will be able to judge one of those proofs in Chapter 3 (Theorem 1) in this book, and forty others in [212, online link]. The consequences on physics will be enormous, and very positive. This book has a lot to do with that finitist and discrete future. Indeed, the inconsistency of the actual infinity will change all.

Experimental sciences have an additional element for their corresponding foundations: the inductive principles or laws. Statements whose veracity is accepted without formal proofs (such as axioms) but confirmed by experimentation and by observation of natural phenomena. This is the case, for example, of the Principle of Inertia, including preinertia. But here, too, it is important to be careful. In this sense, it is convenient to recall Russell's famous metaphor of the chicks [307, p. 31]): the innocent animals who lived happily on the farm in the care of their attentive farmer without suspecting the existence of fried chicken with potatoes. Fortunately, humans are, in general, smarter than chicks and we have discovered that it is wise to be cautious when drawing inductive conclusions about the physical world.

5.7 Infinite regress of definitions

Although the infinite regress has always been discussed for the case of arguments (demonstrations), its extension to other formal elements such as definitions and causes is immediate. And the consequences of Principle 20 and Theorem 23 on these new formal elements are also immediate.

Since concepts are not self-defining (Theorem 20), to define any concept it is necessary to use one or more different concepts; and the same goes for the latter. In this way a recursive sequence of definitions appears to which Theorem 23 can be applied, with the same consequences as in the case of axioms:

Corollary 16 (of Primitive Concepts) *Primitive concepts are inevitable in all sciences and languages.*

Proof: It is an immediate consequence of Theorems 20 and 23. □

Of course, the defined and undefined objects must be legitimized by the axioms or principles of the corresponding theories, or by formal proofs. This is usually the case in the formal sciences, but not always in the experimental sciences. Even in theoretical physics or theoretical biology, which are more formal and mathematical than their corresponding experimental branches. The problem is that when a science pays little attention to its fundamental basis, anything can happen.

The most basic concepts of science, such as set, number, point, force, mass, time, etc., are primitive concepts. Most of them are intuitive: We know what they are even though we have no formal definition of them. This is the case, for example, with set, number, mass, or force. In other cases we may have a false intuition. I think that's the case with point and instant. The intuition we have of point is confused with that of the mark on the paper or board that is trying to represent it. And the intuition of instant is confused with very short intervals of time. And neither is the case.

The concept of point is primitive and fundamental in physics: space would consist of points without extension, points that do not occupy places but define places; and there would be point masses, point charges, point particles, point trajectories, and so on. And as if that were not enough, there are as many points on a line one trillionth of a millimeter long as there are in the entire three-dimensional universe. Obviously, the same could be said of the concept of instant: the same number of them elapse in a microsecond as in the entire history of the universe. This is Cantorian Infinitism!

The effort to define objects, and to establish the axioms and proofs that justify them, may not always have been adequate. Of course, definitions must be formally productive: once legitimized by the axioms, they must be usable in subsequent proofs. A very notorious case is that of Euclidean geometry: it is possible to define a new foundational basis with 29 productive definitions and 10 axioms, in which it is possible to prove, like any other theorem, the statement of the Euclidean axiom of parallels [196, 200].

It is worth clarifying this issue further. The concepts of point, line, and straight line are primitive. So a straight line (a central concept in Euclidean geometry) is something for which we have no formal definition; a straight line belongs to a class of objects (lines) for which we also have no definition; and a line is made up of points for which we also have no definition. Perhaps too many *indefinitions*.

It is possible, however, to give a formal definition of a straight line (although the concepts of point and line remain primitive). It is a formally productive definition that, together with the rest of the new foundational elements of Euclidean geometry, allows us to demonstrate the Euclidean statement of parallels [196, 200, p. 40].

5.8 Infinite regress of causes

In accordance with Theorem 20 of Formal Dependence, physical (and formal) objects and natural phenomena are not self-causing, i.e. they cannot be the cause of themselves. In consequence Theorems 23 and 22 apply to them. We must, therefore, accept the following:

Corollary 17 (of the First Cause) *To explain any physical object or phenomenon, a first cause not explainable in terms of other causes deduced from our knowledge of the observable universe is necessary.*

Proof: It is an immediate consequence of Theorems 20 and 23. □

Taking into account the above results, and those demonstrated in the previous Chapter 4, the following results, which are anything but irrelevant, can be demonstrated.

Theorem 24 (of the Finite Universe) *The universe is finite in extent, age and number of components.*

Proof: The universe is consistent (Theorem 18 of the Consistent Universe). In consequence the number of its space discrete units (qusits) can only be finite. Therefore the spacial extension of the universe is finite. The same applies to the number of time discrete units (qutits) and to number of its physical components. Therefore, the age of the universe and the number of its physical components can only be finite. □

Theorem 25 (of the Origin of the Universe) *The universe must have had an origin whose cause is unknowable.*

Proof: The number of qutits of the observable universe is finite (Theorem 24), therefore, and taking into account that this number grows towards the future and decreases towards the past, there had to be a first qutit in the history of the universe. Therefore, the universe must have had an origin. Moreover, the universe as an object can only be explained by a first cause unknowable in terms of other causes deduced from our knowledge of the observable universe (Corollary 17 of the First Cause). □

Theorem 26 (of the Discrete Universe) *The universe has evolved through a finite and discrete sequence of qutits.*

Proof: The universe has evolved (Principle 1 of Directional Evolution) along a finite sequence of qutits in which there is a first qutit (Theorem 25 of the Origin of the Universe), a last qutit (the current qutit), and each qutit has an immediate predecessor, except the first one, and an immediate successor, except the last one. Therefore, that sequence is discrete (Definition 13) and finite (Theorem 9 of the Discrete Sets). □

As noted above, the consequences of the above theorems and corollaries on the universe are anything but irrelevant. In any case, do not forget that science should be free of prejudices, free even of religious and anti-religious prejudices. And that, on the other hand, without the necessary formal rigor, language cannot be scientific. Without the rigorous use of language, anything can be demonstrated. To affirm, for example, that the universe arose from a fluctuation of nothingness, implies that nothingness is not nothingness but something with the capacity to fluctuate universes. As T. Maudlin would surely say, the importance of the following conclusion of Corollary 17 of the First Cause cannot be exaggerated: The evolution of the universe, as such a natural process, must also have a first cause outside the evolution of the universe itself; a first cause that cannot be explained in terms of other causes deduced from our knowledge of the observable universe, i.e. a first cause that cannot be explained in physical or logical terms.

6. The Formal Scenario

6.1 Introduction

Although all the formal results obtained up to this chapter can be considered as elements of the formal setting for the discussions that follow in the following chapters, not all of them have the same relevance because some of them are intermediate results that are necessary in the proofs of the fundamental results that do need to be taken into account in those discussions. This short chapter is a simple listing of all the formal statements established or proved up to this chapter. Most of those results have to do with infinity and the geometry of space, and are dissenting with the dominant infinitist current in contemporary mathematics and physics. They are gathered here to facilitate their access to the reader.

6.2 Formal elements for a new theory of space and time

The 52 formal elements listed below (1 principle, 13 definitions, 25 theorems and 16 corollaries) were established and, where appropriate, proved in the previous three chapters. Their content can be summarized as follows:

> If one accepts the Principle of the Directional Evolution of the Universe, or put very informally: if one assumes that the gas released from a bottle of champagne when the bottle is uncorked will never spontaneously and naturally return to the uncorked bottle, then the universe is a consistent object that evolves according to a consistent set of laws. Furthermore, it had to have an origin whose cause cannot be established from knowledge drawn from within the universe itself. The age, extent, and number of its components, at any scale (from subatomic particles to galaxies), must be finite. None of its components, including space and time, can be divided into an infinite number of parts or units. Prac-

tically all physical magnitudes have to be discrete, with indivisible minimal units. And the number of universes (if more than one exists), and the number of cycles of possible creations-destructions of the universe must also be finite. Consequently, the universe must be finite and discrete, and it must have had an origin whose cause is not cognizable by science constructed from within the universe itself. It could also be the case that the entire observable universe is a hoax, or a joke by something or someone, in which case there would be no point in continuing to do science.

It is a relate extracted from the following list of formal elements:

Definition 4 of Infinite Set: A set is said infinite if it can be put into a one to one correspondence with one of its proper subsets. (Page 12).

Definition 3 of Complete Totality: A complete totality is a set defined by comprehension in which every element that should be in the set, is in the set. (Page 11).

Definition 5 of Successors and Predecessors: In strictly ordered sets, all elements that, in the ordering of the set, follow (precede) a given element of the set, are its successors (predecessors). If between the given element and one of its successors (predecessors) there is no other element, then this successor (predecessor) is the immediate successor (predecessor) of the given element. (Page 12).

Definition 6 of Actual and Potential Infinity: An ordered collection of elements is infinite if there is no last (first) element that completes (initiates) it. The collection is actual infinite if it is considered a complete totality, and potential infinity if it is not considered a complete totality. (Page 15).

Axiom 1 of Infinity: There exists at least one infinite set. (Page 17).

Definition 7 of the Types of Sets: A set is finite if it has a definite and finite number of elements. A set of elements of a certain type is potentially infinite if it always contains a finite number of elements of that type and any finite numbers of new elements of that type can always be added to it, without the set ceasing to be finite and without it being necessary to change its name. (Page 18).

Definition 8 of the Types of Infinities: The actual infinity is the infinity of the infinite sets. The potential infinity is the infinity of the potentially infinite sets. (Page 18).

Definition 9 of Inconsistent Set: A set is inconsistent if a contradiction can be deduced from the number of its elements, or from the number of elements of at least one of its proper subsets. (Page 18).

Corollary 1 of Inconsistent Sets: A set with the same number of

6.2 Formal elements for a new theory of space and time

elements as an inconsistent set, is also inconsistent. (Page 18).

Definition 10 of Denumerable Set: A set is denumerable if its cardinal is the smallest infinite cardinal $\aleph_0$ of the infinite set of all natural numbers. An infinite set is non-denumerable if its cardinal is greater than the smallest infinite cardinal $\aleph_0$. (Page 18).

Definition 11 of ω-Ordered Sets: A set is ω-ordered if being denumerable, it has a first element, each element has an immediate successor and an immediate predecessor, except the first one which has no predecessor. (Page 18).

Theorem 6 of the Axiom of Infinity: The infinity subsumed in the Axiom of Infinity can only be the actual infinity. (Page 21).

Theorem 2 of Denumerable Sets: It is always possible to define a one-to-one correspondence between any two denumerable sets. (Page 18).

Theorem 3 of Non-Denumerable Sets: Every non-denumerable set has denumerable proper subsets. (Page 19).

Theorem 4 of Indexation: The elements of a denumerable set can be reordered with the same order as the elements of any other denumerable set. (Page 19).

Theorem 5 of the Denumerable Infinity: All denumerable sets are inconsistent. (Page 19).

Corollary 2 of ω-Ordered: All ω-ordered sets are inconsistent. (Page 21).

Corollary 4 of the Inconsistent Infinite Sets: All actual infinite sets are inconsistent. (Page 22).

Corollary 6 of the Inconsistent Axiom of Infinity: The axiom of infinity is inconsistent. (Page 21).

Theorem 3 of the Actual Infinity: The actual infinity is inconsistent. (Page 22).

Corollary 5 of Infinite Divisibility: The actual infinite divisibility of any formal or physical object is inconsistent. (Page 22).

Corollary 6 of Consistent Collections: A set can be either a finite complete totality or a potentially infinite and incompletable totality. Otherwise it is inconsistent. (Page 22).

Definition 12 of Densely Ordered: If no element of a strictly ordered set has an immediate predecessor nor an immediate successor, the set is said to be densely ordered or to define a continuum. (Page 22).

Theorem 7 of the Inconsistent Dense Order: Densely ordered sets are inconsistent. (Page 22).

Corollary 7 of the inconsistency of Q and R: When considered as complete infinite totalities, the set $\mathbb{Q}$ of the rational numbers and the set $\mathbb{R}$ of the real numbers are both inconsistent. (Page 23).

Theorem 8 of the Inconsistent Continuum: The spacetime continuum is inconsistent. (Page 23).

Definition 13 of Discrete Sets: A set is discrete if it has a first element, a last element and each of its elements (except the first one) has an immediate predecessor and (except the last one) an immediate successor. (Page 23).

Theorem 9 of the Discrete Sets: All discrete sets are finite. (Page 23).

Theorem 10 of the Strictly Ordered Sets: Every strictly ordered set is discrete. (Page 23).

Theorem 11 of Discrete Sets: Every set is either discrete or discretely orderable. (Page 24).

Corollary 8 of the Eternal Universe: A consistent universe cannot be eternal. (Page 25).

Corollary 9 of the Finite Number of Universes: In a consistent reality only a finite number of universes could exist. (Page 25).

Corollary 10 of the Finite Number of Cycles: In a consistent reality there can only be a finite number of cycles of creation-destruction of universes. (Page 25).

Corollary 12 of the Finite Universe: A consistent universe cannot contains an actual infinite number of physical objects. (Page 26).

Corollary 13 of the Finite Mass-Energy: The mass and the energy of the observable universe cannot be actually infinite. (Page 26).

Definition 14 of Discrete Magnitudes: A magnitude is discrete if any of its values is an integer multiple of an indivisible and invariant minimum. (Page 30).

Theorem 13 of the Discrete Physical Laws: If the mathematical output of a physical law is a discrete magnitude, its definition cannot contain continuous variables. (Page 31).

Theorem 14 of the Discrete Magnitudes: All physical magnitudes are discrete, except those that are not mathematically related to discrete magnitudes, in which case they could be continuous. (Page 33).

Theorem 15 of the Discrete Space and Time: Space and time magnitudes are discrete, each with an indivisible and invariant minimum. (Page 34).

Theorem 16 of Discrete the Threshold: The laws of physics do not apply in spaces smaller than the minimum unit of space nor in times

6.2 Formal elements for a new theory of space and time

smaller than the minimum unit of time, both being of non-zero extension (duration). (Page 34).

Corollary 14 of the Physical Laws: The laws of physics apply to all regions of space and time, provided they are not less than their respective indivisible minimum units. (Page 35).

Theorem 17 of Adjacency: No space exists between any two successive space minimum units, and no time elapses between two successive time minimum units. (Page 35).

Principle 1 of Directional Evolution: The observable universe always evolves independently of its rational observers and in the same direction of increasing its global entropy. (Page 41).

Definition 15 of the Consistent Set of Laws: A set of physical laws is consistent if under the same conditions it always leads to the same results. (Page 42).

Theorem 18 of the Consistent Universe: The universe evolves under the control of a unique set of invariant and consistent physical laws. (Page 42).

Corollary 15 of the Physical Laws: The laws of physics apply to all regions of space and time. (Page 43).

Theorem 19 of Identicality: All particles of the same type have the same properties and behave the same way under the same conditions. (Page 43).

Theorem 20 of Formal Dependence: No concept defines itself; no statement proves itself; no physical object is the cause of itself; and no cause is the cause of itself. (Page 43).

Theorem 21 of the Reference Frames: The laws of physics are the same in all discrete reference frames. (Page 44).

Theorem 22 of Incompletable Regress: Every recursive sequence S of proofs, definitions or causes in which there is a last element to be proved (defined, caused) and each element has an immediate predecessor that proves (defines or causes) it, is incompletable. (Page 45).

Theorem 23 of the First Element: A consistent sequence in which there is a last element and each element has an immediate predecessor is a complete totality only if it has a first arbitrary element without predecessors. (Page 46).

Corollary 16 of Primitive Concepts: Primitive concepts are inevitable in all sciences and languages. (Page 48).

Corollary 17 of the First Cause: To explain any physical object or phenomenon, a first cause not explainable in terms of other causes deduced from our knowledge of the observable universe is necessary.

(Page 49).

Theorem 24 of the Finite Universe: The universe is finite in extent, age and number of components. (Page 49).

Theorem 25 of the Origin of the Universe: The universe must have had an origin an origin whose cause is unknowable. (Page 49).

Theorem 26 of the Discrete Universe: The universe has evolved through a finite and discrete sequence of qutits. (Page 50).

7. Space in ancient Greece

7.1 Introduction

Prior to early Greece, usually referred to as pre-Socratic Greece, there is little documentation on the use of space-related concepts. In the so-called river Mesopotamian cultures ([308, 33, 317, 266, 329]) units of measurement related to length, extent and volume were introduced, all linked to the interest of their practical use. The only abstraction here would be a practical abstraction, i.e. the conventional use of those practical units of measurement [176]. We have no documented evidence that a philosophical interest in the notion of space existed in these cultures.

The philosophical interest in the concept of space, as in almost all concepts, appears in the early Greece of the first pre-Socratics, especially in the first cosmologists, the Pythagoreans and the atomists. Even then, certain problems related to space were posed, which are still posed in the 21st century, without being solved, except for the final solution given by an important part of contemporary physics: space does not exist, it is not real, and therefore poses no problem. In these first abstract reflections on space, on extension, concepts related to space are already used, such as the concept of emptiness and the concept of place, the place occupied by physical bodies.

At this early stage, some problems related to the limits of space, to the finite/infinite nature of its extension, and to its divisibility are already discussed. In fact, the problems posed by the actual infinity appear already from the first abstractions on space, and continue to be posed, not solved, until our days. As in the case of physicists with space, mathematicians proposed at the beginning of the last century a solution that claims to be definitive, final, although in this case in the opposite direction to that given by physicists to space: now it will be the dogmatic acceptance of the existence of the actual infinity (Axiom of Infinity). A decision that could be wrong for reasons that have already been given in [212], and that will continue to be given in this book.

It is interesting to note at least two aspects of the first conceptions of the universe. One is the proposal of models in which there is a single principle of creation (and destruction) of all things. Although their relationship to Cellular Automata Like Models (CALMs) is currently unclear, they all belong to the same basic model of explaining the world. On the other hand, and also related to CALMs (with their intimate spatial structure), we have to consider the fact that one of the first theoretical conceptions of space was that of a discrete space defined by extended points, points of non-zero extension. Although in the later development of the theory, the irrational numbers (all of them with an actually infinite number of decimal places) made their appearance and there ended the first discrete possibility for space. Plato and his (rebellious) student Aristotle, certainly the two most important figures of classical Greek culture, also devoted part of their thinking to space, much more the latter than the former. Although in both cases, the former more than the latter, their reflections on space were more ontological and metaphysical than physical.

Some of Aristotle's spatial arguments have the historical interest of having been an inexhaustible source of contradictions for a good part of his Arab followers from the tenth century AD. These contradictions motivated the criticism of the great Greek thinker, who despite the criticism in this matter, continues to be the great thinker of reference in the Arab world and in the European medieval scholastic world, nourished by the translations that the Arabs made of the main works of Aristotle, including those in which space is discussed more extensively, for example his *Physics* [16], or his *On Heaven* [14].

Although he is also a Greek classic, like Plato or Aristotle, the importance of Euclid in the content of this book is so outstanding that he deserves a chapter exclusively devoted to his geometry, as is done in Chapter 8 of this book. As will be seen there, Euclid is more mathematician and physicist than metaphysician, and his influence on modern mathematical and physical sciences is the most important of all Greek authors, pre-Socratic and classical. Euclidean three-dimensional space is considered since Newton as the geometrical structure of physical space, real and absolute for Newton, not so real and not so absolute for most contemporary physicists. This question, the reality of physical space, will be a key issue in the content of this book.

7.2 The first cosmologists

As already indicated, it is in early Greece (8th-7th century BC) that space is included as an object of study proper to philosophy. What existed before that, according to historical documentation, was a pragmatic use of this concept: the knowledge of the spatial directions in

7.2 The first cosmologists

which things are located, a kind of system of reference of practical and social interest, and the definition of measurements related to the extension of objects (length, area and volume) for exclusively practical purposes, almost always related to surveying and construction.

Cosmogony, the origin of all things (the origin of the universe in modern terms), appears at the beginning of practically all known philosophies, including those developed in greater or lesser detail by the first Greek authors in what we now generally know as the pre-Socratic philosophers, which include mainly the first cosmologists (Hesiod, Thales, Anaximander, Anaxagoras, Heraclitus) and the Eleatic, Pythagorean and materialist (atomist) philosophies.

The first idea of space to appear in Greek literature is found in Hesiod (8th-7th century BC): the CHASM, a kind of extension in which things can originate and exist [314, p. 18-20]. The chasm, then, was the first thing to exist. (Note that this idea of Hesiod's is not far from the idea of space as the generator and container of all physical objects, an idea not alien to the concept of CALM (Cellular Automata Like Model), which will emerge throughout the book as a consistent model for beginning to explain the physical world. The second thing that must have existed, according to Hesiod, was our Gaia, which gave birth to Ouranos (above) and Tartarus (below). And thus appear the three basic directions of space (six, if one considers the two senses of each of them, as was usual at that time). But Hesiod did not go into details, he did not develop his spatial concepts, which, moreover, are not clearly distinguished from the temporal concepts.

A little later, in the school of Miletus, another idea appears that is also related to the concept of physical space that will be introduced here, although at the moment this relationship may seem somewhat obscure. This idea is that of a basic generating principle (the ARCHÉ) of all things, of the whole universe. The first proposal of such a principle was that of Thales of Miletus (624-546 BC): that generating principle of all that exists in the universe would be *water*. According to Thales, the other elements are derived from this generating principle.

Anaximander of Miletus (610-545 BC) rejected this proposal, considering that water being one of the fundamental elements, characterized by the cold and the wet, could not originate other fundamental elements such as fire, characterized by the opposites hot and dry. Instead he proposed a principle, the APEIRON (the unlimited and indeterminate), because being itself indeterminate and unlimited it could originate the fundamental elements and a multitude of material, spatial and temporal processes [314, p. 23]:

1. The apeiron is that from which the heavens and the worlds arise. Therefore, it appears to be a matter from which everything can

arise. The generation of everything from it is possible, since it itself is not determined or limited as is a piece of matter or a material thing.

2. The apeiron is eternal and ageless and, therefore, temporally infinite. Infinite duration is a necessary condition for the unceasing processes of the phenomenal world.

3. The apeiron surrounds all worlds and is therefore the most encompassing space. It allows all movements and changes to take place in it.

As is also the case of CALMs, Anaximander's apeiron has spatial, temporal and material qualities. Everything is born and dies in it as a consequence of the struggle between opposites (interactions) [33, p. 48-58],[268, p.33-38]. Thus:

> Things perish in the very thing that gave them being, according to necessity [268, p. 36].

But the birth of a thing in the bosom of the apeiron would be an act against the uniqueness of being, the uniqueness of the apeiron. It would be a kind of sin that time makes pay with the disappearance, with the death of that thing. The universe would have, as such an object, an allotted time of existence, at the end of which it is destroyed and returns to the apeiron. This cyclical vision of the universe and of time is not rare in Greek authors.

Anaximenes of Miletus (590-525 BC) did not agree with his teacher Anaximander that the indeterminate, the apeiron, could originate the determinate, the material objects of the physical world, the physis. He then proposed AIR as the generating principle, although it would be an element different from the material air of our days. That of Anaximenes is more complex and not only material: it would contain the generating principle of life. And by different degrees of rarefaction and condensation it would originate other material objects, such as fire, water or earth. Anaximenes therefore proposes a transforming cause that originates the different things of the world from the original principle of all of them: rarefaction/condensation.

Heraclitus of Ephesus (6th-5th centuries BC) also considered a single generating-regulating principle of all things: that without which nothing could be explained. His principle is more relational, more legal and logical than material. It is the LOGOS, the law that binds and holds together the things that constitute physis. It would be present in all things, although it cannot be visualized in any of them, except in their becoming. The physis itself would be giving us (logical) signs of its existence, although humans, according to Heraclitus, tend to ignore those signs led by habit and lack of deep reflection. Only that deep reflection can lead to a thought that reflects the true becoming of physis. A truth that is not to be sought in mysticism but in the simplest objects of the physical world.

Heraclitus also defended what could be called the Principle of Anti-Identicality, as opposed (at least in some respects) to the Theorem of Identicality which will be proved later, in Chapter 5 of this book. According to this Heraclitean principle all things share the same attribute: *being different from one another, at least in the space and time occupied by each thing*. In his deep and obscure style he tells us (quoted in [268, p. 69]):

> Listening not to me, but to the logos, it is wise to recognize that all things are one.

Naturally, both the Anti-Identicality Principle and the Identicality Theorem also apply in the world of CALMs, and in practically all scientific models that try to explain the physical world.

7.3 Parmenides and Zeno of Elea

If Heraclitus is the philosopher of becoming, of continuous change, Parmenides (530(515)-? BC) is the philosopher of the permanent being: being is incompatible with non-being; being cannot not be; being is the opposite of nothingness, and nothingness is not even something that can be considered or conceived, (quoted in [268, p. 112]):

> ... so that you could never cut so that being does not follow with being.
>
> It must be what can be said and conceived. Because there is being, but nothing, there is not.
>
> And it is that such a thing will never be violated, so that something, without being, is.

Parmenides' philosophy contains, then, the two great principles that will end up founding logic: the Principle of Identity (a thing is what it is, and it is not what it is not), and the Principle of Non-Contradiction (it is not possible to be and not to be at the same time). It would not be out of place to consider Parmenides as one of the fathers of Classical Logic.

The philosophy of Parmenides, surely one of the most closed, challenging and difficult in the history of human thought, can also be opened and explained by admitting, as will be seen later in this book, that, indeed, that which is, can only be; but both as what it was, or as something else into which it has been transformed. It is the complete being, it is the being that includes both what it is and what it can be transformed into. We shall also see in this book that for the same reasons that a first cause is necessary to explain being (Corollary of the First Cause, Chapter 5), a final cause is also necessary to explain complete non-being (Chapter 5). The universe is closed, everything is transformed but nothing disappears from the universe, it cannot disappear without the final inexplicable cause, in that Parmenides is right. And it is an important detail that all cosmology should take into account.

Like his teacher Parmenides, Zeno of Elea (490-430 BC) was more interested in ontology than in physics, but he developed part of his famous arguments by making use of a mathematical property of space. A property that, both in his time and in ours, is still a hypothetical property of space: its supposed infinite divisibility (actual infinity) and its consequent modern dense order: between any two points of a simple straight line of 1mm length there exists a non-numerable infinity ($2^{\aleph_0}$) of distinct points; and between any two points of those infinite points there exists another non-numerable infinity of distinct points; and between any two points of those infinite points there exists another non-numerable infinity of distinct points; and so on and on. In Zeno's words [33, p. 177]:

> If there are many beings, beings are infinite, for there are always others in the midst of beings, and in turn others in the midst of these, and thus beings are infinite.
>
> What moves, does not move where it is or where it is not.
>
> If everything there is is in a space, it is evident that there will be a space of space, and that will go on to infinity.

By making use of the supposed infinite divisibility of space, Zeno proves, for example, the impossibility of motion:

> To go from A_1 to A_2, it is first necessary to go from A to the middle A_3 of $A_1 A_2$.
>
> To go from A_1 to A_3, it is first necessary to go from A to the middle A_4 of $A_1 A_3$.
>
> To go from A_1 to A_4, it is first necessary to go from A to the middle A_5 of $A_1 A_4$.
>
> To go from A_1 to A_5, it is first necessary to go from A to the middle A_6 of $A_1 A_5$.
>
> and so on to infinity.
>
> Therefore, the motion from A_1 to A_2 cannot be started.

Already here appears the fascination of humans with the strange and bizarre, as if the strange and bizarre added scientific value to theories. We like to prove things like the impossibility of motion despite its overwhelming evidence. It seems clear in this case that, in the face of such overwhelming existence, Zeno would have to have considered the possibility of some flaw in his argument. And the flaw, as we shall see, is the Hypothesis of the Actual Infinity which, briefly stated, considers that the ordered list of natural numbers exists as a complete totality, even though there is not a last natural number completing the list. As if the incompletable could exist as completed, as Aristotle would say (recall that a complete totality is a set defined by comprehension in which every element that should be in the set, is in the set). The infinite will accompany us throughout the book, meanwhile the reader can analyze some demonstrations of its inconsistency in [204, 210, 211] and especially in [212] and [213, Link].

Indeed, in a discrete physical space, with minimal indivisible units (qusits) there exists immediate successiveness, so that between a qusit and its immediate successor there is no other qusit. Under these conditions it will be proved that between any two qusits there always exists a first qusit, a last qusit and between them a finite number of qusits. All Zeno paradoxes are immediately dissolved in this finite and discrete scenario.

7.4 The Pythagoreans

The abstract notion of extension and place appears in the first Pythagoreans. It is a spatiality linked to the natural numbers, which for them were more real than sensible reality itself. Indeed, according to Aristotle [16, 213b, p. 230] the Pythagoreans considered the natural numbers with a certain spatiality, necessary to guarantee their discrete character: between two successive natural numbers no other natural number can exist. The void delimits the natural numbers. They also considered the void as a kind of division or separation between objects. This primitive space (pneuma apeiron) had no physical implications, except that of separating things [176, p. 9]. Some pre-Socratics identified it with the limitless, with the void, with air, and even with night [47, p. 433-434]. It is only the beginning of the abstract conception of space.

Already in this epoch, the earliest of abstract thought, the first idea of a discrete space appears. Indeed, the early Pythagoreans believed in the existence of a geometric space formed by indivisible points with a length δ greater than zero. Consequently, all lengths would have to be commensurable: the ratio between any two of them, say L_1 and L_2, would be the ratio between two natural numbers, i.e. a rational number [231, pp. 11-16]l:

$$L_1 = n_1\delta; \; L_2 = n_2\delta \tag{1}$$

$$\frac{L_1}{L_2} = \frac{n_1 \delta}{n_2 \delta} = \frac{n_1}{n_2} \tag{2}$$

But, as is well known, they themselves discovered the incommensurability of two well-defined lengths: the length L of the side of any square and the length L_d of its diagonal. For example, for a square whose side has length 6δ we would have:

$$L_d = \sqrt{6^2 \delta^2 + 6^2 \delta^2} \tag{3}$$

$$= 6\delta\sqrt{2} \tag{4}$$

$$\frac{L_d}{L_s} = \frac{6\delta\sqrt{2}}{6\delta} = \sqrt{2} \tag{5}$$

where $\sqrt{2}$ is an irrational number: a number with an infinite, non-periodic number of decimal places. We will deal with numbers with infinite decimal places later in this book. For the moment we regret that the Pythagoreans did not discover discrete (integer) division, the only division compatible with discreteness, for example:

$$L_d = \lfloor \sqrt{6^2 \delta^2 + 6^2 \delta^2} \rfloor \tag{6}$$

$$L_d = \delta \lfloor \sqrt{6^2 + 6^2} \rfloor \tag{7}$$

$$= 6\delta \lfloor \sqrt{2} \rfloor \tag{8}$$

$$= 6\delta \tag{9}$$

$$\frac{L_d}{L_s} = \frac{6\delta}{6\delta} = 1 \tag{10}$$

where $\lfloor x \rfloor$ stands for the integer part of x. As we will see later in this book, equations (6)-(10) represent the discrete version of Pythagoras theorem, which we will also deduce later on. And naturally, the Pythagorean metric, based on the Pythagorean theorem, will be key in the geometries and theories of space to come.

The newly discovered incommensurability between certain lengths led the Pythagoreans toward the notion of continuous space [231], a precedent of the relativistic spacetime continuum. Perhaps due to the enormous influence of our sensory perception of the physical world as a continuous space-time scenario, discrete (discontinuous) arithmetic was not developed in Greek culture, and is yet to be developed in ours in formal and universal terms.

In any case, one of the consequences of the Pythagorean discovery of incommensurable lengths was the abandonment of extensive points in favor of non-extensive points, which are the same ones we still use today in all continuous geometries, Euclidean and non-Euclidean, practically the only geometries in contemporary physics. But the story is not over, as we will see throughout this book.

The Pythagorean Archytas (435/410-360/350 BC) seems to have written a book on space, although only a few fragments have survived. He is one of the first authors to consider the problem of the limit of space. He did so with a well-known and recurrent argument (his discussion is repeated at least until 1690, when J. Locke uses it in his famous text *Essay concerning human understanding* [217, C. XIII, 21, p 102]): Archytas wonders whether placed a man at the boundary of space he may, or may not, extend his arm beyond that boundary.

The same question we can ask ourselves today: what will we find if we travel from the center of the Universe in a straight line 46 billion light years and stand at its boundary? Does that boundary exist? Does the outside of the Universe exist? What could happen if we emit a visible laser beam from that supposed boundary in the direction of the supposed outside of the Universe? And if the outside of the universe does not exist, does it not exist because it is infinite, or does it have some kind of physical limit? If the universe is consistent, and we will prove that it is, and infinity is inconsistent, and we will prove that it is, then it must be finite, and therefore could have a physical limit. But, in addition to asking questions similar to those we have just asked about the limits of the universe, Archytas reflected on other aspects of space, especially in relation to the objects contained in space:

1. He distinguishes between space and matter, and considers space to be independent of matter.

2. Every object occupies a place and the object cannot exist if the place does not exist first. The place must exist before all things.

3. A salient feature of space is that it contains all things, but space is not contained in anything else.

4. Space determines the volume of all bodies: it exerts a kind of pressure on them, preventing them from reaching an infinite size.

5. Space is thus a kind of primitive atmosphere with pressure and tension.

6. Beyond space lies the infinite void.

7. Since there is nothing after all things, there is no outside, and therefore space is without end and without limit. And no matter in which region it is situated, it will have the same infinitude in all directions.

7.5 The atomists

As it is well known, and its name indicates, for the atomists Leucippus (460 - 370 BC) and his disciple Democritus (460 BC - 370 BC) matter was discrete, it was formed by indivisible units: the atoms. With non-zero shape and size, the atoms existed in an infinite number (actual infinity) and therefore would occupy an infinite space, an empty extension without influence on the motion of the atoms, incessant motion due to their continuous collisions. The existence of empty space had been rejected by other pre-Socratic thinkers such as Melissus (quoted in [176, p. 11]):

> Nor is there anything empty, for the empty is nothing and that which is nothing cannot exist

But for the atomists it was a logical necessity, according to their atomic theory of matter. The disciples of Leucippus and Democritus added weight to the atoms as the cause of their upward and downward motion, which added to space a preferred directionality: the vertical. Space was then homogeneous but anisotropic. There remains the doubt as to whether the unlimited space was for the atomists something that penetrated all bodies and was penetrated by all bodies, or was only the sum of all the gaps between all atoms and between all bodies.

Lucretius, a late atomist (99 BC - 55 BC) expressed the foundations of atomic materialism in a long and famous poem that was lost for several centuries until it was found in 1418: The Nature of Things. With respect to space we can read [221, p. 108]

> Let us return to our reasoning:
> all nature, then, is based
> on two principles: bodies and void   (420)
> in which they swim and move:
> that there are bodies, common sense
> proves it; an irresistible principle
> without which reason, abandoned
> from error to error would be lost.
> If there were not, therefore, that space
> which we call emptiness, there would not be
> bodies would not be seated, nor would they move
> could, as I have just told you

And also: [221, p. 129]

> If, in addition, space is limited

7.5 The atomists

> and someone stands at the end
> and shoots a flying arrow,
> do you want it to be shot with great strength
> it flies lightly to reach the target,
> or do you think that impeded by some hindrance
> its flight does not let it go forward?
> One or the other you must confess. (1220)
> Whichever one you choose, you must forcibly
> you must remove the limits to the whole:
> For it may well be an obstacle that hinders
> and hinders the arrow from reaching the target,
> or else it passes it, here there is no end:
> where you set limits, I will at once
> I will ask what has become of the arrow:
> you will never find the end like this;
> its immensity always leaves a space
> for the fugitive arrow to cross.

Space becomes according to Lucretius an infinite receptacle of all things (bodies). On this infinity, Lucretius gives a new argument, invoking, like other disciples of Leucippus and Democritus, the weight of atoms and a directional preference in space (above and below). [221, p. 129]:

> Moreover, if nature (1231)
> had set limits to the whole,
> already the matter with its own weight
> would be gathered in the deepest places;
> beneath the vault of heaven
> nothing would be produced.

The atomists, the first materialists in history, were also the first to admit that something immaterial like the unlimited void would have an existence as real as that of material objects. Real but different in their essence, as also stated by Gorgias (460-380 BC), who also gave one of the first proofs of the finiteness of space (reconstructed in [176, p. 14]):

> The first clear idea of space and matter as belonging to different categories is to be found in Gorgias. Gorgias first proves that space cannot be infinite. For if the existent were infinite, it would be nowhere. For were it anywhere, that wherein it would be, would be different from it, and therefore the

existent, encompassed by something, ceases to be infinite; for the encompassing is larger than the encompassed, and nothing can be larger than the infinite; therefore the infinite is not anywhere. Nor on the other hand, can it be encompassed by itself. For in that case, that wherein it is found would be identical with that which is found therein, and the existent would become two things at a time, space and matter; but this is impossible. The impossibility of the existence of the infinite excludes the possibility of infinite space.

7.6 Space according to Plato

According to Aristotle, Plato was not very satisfied with the explanations given by his predecessors about the existence of space, so he tried to explain it [16, 209b]. And he did so in his Timaeus, a dialogue between Socrates, Critias and Timaeus of Lycritus (an old Pythagorean of dubious historicity). The dialogue is sometimes obscure, but it is undoubtedly one of the most influential works in the history of philosophy and science. The Timaeus is a cosmogony that includes reflections on matter and on living beings [278]. For the reasons that will be given below, Plato's text will be very significant for the physical discussion of physical space proposed in this book. In the Timaeus we can read [278, Pos. 520-536] (the texts in straight brackets are mine):

> The same reasoning also applies to the nature that receives all bodies. We must say that it is always identical with itself, for it does not change its properties at all. Indeed, it always receives everything without adopting in the least any form similar to anything that enters it, since by nature it underlies everything as a mass which, because it is changed and shaped by what enters, appears diverse at various times; and both what enters and what leaves are always imitations of beings, imprinted from them in a difficult to conceive and admirable manner which we shall investigate later. Certainly, now we need to conceptually differentiate three genres:
>
> (i) That which becomes, [the objects that are formed in the imperfect material world],
>
> (ii) that [the medium] in which it becomes, [the receptacle in which material objects are formed]
>
> (iii) and that through the imitation of which that which becomes is born. [The Ideas or Perfect Forms]
>
> And one can also liken the vessel to the mother, that which is imitated to the father, and the intermediate nature to the

7.7 The Aristotelian space

son, and think that, similarly, when a relief is to be of a great variety, the material on which the engraving is to be made would be well prepared only if it lacked all those forms which it is to receive from somewhere. If it were similar to anything of what goes into it, by receiving the opposite or what is in no way related to that, it would imitate it badly because it would manifest, in addition, its own appearance. It is therefore necessary that that which is to take all species in itself should be exempt from all forms. As happens in the first instance with artificially perfumed oils, the liquids that are to receive the perfumes are made to be as odorless as possible. Those who attempt to print figures on some soft material do not allow any figure at all, but flatten it first and leave it completely smooth. It likewise corresponds that what is to receive often and well to its full extent imitations of eternal beings should by nature lack all form. Therefore, let us conclude that the mother and receptacle of the visible become and completely sensible is neither earth, nor air, nor fire, nor water, nor whatever is born of these, nor that from which they are born. If we affirm, on the contrary, that it is a certain invisible, amorphous species, which admits everything and which participates in the most paradoxical and difficult to understand way of the intelligible, we will not be mistaken.

Consequently, and according to Plato, in addition to that which becomes (i), and that through whose imitation that which becomes (iii) is born, it is also necessary to consider that in which it becomes: the receptacle (ii). Note that this Platonic receptacle is material, but of a different matter from that which forms the material objects of the physical world, and while these are in continuous change, the same is not true of the receptacle, which remains unalterable. Plato will also end up calling the receptacle space. Plato then explains how the universe was set in motion, starting from a receptacle that already included traces of the four fundamental elements (fire, air, water and earth) in initial chaotic movement, but that story is already alien to our objectives.

Before leaving the Platonic receptacle, let us think about the matrix of cells of a CALM (Cellular Automata Like Model) and the functioning of the CALM. It is in this matrix that CALM objects are formed and evolve, resulting from the CALM laws and the state of the cells. CALM objects, as such, are different from the cells and can move through the CALM while the cells remain immobile. It seems appropriate, then, that from now on we also refer to the array of discrete elements (cells) of a CALM as a receptacle.

7.7 The Aristotelian space

Aristotle's texts have the reputation of being unfinished, of being drafts for future manuscripts that, although mentioned by his first commentators, do not really exist [164, p. 72]. This situation is conducive to different interpretations of some Aristotelian texts, which is what happens with those devoted to space, as we will see here. The situation is complicated because space could be a primitive concept (indefinable in terms of other more basic concepts) that Aristotle tries to define using twisted dialectical means. Without success, of course. As indicated in the presentation of this book, until now no one has succeeded. Neither the authors we have already remembered nor those we will remember. In such situations I always remember Newton's words, which are almost a joke. I repeat them again here [259, p. 77]:

> I do not define time, space, place and motion, as being well known to all.

In his Categories [15, Part 6, p. 15], Aristotle distinguishes between discrete and continuous magnitudes according to whether there are discontinuities or jumps between their successive values (discrete magnitudes), or not (continuous magnitudes). Between any two values of a continuous magnitude there are always intermediate values (dense order), which does not occur with discrete magnitudes (discrete order). For Aristotle, the quantities of space and time are of continuous type, since there are no discontinuities between their respective parts. The intervals of space and time are infinitely divisible, although the infinite parts do not all exist in act (actual infinity) but in potency (potential infinity).

To reconcile the indivisibles with the divisibles ad infinitum, Aristotle defined three types of relations [16, Book V, 228a]: successiveness, contiguity, and continuity, establishing the conditions to be met by continuous, contiguous, and successive elements. Between elements of one type there could be elements of the other types. Successiveness implied neither contiguity nor continuity, but if two successive elements touched, then they were contiguous (juxtaposed), as occurs, for example, when air touches the surface of a glass of water. If adjacent elements had coincident boundaries, then they would be called continuous and would be a single reality (half of a stick is continuous with the other half). Physical contiguity is mathematical continuity when all contiguous elements are of the same type, homogeneous. In the case of different physical objects, there can be contiguity but not continuity, because they are qualitatively different elements.

It is mainly in his Physics that Aristotle analyzes the concepts of space and place, although he hardly uses the word space, probably trying to avoid confrontation with a predictably primitive, indefinable

7.7 The Aristotelian space

term. In fact, the word "*space*" appears only 10 times in Aristotle's Physics, 9 of which, according to other authors, refer to space without giving a definition of it. Only once is the word place used in the sense of space (chôra). On the contrary, the word "*place*" appears 383 times because it gives a definition of place, which, as we will see here, is a pseudo-definition. Recall that the universe, according to Aristotle, is a finite sphere composed of several concentric layers with a center at the center of the Earth. The outermost shell, which limits the size of the universe, is formed by the fifth Aristotelian element (quintessence) called ETHER, which is distinct from the other four elements (earth, water, air, and fire) that form the inner sphere or sublunar world.

The outer layer or envelope, the heavens in Aristotelian terms, would be in continuous rotation, which leads to certain contradictions with Aristotelian mechanics that were discovered by the Arabian scholars of the ninth to twelfth centuries. We will analyze them in Chapter 9. For its part, and in an ideal state, the internal sphere would be formed by four concentric layers that from the exterior to the interior would be: fire, air, water and earth. But in its real state the four elements are mixed, although conserving, as one of their essential properties, their tendency to move towards their corresponding natural places, that is to say towards each of these four layers ordered in the way that has been indicated. These movements would be natural, not forced.

The natural motion is, therefore, teleological: the fundamental elements would move with the purpose of occupying their natural places in their corresponding layers. In addition to these natural movements towards natural places, there would also be forced movements of objects towards other places. Aristotle then tries to characterize the concept of place in order to then try to deduce what the place of a thing must be [16, Book 4]. Among these characteristics of place he highlights:

1. Places have a real existence although not independent of the things that occupy those places.
2. The place of a thing is that which embraces it.
3. The place of a thing is not part of that thing.
4. The place of a thing is neither larger nor smaller than the thing.
5. The place and the thing it contains can be separated.
6. Each location implies absolute directionality up and down.
7. The place has no place.
8. The place is different from its changing content, then it is real.

Aristotle then proposes four possible definitions of place, to prove that

three of them are impossible:
- (i). The place is the form.
- (ii). The place is the matter.
- (iii). The place is a certain extension between the extremes of that which contains the thing.
- (iv). The place is the end that contain the thing.

The first two alternatives are impossible because neither form nor matter are separable from the thing, while the place of the thing is separable from the thing. The proof that the third alternative is also impossible is much more obscure [16, 211b, p. 224-225]. Therefore, the only possible definition left is the fourth (iv) of the above alternatives. [16, 212a, p. 225]:

> The place of a thing is the limit of the containing body that is in contact with the contained body.

And as the reader will have guessed, the above definition does not define anything unless it is indicated what the containing body is, which is nowhere to be found in the Aristotelian text, which at most refers to vessels and liquids contained in the vessels, insisting that the containing body must be in contact with the contained body, leaving no gaps between the two. So, after all and as expected, the place of a thing is the boundary of something that is not known what it is and that is in contact with the contained body. The place of an object is a kind of shell that envelops the object, but a shell of an unknown nature. That is the Aristotelian (pseudo) definition of place, because the definition of space does not exist. Only in [16, 208b, p. 213] one can read:

> Hence, it may appear that the place or space [chôra] to which or from which the bodies have changed is distinct from them.

where Aristotle equates space with place. But he does not give an explicit definition of space. Some authors consider that Aristotelian space would be the set of all the places of all objects [176, p. 20], that is to say the set of all the external shells of all material objects. These external shells would be surfaces without thickness, with the same morphology as the external morphology of the enveloped body [164, p. 77].

Some authors give a definition of Aristotelian place different from the one given above. For example, in [314, p. 42] the following definition of Aristotelian place is given (which appears written in both Greek and English):

> The first/immediate unmoved limit of that which surrounds- that is topos.

7.8 Space in post-Aristotelian Classical Greece

The place would be, in addition to the immediate boundary or first, that which surrounds and is immovable. For example, the place of a boat sailing down a river would not be the outer shell of the boat, but the immovable boundary of the riverbank, since the container object must be immovable. It would be something like a material (physical) reference frame in which it is possible to describe motion (in this case of the boat), because a boundary of the contained body of the same size as the contained body would not allow motion, and motion is one of the fundamental concepts of Aristotle's physics.

As M. Jammer points out, an interesting aspect of the Aristotelian concept of place is its similarity to a field of forces [16, 208b, p. 213]:

> Moreover, the displacements of simple natural bodies, such as fire, earth, and the like, not only show us that place is something, but also that it exerts a certain power. For each of these bodies, if nothing prevents it, is carried toward its proper place, some upward and some downward.

In the same sense, other authors [314, 233] emphasize the dynamic role of the "container body" as a transmitter of forces and as a cause of the maintenance of the directions of motion of the "contained bodies", reminiscent respectively of preinertia (which will be discussed in Chapter 19 of this book) and Newtonian inertia.

But the fundamental problem of a definition of space remains unresolved. And as will be seen throughout this book, it remains so today. Consequently, the main questions raised by the physical nature of space remain open:

- Does space have substance?
- If it does, what kind of substance is it? because it must be different from ordinary matter.
- How does it interact with ordinary matter?
- Does it penetrate all objects?
- Does it allow itself to be penetrated by all objects?
- What is space the cause of? or is it the cause of nothing?

Although in our days, all of them are solved by the expeditious way of the negation of space: space does not exist, it is not real. However, space can expand, deform, vibrate and be the transmitting medium of its own vibrations and of other vibrations such as electromagnetic waves. And one wonders how something that is not real can do all that? How can the vibrations of an object that does not exist be recorded experimentally? We will return to these questions and propose some answers at the end of the book.

7.8 Space in post-Aristotelian Classical Greece

Although Plato and Aristotle were the fundamental pillars of metaphysical and physical thought until at least the Scientific Revolution of the 16th-17th centuries [182, 103, 320], their approaches to space were already disputed by Aristotle's own disciples and later by his Arabic translators [352, 357]. Thus, his disciple and successor in the direction of the Lyceum, Theophrastus (371 BC-287 BC), came to the conclusion that space is not an entity of its own but a system of relations between bodies that determines their relative positions. Here appears already a theory of the group of relational theories as a counterpoint to the substantival theories. Naturally, Euclidean space should also be included in this section, but, as already indicated, given its relevance to the central theme of this book, chapter 8 will be devoted to it.

The Stoics modified the Aristotelian definition of place: instead of the bounding surface of the first immobile container body, they used the alternative of the interior volume defined by that bounding surface, a concept closer to the intuitive notion of space occupied by an object. The purely geometrical continuity in Aristotle becomes a physical principle in the Stoics, allowing the propagation of physical phenomena and physical interactions between objects throughout the universe, even beyond the sublunar world: for example, Posidonius (135 B.C. -51 B.C.) discovers the influence of the Moon on terrestrial tides.

The universe according to the Stoics would be formed by the set of material objects physically interrelated through the physical space in which they are included, all surrounded by an undifferentiated vacuum, externally unlimited, infinite, and without influence on the underlying material world. According to the Peripatetics, the material universe of the Stoics would eventually have to dissipate into the infinite void, but the Stoics argued that such dissipation into the external void would not occur because of the tension and interactions between the material parts of their universe [176].

Strato of Lampsacus (335-268 B.C.) was a Greek peripatetic philosopher who succeeded Theophrastus in the direction of the Lyceum founded by Aristotle. He was especially devoted to the study of nature, including new natural elements in the explanation of the world, always seeking an agreement with daily experience, to the point that the intervention of the gods in the creation of the universe was unnecessary. Strato defined space as the container of all things. A container that would exist even if there were no things inside it. So, for Strato, the vacuum was not entirely impossible, it could exist in the interstices of material particles. The Alexandrian engineer Heron (10?-70? B.C.) used the penetration of material rays of light and heat in water as proof of the existence of such interstices.

7.8 Space in post-Aristotelian Classical Greece

The first indication of a connection between space and God appears in Palestinian Judaism of the first century. In Greek philosophy the use of the word place as a reference to God does not occur. In Sextus Empiricus (160 - 210 B.C.) an empiricist Greek physician and philosopher, we can read a hint of this usage (quote taken from [176, p. 29]):

> And so far as regards these statements of the Peripatetics, it seems likely that the First God is the place of all things. For according to Aristotle the First God is the limit of Heaven. Either, then, God is something other than the Heaven's limit, or God is just that limit. And if He is other than the Heaven's limit, something else will exist outside Heaven, and its limit will be the place of Heaven, and thus the Aristotelian will be granting that Heaven is contained in place; but this they will not tolerate [...] And if God is identical with Heaven's limit, since Heaven's limit is the place of all things within Heaven, God -according to Aristotle- will be the place of all things; and this, too, is itself a thing contrary to sense.

Whereas in the Jewish theology of the time, and probably earlier, they wrote things like (quoted in Latin in [176, p. 29]):

> The Hebrews do not doubt God, because no one contains Him, but He Himself, by His immense power, contains all things, having to be called "makom" or place, as is often done in the booklet of the Paschal rites published by Rittangelius.

8. Euclidean space

(Content partially taken from [200])

8.1 Introduction

This chapter is entirely devoted to Euclid's Elements, a fundamental work in the history of science, written more than 2300 years ago by Euclid (~325 BC - ~265 BC), and still valid today, at least in its most basic aspects. Euclid's Elements have the added interest of providing a possible geometric model for physical space, which is the central theme of this book.

As we shall see, and despite the fact that physical space is not real for most contemporary physicists, physical space could be real and discretely Euclidean[1], although massive objects could locally deform it and transform its geometry from Euclidean to non-Euclidean. In Chapter 25 of this book we will see that there are Euclidean alternatives to explain these non-Euclidean deformations.

After a brief presentation of Euclid and his Elements, one of the formal shortcomings of the Euclidean text is analyzed: the absence of a functional definition of a straight line, i.e., the absence of a definition that, realized in terms of the properties of lines, characterizes straight lines exclusively and that can be used explicitly in formal proofs. It is also explained why this deficiency is important, although the matter is treated in detail in Appendix C of this book. Finally, the physical reasons why it is still believed that physical space is Euclidean are discussed.

8.2 Euclid

Euclid is the name of a Greek mathematician, the author of the Elements, a book that laid the foundation for a mathematical discipline now known as Euclidean geometry. As with many other great thinkers

---

[1] As we will see later in this book, physical space cannot be continuous, as in Euclidean geometry, but discrete.

of ancient Greece, almost nothing is known about the man. In fact, all we know about the author of the Elements comes from two texts, one by Proclus Diadochus (412-485 AD) and the other by Pappus of Alexandria (290-350 AD). From Proclus' text [283] we infer that Euclid lived in the time of Ptolemy I Soler (367-283 BC), and that he was "younger than the pupils of Plato, but older than Eratosthenes and Archimedes". It seems reasonable to conclude that he flourished around 300 BC and that he received his mathematical education in Athens, from the students of Plato.

In the same passage of Proclus's text we can read the well known anecdote on Euclid (quoted from [150, p. 1]):

> ...Ptolemy once asked him [Euclid] if there was in geometry any shorter way than that of the Elements, and he answered that was no royal road to geometry.

From Pappus' text [265] it can be inferred that Euclid *taught and founded a school in Alexandria* because Pappus wrote about Apollonius of Perga (262-190 BC) that *he spent a very long time with the pupils of Euclid in Alexandria, and it was thus that he acquired such a scientific habit of thought*. In the same text, Pappus wrote a favorable comment on Euclid as a response to the less favorable opinion of Apollonius on Euclid's work on conics.

From 1332 to 1493 Euclid was believed to be the philosopher Euclid of Megara who lived about 400 BC. In 1493, Constantinus Lascaris resolved definitively the error. Other misunderstandings and questionable anecdotes related to Euclid and his Elements come from the Arabian authors, some of which defended the theory that it was Apollonius, not Euclid, the author of the Elements. Euclid not only wrote the Elements, at least half a dozen of other scientific works were surely authored by Euclid. Among them:

- The Data: an introduction to higher analysis.
- The Phenomena: on theoretical astronomy.
- The Optics: on the (rectilinear) propagation of light.
- Elements of Music: on harmony and Pythagorean theory of music.
- The Porisms: three lost books of very controversial content (probably advanced mathematics).

8.3 Euclid's Elements

In my opinion, the 13 books that make up Euclid's Elements are the first great scientific work in the history of science. It is certainly the most edited work and the most widely used source of scientific knowledge by people of all times and places. The word "Elements" in the title of a scientific work (Elements of Chemistry, Elements of Geology, etc.) usually indicates that it is an introductory text to a discipline, the aim of the work being to provide the reader with the basics to get started in a science. Generally, these works contain the definitions, principles, and axioms upon which the science in question is built. This is a tradition that dates back to ancient Greece. Such is the case with Euclid's Elements, although here there is something more. In fact, Euclid's Elements represent:

a) A model of how to proceed in the development of a mathematical theory.

b) A model of mathematical reasoning.

c) A prototype of the axiomatic method (scientific method of the formal sciences).

d) A compendium of the main geometric results known in Euclid's time.

e) The creation of a new science.

But Euclid's Elements were neither the first nor the last Elements of Geometry. Among other authors of this type of works we can mention Hippocrates of Chios (not to be confused with physician Hippocrates of Kos), Leon, Theudius of Magnesia, Amyclas of Heraclea, Cyzicenus of Athens, Philippus of Mende or Aristaeus. The success of Euclid's Elements, perhaps the most read and studied book ever, made practically disappear the other *Elements*. According to T.L. Heath [150, p. vii], Euclid's work is "*one of the noblest monuments of antiquity*". I fully agree.

Naturally, Euclid's text was built on the basis of the geometry known at the time. Among the authors on whose experience and achievements Euclid built his Elements we must mention the followings [136, p. 9]:

a) Pythagoreans: Books I, II, III, IV, VII and IX.

b) Archytas: Book VIII.

c) Eudoxus: Books V, VI, and XIII.

d) Theaetetus: Books X and XII.

The thirteen books include 5 general axioms, 5 geometric axioms, 131 definitions and 465 propositions. The propositions proved in one of the books can be used to prove other propositions in the same or in other subsequent books, so that between them there exists a complex network of formal relations that are now being analyzed with the aid of graph theory and computer programming [316]. These types of analyses allow to calculate the number of formal connections between any two propositions as well as the number of logical paths connecting two propositions. The Book I, which has been always considered as the most perfect of the thirteen books, is the richest regarding the number of formal connections between its propositions. For instance, between Proposition 1 and Proposition 45 there is a formal path composed of 20 different propositions, and 558 different logical paths connecting them [316, p. 25]. It is also the book that poses Euclid's enigma we will examine in the next section.

8.4 The enigma of the parallel straight lines

To avoid the infinite regress of arguments and circular arguments, all sciences, whether formal or experimental, must be built on assertions whose veracity must be accepted without proof. In the formal sciences these assertions are known as axioms. Ideally they should be short in number and highly self-evident. If we construct a science on an excessive number of axioms the output could result excessively speculative. If the axioms are not self-evident the output would be excessively abstract. For these reasons the set of axioms selected to found a formal science should be carefully examined. In the case of the experimental sciences, biology, geology, physics and chemistry, it is the inductive knowledge (that of Russell's chicks (see Chapter 5) which guides the choice of axioms, which are usually called principles or fundamental laws.

Euclid's Element are based on five general axioms (that apply to all sciences) and five geometric axioms (Euclid's Postulates). It is this group of axioms, or postulates, which poses the problem of Euclid's enigma, also known as the parallel enigma. A simple reading of these five axioms suffices to understand from were the problem arises.

Let the following be postulated [150, p. 154-155]:

1. To draw a straight line from any point to any point.
2. To produce a finite straight line continuously in a straight line.
3. To describe a circle with any centre and distance.
4. That all right angles are equal to one another.
5. That, If a straight line falling on two straight lines makes the interior angles on the same side less than two right angles, the two straight lines, if produced indefinitely, meet on that side on which are the angles less than two right angles.

The first four postulates are short and self-evident assertions. The fifth one is neither short nor self-evident. It has rather the aspect of a typical Euclidean proposition or theorem. For both reasons it was

8.4 The enigma of the parallel straight lines

put into question, as such a postulate, from the very beginning of the history of Euclidean geometry. For centuries, the same questions have been being asked: can this Fifth Postulate be derived from the other four? can Euclidean geometry be built without the Fifth Postulate? what differences would there be between a geometry with the Fifth Postulate and one without the Fifth Postulate? These questions summarize the enigma of the Fifth Postulate. In 1868 E. Beltrami proved the Fifth Postulate cannot be deduced from the other four [26], so it is not necessary to include it in the foundational bases of other types of geometries (see next section).

The first known attempt to resolve the enigma of the Fifth Postulate dates from the 2th century AD. And the attempts continued until the end of the 19th century, even after the birth of non-Euclidean geometries. So, the accumulated literature on the Fifth Postulate is enormous (see [294]). Among the main authors that tried to solve the problem of parallels we found: C. Ptolemy (2nd century), Proclus (5th century), al-Gauhary (9th century), Omar Khayyam (11th century), Nasir ad-Din at-Tusi (13th century), John Wallis (1616-1703), Gerolamo Saccheri (1667-1733), J. H. Lambert (1728-1777), J. L. Lagrange (17-36-1813), or A. M. Legendre (1752-1833).

Euclidean geometry is intuitive because it is closely and unequivocally related to our interactions and experiences with the physical world, in which we perceive space and objects arranged in that space. For that reason, Euclidean geometry is easy to understand. Although, for that very reason, it is not uncommon to take for granted what cannot be taken for granted. Or in other words, it is very easy to assume hypotheses implicitly, without realizing that one is assuming an implicit hypothesis, i.e. an hypothesis that is not included in the initial basis of assumed hypotheses (axioms) that should be the only hypotheses used in demonstrations. This type of error has always been present in the history of Euclidean geometry, particularly in the history of the Fifth Postulate. A matter on which, after centuries of discussions, the only thing that could be found were alternative statements for Euclid's Fifth Postulate. Most of them are in themselves problems of great geometric interest. These include the following ones (taken from [150, p. 220] and [257]):

1) Through a given point only one parallel can be drawn to a given straight line (Proclus and Playfair).

2) If a straight line intersects one of two parallels, it will intersect the other also (Proclus).

3) Straight lines parallel to the same straight line are parallel to one another (Proclus).

4) Parallels remains, throughout their length, at the same finite distance from one another (Proclus).

5) There exist straight lines everywhere equidistant from one another (Posidonius and Geminus).

6) Non-equidistant straight lines converge in a direction and diverge in the other (Thabit ibn Qurra).

7) If in a quadrilateral figure three angles are right angles, the fourth angle is also a right angle (Clairaut).

8) Two perpendiculars of the same length to the same straight line defines a rectangle (Farkas Bolyai).

9) There exists a triangle in which the sum of the three angles is equal to two right angles (Legendre).

10) A straight line perpendicular to a side of an acute angle cuts also the other side (Legendre).

11) Through any point within an angle less than two-thirds of a right angle, a straight line can be drawn which meets both sides of the angle (Legendre).

12) There exists no triangle in which every angle is as small as we please (Worpitzky).

13) Given any figure, there exists a figure similar to it of any size we please (Wallis, Carnot and Laplace).

14) There exist two unequal triangles with equal angle (Saccheri).

15) A rectilineal triangle is possible whose area is greater than any given area (Gauss).

The positive side of all this work is that, thought Euclid's enigma could not be resolved, Euclidean geometry was enriched and extended with an increasing collections of new and exciting problems. But the question that interests us here is: Can the Fifth Postulate be proved on a foundational basis DIFFERENT from Euclid's? Ockham's razor suggests an affirmative answer. And Ockham's razor is not usually wrong.

8.5 The definition of straight line

The infinite regress of definitions makes the use of primitive (indefinable) concepts inevitable in all sciences, which is not always explicitly recognized. Euclid's Elements is not an exception. Book I includes 23 definitions, some of which are rather confusing. Among the first definitions are those of point, line and straight line:

1. A point is that of which there is no part.

2. A line is a length without breadth.

8.5 The definition of straight line

3. A straight-line is a line which lies evenly with the points on itself.

The definition of point contains a generic, unspecified "*that*" which raises more questions than answers: does it have no parts because it has no size or because it is homogeneous (there are many different homogeneous objects)? Can something exist that has no size? Can extended Euclidean objects be composed of non-extended Euclidean objects? In the case of the definition of line it is clear that a line is not a length. And the definition of straight line does not specify what it is to lie evenly, unless it means to extend in a straight line, in which case the definition is circular. As expected, Euclid did not use any of these definitions in the formal proofs of his propositions. And the same is true in later Euclidean geometries, in which the three concepts are often used as if they were primitive concepts.

The Euclidean definitions of point, line and straight line are thus formally non-productive: a productive definition of an object establishes properties of the object which, once the object is axiomatically legitimated, are explicitly used in demonstrations. Surely the concepts of point and line can only be primitive concepts; if they were not, they would have to be defined by other more basic concepts, which would be the new primitive concepts. There is no escape from this potentially infinite regress. Another thing is the definition of a straight line. In this case it is possible to give a productive definition. Indeed, let us consider the following definitions of straight line that followed one after the other in post-Euclidean time:

- Definition by Heron of Alexandria (10-70 DC) [150, p. 168]: [a line such that] all its parts fit on all (other parts) in all ways.

- Definition by Proclus (412-485 DC) [150, p. 168]: that line which with one another of the same species cannot complete a figure.

- Definition by J. Playfair (1748-1819 DC) [279, p. 8]: if two lines are such that they cannot coincide in any two points without coinciding altogether, each of them is called a straight line.

- Definiton by E. Beltrami (1835-1900 DC) [26, p. 2]: [a line whose] specific character is to be completely determined by only two of its points, because two [straight] lines cannot pass through two given points of space without coinciding in all their extension.

These definitions point to a unique property of straight lines: if two straight lines have two common points, all points between those common points are also common points. The New Elements of Euclidean Geometry [200] is built on a new foundational basis that includes 29 definitions, 10 axioms and 45 corollaries. In that new foundation straight lines are defined as follows:

1. Points and segments that do not belong to the same line are said

non-collinear. Non-collinear lines with at least one common segment are said locally collinear.

2. Lines whose segments have the same definition as the whole line are said uniform. Two or more uniform lines are said mutually uniform iff any segment of any of them has the same definition as any segment of any of the others.

3. To extend a given line by a given length is to define a line, said extension of the given line, that is adjacent to the given line, has the given length, and the extension and the extended line are lines of the same class as the given line. Lines that can be extended from each endpoint and by any given length are called extensible lines.

4. **Definition 16 (of Straight Lines)** *Straight lines: Extensible and mutually uniform lines that can neither be locally collinear nor have non-common points between common points.*

Therefore, it is possible to give an exclusive definition of straight line not based on metric concepts alien to the nature of lines[2] but on concepts proper to the topological nature of lines. It is also a functional, productive, definition. By the way, the new foundational base of Euclidean geometry allows to prove as a theorem the Euclidean postulate of the parallel straight lines, see Appendix C for a summary and [200] for a full discussion

8.6 Physical space is Euclidean

Until the beginning of the 19th century, the axioms of Euclidean geometry had enough evidence to consider that the geometry of physical space was Euclidean. Recall Gauss's famous fictitious experiment that proved the physical reality of that geometry: the angles of a triangle were 180º when the sides of that triangle were three rays of light properly emitted from the tops of three mountain peaks. Although the experiment is only a fiction, it illustrates well the conviction that the geometry of physical space was (and still is) truly Euclidean. What is not a fiction, as will be seen below, is the estimate of the critical density of the universe. However, we must first remember the birth of non-Euclidean geometries, around the same time as Gauss's fictitious experiment.

In the early years of the 19th century, all attempts to deduce Euclid's Fifth Postulate from the other four had failed. For that reason, the parallel's problem was called at that time the shame of mathe-

---

[2]As is the case of the non-Euclidean definition of straight line as the line that minimizes the distance between any two given points.

matics [264, p. 9]. Frustration with parallel lines led to the birth of non-Euclidean geometries, which occurred in the first half of that century [297, 136, 264, 246, 135, 311, 257, 134, 275, 348]. The axioms of these non-Euclidean geometries no longer include Euclid's Fifth Postulate. For that reason, these geometries lead to results very different from the classical results of Euclidean geometry. And much less intuitive, more stranger to our daily experience with forms and with their spatial relationships. At the end of the 19th century, E. Beltrami demonstrated the formal consistency of non-Euclidean geometries [25], which implies that, as suspected, Euclid's postulate number 5 cannot be deduced from the other four Euclid's geometric postulates.

But why Euclid's Fifth Postulate should be a theorem and not a postulate? As shocking as it may seem, the answer is related to the role that simplicity and beauty play in the construction of scientific theories, both in the formal sciences (such as geometry) and in the experimental sciences (physics, for example). In this sense, Ockham's razor has always been a good aesthetic reference based on simplicity. And from the aesthetic point of view of simplicity, Euclid's Fifth Postulate lacks the simplicity and self-evidence expected from an axiom or postulate. And if everything indicates that it should be a theorem, why has it been impossible to prove that it was? The answer now has to do with the servitudes of human knowledge. The contrast between Euclidean and non-Euclidean geometries is quite clear:

The Hyperbolic Axiom reads:

> There exists a line l and a point P not in l such that at least two distinct coplanar lines parallel to l pass through P.

The Elliptic Axiom states:

> Through a point exterior to a given line, there is no line parallel to the given line.

While Playfair's Axiom (a variant of Euclid's Fifth Postulate) reads:

> Through a given point one, and only one, parallel can be drawn to a given straight line.

Apart from the non existence of parallels, another notable difference between Euclidean geometry and Riemann elliptic geometry is that in the latter there are infinitely many different straight lines passing through the same two points, which contradicts the strong version of Euclid's First Postulate, according to which there is only one straight line between any two points. Euclid's original statement (weak version of the First Postulate) establishes the existence of AT LEAST one straight line between two points. Hence, his statement is compatible with the existence of more than one straight line between two points. Although it does not seem probable that this was Euclid's belief, nor that of the majority of the subsequent Euclidean authors. In Appendix C, it will be proved that any two points can be the endpoints of one, and only one, straight line (according to the Definition 16 of straight line proposed above). Other abuses of language in non-Euclidean geometries, all of them related to the definition of a straight line, will be discussed in Chapter 14.

As is well known, the general theory of relativity states that, depending on its energy density at the time of its formation, the universe could be closed (elliptical geometry with curvature greater than zero), open (hyperbolic geometry of curvature less than zero) and plane (Euclidean geometry of zero curvature). In the first two cases, the al-Tutsi-Legendre version of Euclid's 5th Postulate [200] is not verified (in the elliptic case the internal angles of a triangle add up to more than 180º and in the hyperbolic case less than 180º). Also in the case of the closed universe, the universe would collapse gravitationally, while in the case of the open universe the expansion would be forever [340, 113, 233, 291, 41, 95, 91, 93, 88, 90, 367, 224, 149, 131, 290, 334, 341, 189, 40].

There is a single value for the initial energy density of the universe that separates closed universes from open universes, the value that corresponds to the flat universe. This unique density is called the critical density ρ_{crit}. The energy density of the present universe has been calculated from astronomical observations in three independent calculations: the energy density due to ordinary matter ($\rho_{om} = 0.049\rho_{crit}$), the energy density due to dark matter ($\rho_{dm} = 0.268\rho_{crit}$) and the energy density of dark energy ($\rho_{de} = 0.683\rho_{crit}$). As can be seen, the sum of the three independent measurements is precisely the critical density ρ_{crit}, which corresponds to that of the flat universe:

$$\rho_{om} + \rho_{dm} + \rho_{de} = \rho_{crit} \tag{1}$$

It should be noted that in the case of a closed universe and in the case of an open universe, the formation of structures such as galaxies would be compromised, at least in the long and very long term. And it should also be noted that the initial energy of the universe had to be such that its initial energy density could not differ from the critical density ρ_{crit} by a factor greater than 10^{-62}, an extremely small factor. Explaining this coincidence is one of the biggest problems facing cosmology today. It also has a great interest in the discussion about the origin of the universe. Obviously, equation (1) can be considered, or not, as a mere random coincidence. But in any case it would be proving that the geometry of physical space is Euclidean.

It is also well known the relativistic explanation of gravity by the general relativity: instead of a force, gravitational attraction would be caused by the local deformation of spacetime, in turn caused by the local presence of massive objects. The geometry of that deformed space would no longer be Euclidean but Riemannian. As will also be seen in Chapter 25 of this book, other physical explanations based on preinertia are possible that do not need to deform neither space nor time.

9. Space light and Gold

9.1 Introduction

This chapter examines some of the ideas about physical space that developed between the first century and the Scientific Revolution of the 16th-17th centuries. During this period the Aristotelian theory of space and place is discussed, discussions in which theology played an important role. Indeed, the words "God" and "place" were used in Alexandrian Jewish theology as practically synonymous words. An equivalence that will be introduced also in Christian theology, which became an important factor in the development of theories about space from the time of Philo (20 B.C.-50 A.D.) to Newton (1642-1727), or even later; that is, at least from the 1st century to the 18th century. With the development of these scientific theories, their corresponding authors also tried to prove in formal terms the existence of God.

9.2 The Neoplatonics

By 146 BC, Greece was already a Roman province, as was Egypt a little later. The two great centers of Greek culture, Athens and Alexandria, came under the political control of Rome. The Roman Empire (29 BC - 476 AD) came to occupy a large part of Europe, from Spain to the Rhine and from North Africa to Persia. More focused on the efficient administration of its vast empire, Rome devoted most of its efforts to the development of law, public administration, and great civil works. Neoplatonism is part of the late Hellenism that developed in Athens and Alexandria during the Roman period. It is a synthesis of Platonic elements, enriched with contributions from other great Greek authors such as Zeno and Aristotle, and Eastern mystical influences from both Hinduism and Judaism of the time. Neoplatonism, in turn, had a great influence on medieval Christian mysticism.

Plotinus (204-270), disciple of Ammonius Sachas (175-242), is one of the most significant authors of Neoplatonism. According to him, there would be a First Principle as the cause of all being, which is why that First Principle, the One, cannot be described as being; it must be understood as beyond being, as something completely indeterminate. This inevitable indeterminacy of the One is the consequence of what we will here call the potentially infinite regress of causes and the logical necessity of a first cause not explainable in terms of other causes, contained in the Theorem of the First Element, deduced from the Principle of Directional Evolution (Chapter 5). For Plotinus, light has the highest degree of existence. It is the medium that maintains the universal order and permeates the space of the whole universe. In its purest reality, light is God. According to the Cabala, the light of the Holy and Infinite One originally occupied the entire universe and then was withdrawn and concentrated into its own substance, creating empty space.

But the first unequivocal redefinition of neoplatonic space is found in the work of the Greek neoplatonic philosopher Proclus (412-485). Indeed, in Proclus we find a definition of space similar to that of the Stoics:

Space is the interval between the limits of bodies.

Consequently, it must be a magnitude commensurable with corporeal objects, although it must also be immaterial and immobile. Space contains the whole material world but is not contained by the material world, being therefore coextensive with the domain of light. This neoplatonic metaphysics of light and space will spread through Jewish philosophy and mysticism and will exert a great influence on most of the natural philosophers of the Renaissance.

The neoplatonic Damascius (458-538?) conducted a profound investigation on the nature of space. A key concept of that research was the position or location of an object. For Damascius, position is an inseparable attribute of every object, and has a double meaning: on the one hand it denotes the relative location of the different parts of the object, and on the other it signifies the position of the whole object in the universe as a whole. If position is a quality of each object, space makes it possible to determine that quality in quantitative terms. In this sense, space is the numerical measure of position.

Space is different from position in the same sense that time is different from motion. Position is an inseparable quality of the object, even when it is in motion. Position is not transferable from one object to another, although always changing it never becomes the position of another object, it simply ceases to exist when the object acquires a new position. One could say that for Damascius space functions as a kind of absolute reference frame.

Damascius also discussed another important issue related to space: the first relativistic question of whether all motion requires the existence of a body at rest. Euclid had already weighed in, saying that an object can appear to be at rest to an observer walking toward it, or it can appear to be in motion to the observer if the observer considers himself to be at rest. But in general, Aristotelian scientists and philosophers thought that motion required the existence of an immovable object (which might imply the immobility of the Earth).

For Damascius, on the contrary, motion does not presuppose the existence of an immobile object, only our perception of motion requires that existence. As Galileo would later say, humans do not have sensors to perceive uniform rectilinear motion [126, p. 529] (as we do, for example, to perceive temperature). Although Damascius still maintains the traditional doctrine of natural places, which remain fixed and motionless, i.e., independent of the actual motion of the concrete parts of the universe. The natural place is, for Damascius, the directing force towards perfection.

J. Philoponus, also called John the Grammarian (490-566), discovered an inconsistency in Aristotle's theory of space: what is the (Aristotelian) place of the sublunar world? According to Aristotle it is the concave surface of the first celestial sphere, which is the orbit of the moon. That sphere is in rotation, but the rotation of the sphere was not considered motion because the sphere, as such a sphere, remains always in the same place. Philoponus disagreed: the parts of the sphere move because they occupy different places over time. Therefore, if the place of an object has to be the first immobile envelope of that object, the place of the sublunar world could not be the concave sphere of the first celestial sphere. Moreover, there is the problem of the place or space in which the outermost celestial sphere moves, because there is no space outside.

For Philoponus, therefore, a new definition of place and space was necessary: space would be a three-dimensional incorporeal volume, different from the objects contained in it. Space and emptiness would be identical. Any region of space could successively receive different material bodies, but space would not intervene in the movement of objects. If objects move toward their natural places it is not because of the intervention of space but because of their tendencies to reach the places that the Demiurge has assigned to them. Changing the Demiurge for the laws of physics, the story of Philoponus acquires a certain relevance. At this point it seems appropriate to recall the words of Copernicus (quoted in Latin in [176, p. 57]):

> In fact, I believe that gravity is nothing other than a certain natural desire given by divine providence to the parts of the

universe to come together in unity and integrity in the form of a group.

9.3 Arab and Judeo-Christian ideas about space

This section briefly analyzes the path followed by classical Greek science, together with Indian and Arabic science, towards the Christian world. An essential first step for the birth of universal modern science from the so-called Scientific Revolution of the sixteenth and seventeenth centuries.

SCIENCE IN THE ROMAN EMPIRE

As noted above, by 146 BC, the Roman Empire (29 BC - 476 AD) came to occupy a large part of Europe (from Spain to the Rhine and Persia) and North Africa. More focused on the efficient administration of its huge empire, Rome devoted most of its efforts to the development of law, public administration and great civil works. In 320 AD it recognized the Christian religion, which gradually became the majority religion.

The pressure of the "barbarbarian" peoples of the north eventually led to the fall of the western part of the Roman Empire in 476. The eastern part of the Empire held out until 1453, when the Turks conquered Constantinople. But the barbarian peoples settled in the former western part of the Empire quickly became Romanized, and also Christianized (even before they became Romanized). The common language of the Empire, Latin, remained the common cultured language throughout western Europe until the 18th century, which facilitated the spread of culture throughout the different European regions.

Philosophy aroused less interest in Rome than in Greece. Few Greek authors were translated into Latin in the Western Roman Empire: Plato's Timaeus, Nicomachean Arithmetic and some of Euclid's books. Scientific knowledge was mainly oriented towards practical applications related to surveying and major public works. In this regard, Cicero writes (quoted in [352, p. 68]:

> The Greeks gave the geometrician the highest honor; according to this nothing had a more brilliant progress than mathematics. But we have set as the limit of this art its usefulness for measuring and counting.

Among the authors of scientific-practical works and encyclopedias of the Roman period, the following stand out:

1. Titus Lucretius Carus (198-55 BC)): *De rerum natura*, a work written in verse on Epicurean atomic theory. The work begins with a

very significant principle for contemporary science[1] [221, p. 98-101]:

> We will begin with a principle of his:
> no thing is born from nothing;
> the divine essence cannot do it:
> though it represses all mortals
> fear so that they are inclined
> to believe produced by the gods
> many things of heaven and earth,
> because they do not understand their causes.
> [...]
> To this is added the fact that nature
> annihilates nothing, but reduces
> everything to its primitive bodies;

Note the second verse and compare it with the last statement of the following theorem which will be formally proved later (Chapter 5):

Theorem of Formal Dependence: No concept defines itself; no statement proves itself; no physical object is the cause of itself; and no cause is the cause of itself.

Among many other things, Lucretius' work denies action at a distance and infinite division; gives an explanation of colors, sounds and atmospheric phenomena; and expounds a corpuscular theory of light and heat.

2. Marcus Vitruvius Pollio (s. I A.C.): *De architectura*, in which a great variety of problems related to architecture and some strictly scientific problems collected from Greek authors, for example the theory of sound due to waves through the air, are dealt with.

3. Marcus Terentius Varro (116-27 BC): *Disciplinae*, first encyclopedia written in Latin.

4. Marcus Tullius Cicero (106-43 BC): *Somnium Scipionis*, which includes a description of Greek geocentric cosmology.

5. Sextus Iulius Frontinus (40-103 AC): *De aquis urbis Romae*, on aqueducts and water conduction, including some general laws of hydraulics of Greek origin.

6. Lucius Annaeus Seneca (4 BC-65 AC): *Questions*, 7 books on natural phenomena taken from Greek books on meteorology.

7. Gaius Plinius Secundus (Pliny the Elder) (23-74 AC): *Historia naturalis*, 37 books covering most of the knowledge available at the time.

---

[1] Theorem of the First Element, Corollary of the First Cause and Theorem of the Arrow of Time [206].

8. Martianus Minneus Felix Capella (365-440 AC): *De nuptiis Mercurii et Philologiae*, includes the theory of Heraclides according to which Venus and Mercury revolve around the Sun.

Indian science in the 5th-13th centuries

Between 3000 and 2000 BC, the Indo-Aryan peoples settled in India, and with them began its cultural development. The oldest books (the Vedas) date back to 1500 BC. It was not an isolated culture but maintained cultural contacts with Babylon, Persia, Greece and the Roman Empire. Its cyclical conception of time is well known, linked to its religion, in which there was no real separation between the divinity (Brahma) and the physical world. In contrast to Greece, India paid considerable attention to arithmetic and algebra, focusing more on calculation than on proving, which was undoubtedly due to its excellent numbering system, which would eventually become universal (albeit with Arabic symbols).

The positional (sexagesimal) number system originated in Babylon by the Sumerians, although it was eventually lost. The Indian decimal numbering system was also positional and probably originated in the 7th century. The Arabs copied it, incorporated their own symbols for the first ten numbers (Arabic numerals: 0, 1, 2, 3,... 9), and exported it to the territories of their empire. Over time it spread to the rest of the world. Today it is a universal numbering system, with some variations (binary, octal, decimal and hexadecimal). As is well known, in the decimal numbering system the value of each digit in the expression of a number depends on its position in that expression, the value of each position being 10 times greater than that of the previous position, to the left of its writing. The decimal numbering system allows numbers of any size to be represented. Although the latter is just a figure of speech. Indeed, one can define natural numbers (the counting numbers, all of which are finite) so enormous that there is not enough matter in the Universe to represent them graphically at a standard scale of for example 5mm/digit (see Chapter 16).

In addition to the decimal numbering system, including the use of zero, Indian mathematicians of the time also developed algorithms for the four basic arithmetic operations, both with positive and negative numbers, as well as basic notions of trigonometry and methods of solving first-degree and second-degree equations and systems of equations. They also knew some basic theorems of plane geometry. All this Indian mathematical knowledge would pass to the West through Arab authors, including their own contributions.

Arab science in the 7th-13th centuries

Islam as a religious doctrine was established in the Koran around the year 640. From that date, the Umayyads spread throughout North Africa and entered southern Europe, invading part of the Iberian Peninsula. In Cordoba (Spain), an independent caliphate is created. Baghdad and Cordoba become the most important cultural centers of the time. Astronomy and mathematics are the most developed disciplines in both centers. From the year 800 onwards, translations of the great Greek and Indian authors began, which, in addition to being translated into Arabic, were widely commented on. This work was completed in approximately 100 years. From the 9th to the 12th centuries, a period of great splendor developed, especially in mathematics and astronomy. The incursions of the Turks in the 14th century caused the beginning of the decline of this period of splendor in Arab scientific culture, not sufficiently recognized in our arrogant western world.

Among the most noteworthy aspects of this period of Arab scientific splendor, the following can be highlighted:

1. Introduction and use of the Indian decimal numbering system, with its own symbols, including zero.
2. Calculation procedures for square and cube roots.
3. Modern use of fractions.
4. Development of the algebra of first and second degree equations.
5. Method for calculating areas and volumes precursor of integrals.
6. Development of trigonometry applied to astronomy.
7. Important development of astronomy with the construction of astronomical observatories and the perfection of instruments such as the astrolabe.
8. Thabit ibn Kurra (836-901 AC) raises a very significant numerical paradox, antecedent of Galileo's infinitist paradox [318, 78, 212].
9. Development of applied mathematics, especially in optics and mechanics.
10. Interpretation of light as particle rays with finite velocity and different for each transparent medium.
11. Establishment of the laws of the reflection of light.
12. Study of the refraction of light, establishing some of its laws, although not the law of sines.
13. Interpretation of the rainbow as a phenomenon caused by the interaction of light with water droplets in the air.
14. Translation and commentary of all the great philosophical and

mathematical works of the Greeks. In this regard, it is appropriate to recall the following words of Alhazen (965-1040), quoted in [352, p. 79]:

> The seeker of truth is not one who studies the writings of the ancients and following their natural disposition puts his trust in them, but, rather, one who suspects his faith in them and questions what they present [...] Thus the obligation of one who investigates the writings of the philosophers, if he seeks to learn the truth, is to make himself an enemy of all he reads, and applying his mind to their contents, to attack them from all their angles[2].

15. Critique of Aristotle's theory of motion, proposing ideas similar to the theory of momentum (linear momentum).

16. Studies of static equilibrium and analysis of the centers of gravity.

17. Averroes (1126-1198 AC), the great commentator of Aristotle, maintains that there are two ways to reach the truth: reason and the revelation of the Koran.

18. According to Averroes himself, nothing in the world is born or destroyed, but is transformed, an idea that underlies the later statements of the Principle of Conservation of Matter.

Latin translations of the Greek authors

After its reconquest in 1085 by Alfonso VI of Castile, the city of Toledo (Spain) became the center of contact between Christian, Arab and Jewish cultures. At the same time, translations into Latin of Greek works, previously translated into Arabic, and of Arabic works began. At the end of the 11th century, Sicily and southern Italy also became centers of translation of Greek works. In both cases, the translated works were almost exclusively scientific. It is not possible to understand the subsequent history of European science, since the Scientific Revolution, without these translations. They occupy, therefore, a relevant place in the history of physics, including our debate on the reality of physical space.

With the Greek works were also translated the commentaries of their Arabic translators, and some scientific works of the Arabs themselves, including their Arabic numerals, the decimal system of numeration, and Indian arithmetic and algebra, as they are still used today throughout the world. This, together with the translation of Aristotle's Logic, Euclid's geometry and optics, Archimedes' mechanics and the Arabic works on optics and mechanics (by Alhazen and Al Farisi (1260-1320 AC), for example), led to a true scientific revolution in the twelfth and

---

[2] A strategy that should be applied today by all future scientists.

thirteenth centuries. The religious and anti-religious intolerances of our time should look to this wonderful light of knowledge instead of their present dark prejudices.

From the same period are the monastic schools and the European cathedral schools, which from the 12th century onwards will become universities, as is the case of Paris, Oxford, Salamanca, etc. Their role in the development of science will be fundamental up to the present day, including what in my opinion is still a very negative aspect for scientific progress: the intolerant nature of the main currents of thought, which emerged, and continue to emerge, in university centers all over the world. It is worth recalling at this point the words of Adelard of Bath (1080-1152 AC) (quoted in [352, p. 93]):

> One thing is what I have learned from the Arabian masters, under the guidance of reason, and another thing is what you, seduced by the mask of authority, are tied to like a yoke. For what other name but yoke is there to give to authority? You allow yourselves to be led by authority like animals that do not know where they are being led or why.

Fortunately, there are no longer prison sentences (or worse) for intellectual dissidence, but there are sentences of insult and ostracism. I attest to this.

ATOMIC THEORY OF SPACE IN KALAM

Once Greek philosophy became familiar in the Arab world, particularly in the case of the Umayyads, the authority of Aristotle prevailed in almost all physical and metaphysical matters. One of the few exceptions was the Aristotelian theory of space, against which a discrete (atomic) alternative was constructed within the philosophical and theological current known as Kalam (9th-10th century), perhaps comparable to medieval Christian scholastic mysticism. Here, too, the dialectical method was used as a support for theological speculations. The atomic theory of Kalam did not originate in religious speculations, although it was from the religious background that it drew its emotional force of conviction. Kalam was defined as the science of the fundamentals of faith and intellectual proofs in support of theological truths.

According to Kalam, matter is formed by indivisible particles, atoms, equal to each other and devoid of spatial extension. The extension arises, in the three spatial directions, from the establishment of relations between different atoms. The existence of these atoms would be transient, of a very short duration, which requires the continuous divine intervention to maintain the coherence and continuity of the universe. Another important characteristic of the atomic doctrine of

Kalam is the necessary existence of empty space, which, like matter, must also be formed by indivisible units, and the same must apply to time.

Consequently, motion must also be discrete, in jumps; it must consist of a discontinuous succession of jumps in each of which successive positions in space will be occupied. The successive jumps would be separated by one or more discrete units of time in which the corresponding atoms remain at rest, in the same units of space. Slower objects separate their jumps by a greater number of units of time in which they remain occupying the same atoms of space. Interestingly, the theory of physical space proposed at the end of this book, although deduced exclusively in physical and mathematical terms, is reminiscent in many respects of the theory of Kalam.

JUDEO-CHRISTIAN IDEAS ABOUT SPACE

There is sufficient evidence that the Judeo-Christian religious tradition exerted a remarkable influence on the development of physical theories of space from the first to the eighteenth century. Space was nothing but an attribute of God, even the same thing as God. In Palestinian Jewish culture the word "place" was frequently used to denote God. And the divine omnipresence, God occupies all space, is the consequence of a long process of theological thought (which did not occur in the polytheistic religions). As we shall see later in this section, for H. More space is the divine extension. And remember that for Newton space is the divine sensorium.

Another trend in the history of theories of space, very similar to its mystical-theological character and the association of God with space, was the identification of space with light. Light is the medium in which God becomes visible to man: *Ego sum lux mundi*. As noted above, the Infinite Sacred Oneness, whose light originally occupied the entire universe, withdrew its light and concentrated on its own substance, thereby creating empty space. This apotheosis of light became a fundamental feature of late Neo-Platonism and medieval mysticism.

The Franciscan R. Grosseteste (1175-1253) was one of the first scholastics who defended the neoplatonic metaphysics of light: he assumed that light was the first corporeal form and the first principle of motion. The creation of space in the universe was the self-diffusion of light, which according to Grosseteste propagates instantaneously, as can be proved, according to Grosseteste himself, with visible light, which is the basis of spatial extension. Hence the importance of the study of optical geometry and the great interest in mathematics and optics in the 13th century. Light was the means by which universal order was maintained. In its pure reality light was God. In the words of St.

Bonaventure (1221-1274), God was the spiritual light that is actualized in all the senses. The Polish friar Witelo (1230-1314), clearly influenced by Grosseteste, is another Neoplatonic advocate of the identity of light and space: light is the source of all existence, the all-pervading power. Space and light are one.

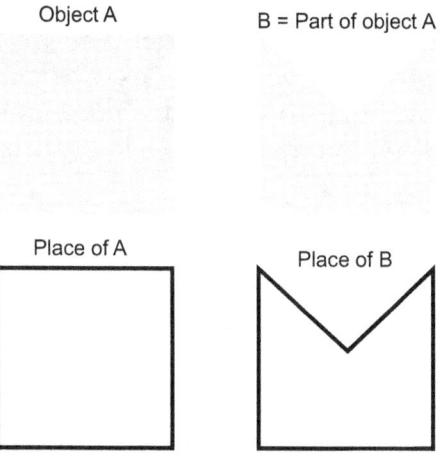

Figure 9.1 – Crescas' Paradox: The aristotelian place of the whole (Left) is less than the aristotelian place of one if its parts (Right).

The Jewish philosopher and jurist Hasdai Crescas (1340-1412) discovered some inconsistencies in the Aristotelian theory of place, among them the so-called Crescas Paradox: the Aristotelian place of the whole can be less than the place of one of its proper parts (Figure 9.1). Consequently, he proposed to change the Aristotelian definition so that the place occupied by any object would always be equal to the sum of the places of its parts, whatever the division of the object into parts.

The theologian and scientist Nicholas of Cusa (1401-1464) offered another solution to the Aristotelian problem of the definition of place: The circumference and the center of the universe could only be God, but from the physical point of view it is absurd to be at the same time the circumference and its center. There is, then, neither the circumference nor the center of the universe. Consequently, the Earth is not the center of the universe or of space, nor is a body at rest necessary for motion to exist, which eliminates the possibility of absolute motion. It would therefore be a relativistic theory of position.

For Tomaso Campanella (1568-1639) space was an absolute spiritual entity characterized by divine attributes. Space was homogeneous and undifferentiated, immobile and incorporeal, penetrated by matter and penetrating matter, destined for the placement of mobile entities. Campanella states that space is in God, but God is not limited by space, space is His "divine creature". It is important to note that P. Gassendi (1592-1655) was in contact with Campanella, and Gassendi was in contact with Newton (1642-1727) and Leibniz (1646-1716).

R. Descartes (1596-1650) identified matter with extension, its key quality. There would be three kinds of matter: ether, the most subtle, identified with space itself; luminous matter, which forms the Sun and the stars; and denser matter, which forms the Earth and the planets. *Empty space* does not exist: a vessel from which all air is extracted must collapse. Space is thus matter, not a separate entity containing matter. Euclidean geometry correctly describes space, and was used by Descartes (and Leibniz) to introduce coordinate systems. As for motion, Descartes explains it on the basis of three fundamental laws [79, p. 421-422]:

1. Each thing remains in the state in which it is as long as nothing modifies that state.
2. Every body that moves tends to continue its movement in a straight line.
3. If a moving body collides with another body stronger than itself, it does not lose any of its movement, but if it meets another weaker body that can move, it loses as much movement as it communicates to the other.

Naturally, the Cartesian theory of space has to overcome the Aristotelian objection that the place occupied by matter and matter itself must be separable. Descartes proposes two ways of conceiving the problem of the extension occupied by a moving object:

1. The extension of a moving object moves with the object: particular non-separable extension.
2. The moving body moves with respect to a fixed extent defined by its position relative to other objects: generic separable extent.

But although it can be thought of as separable and non-separable, in reality this is not possible at the same time. When an object moves, does it remain made up of the same matter, or does the matter remain for the properties to move and attach to successive pieces of matter? The solution Descartes seems to give is that although the matter and place of an object are made of the same substance at any given time, its place is to be identified by its form and relative location.

An outstanding example in which a strong religious bias can be seen in the conception of space is the theory of Henry More (1614-1687). More considers it necessary to complete Descartes' science with cabalistic and Platonic concepts. As for his theory of space, More refers to the cabalistic doctrine as explained by Cornelius Agrippa (1486-1535) in his *De occulta philosophia*, in which space is specified as one of the attributes of God (quoted in [176, p. 42]):

9.3 Arab and Judeo-Christian ideas about space

> And so through that same door through which Cartesian philosophy seems to want to exclude God from the world, I, on the contrary, enter again and strive to introduce him.

The great motive behind More's preoccupation with the problem of space, like that of his whole philosophy, is to find a convincing demonstration of the indubitable reality of God, spirit and soul. He rejects then the Cartesian identification of matter with extension. To demonstrate the existence of spirit it is enough to show that extension is spiritual, provided that extension is real. On the basis of this reasoning, More's treatment of space can be divided into three parts:

1. Extension is not the distinguishing attribute of matter. Matter has the property of impenetrability; impenetrability is the distinguishing criterion between matter and extension. Space is the common ground between the world of matter and the world of spirit. Extension is not an attribute of matter alone but of matter and spirit.

2. Space is real, with real attributes. *Empty space* does not exist for More. But if space is empty of matter it will be filled with spirit. The existence of space is guaranteed by its own measurability. Since there is no accident (measurement) without cause, its measurement proves its existence. Space is incorporeal because it is penetrable, which proves that it is different from matter.

3. Space is of divine character. The necessary existence of space, even without matter, leads More to the final identification of space with God. Space and God both have the property of necessary existence, they are thus the same thing.

More had a great influence on J. Locke (1632-1704), I. Newton (1643-1727) and S. Clarke (1675-1729) and on the eighteenth-century philosophy.

Some Arabic translators and commentators of Aristotle's work had updated the old contradiction between the Aristotelian definitions of place and motion: the ultimate celestial sphere would be in motion without being able to be in motion, since there is nothing beyond the ultimate sphere itself, that ultimate sphere would have to move without a place in which to move. It was necessary to change at least one of the two definitions. William of Ockham (1287-1347) proposed to use the notion of distance from an object to another object of reference to define the object's position. The immobility of a given place was reduced to the constancy of its distance to a given reference body (quoted in [176, p. 72]):

> ... If you are at rest, and even if all the air around you, or any body which surrounds you, is moving, your are always at the same place; for you are always at the same distance

from the center of the poles of the universe. With regard to these the place is therefore called immobile.

On the contrary, N. Copernicus (473-1543) was in favor of eliminating the rotational motion of the outer celestial sphere [72, Book 1, Chapter 5, P. 23]:

> Moreover, since the heavens, which enclose and provide the setting for everything, constitute the space common to all things, it is not at first blush clear why motion should not be attributed rather to the enclosed than to the enclosing, to the thing located in space rather than to the framework of space. This opinion was indeed maintained by Heraclides and Ecphantus, the Pythagoreans, and by Eficetas of Syracuse, according to Cicero. They rotated the earth in the middle of the universe, for they ascribed the setting of the stars to the earth's interposition, and their rising to its withdrawal.

The consequences of these ideas of Copernicus are well known, and will be developed in the rest of the book.

10. Newton absolute space

10.1 Introduction

Newton's Mathematical Principles of Natural Philosophy (Principia) [260] is rightly considered one of the essential works in the history of human thought. First published in 1687, it is still being published and its study (direct or indirect) is still mandatory to know the foundations of classical mechanics, and even of modern mechanics. Newton's Principia represent for mechanics what Euclid's Elements represent for geometry: they contain the first hypothetico-deductive model for rational mechanics. The model of space used in the Principia is, moreover, the Euclidean three-dimensional model.

As is well known, Newton defended the absolute nature of space, time and motion. Or in other words: for Newton, space, time and motion were real, not fictitious or relative. This chapter is devoted to these three absolute (real) notions of Newton, which, moreover, are also closely related to those proposed in this book, although here the general perspective of the continuous will be exchanged for the discrete and finite perspective. By way of dialectical contrast, the chapter also includes the relational view of space defended by Leibniz during Newton's own time, Although Newton's absolute space prevailed over Leibniz's relational space, the latter will be the only one considered by modern physics, which, as everyone knows, is essentially relativistic. A relational view that will eventually prevail until it becomes the practically unique view of space, time and motion in contemporary physics. Dissent also exists, but in a very marginal way, without echo in the "orthodox" scientific community of our days (2024).

10.2 The formal language of Newton's Principia

As the first hypothetico-deductive construction of mechanics, the Principia are compatible with more than one interpretation of the physical world, which among other attributes could be infinite or finite, continuous or discrete, although those matters are not discussed in Newton's text. The Principia consist of three books:

Books I y II: On the motion of bodies.

Book III: On the world system.

The formal content of the three books is richer than that of Euclid's Elements, and includes: definitions, laws, lemmas, propositions, theorems, corollaries, problems, additional hypotheses, additional definitions, rules for philosophizing, phenomena, scholia and examples; distributed as follows:

1. Book I: 8 definitions; 3 laws with 6 corollaries; 29 lemmas; 98 propositions, of which 50 are theorems; 48 problems.

2. Book II: 1 additional definition (of fluid); 1 additional hypothesis; 7 lemmas; 53 propositions, of which 41 are theorems, and 12 problems.

3. Book III: 4 rules for philosophizing; 2 additional hypotheses; 6 phenomena; 11 Lemmas; 42 propositions of which 20 are theorems and 22 problems.

   Some of the lemmas, propositions and theorems also include corollaries and scholia (commentaries).

Geometrical representations and proves are very abundant in the Principia, where geometry (and mathematics in general) have a very realistic, or naturalistic, use: they are not a simple set of logically related formal elements but in a certain way a part of mechanics. The laws (axioms) and hypotheses of the Principia are not arbitrary statements disconnected with the real world; for Newton they are an inevitable consequences of the immediate experience with the physical world. Although he makes a distinction between science and metaphysics (quoted in [176, p. 97]):

> I consider here mathematical quantities, not as composed of the smallest possible parts, but as described by continuous motion. These genes have a real place in nature and are seen in the motion of bodies every day.

Newton sought to separate science from the religious and transcendental, with the sole exception of space, one of the four fundamental concepts for the study of nature. The other three are time, mass, and force, which are as real as space. Although, as is always the case with the most basic concepts, their definitions are imprecise or circular. Even Newton's definitions. This inevitable limitation of human knowledge is hardly ever acknowledged. Not even in the Principia. Here we will devote a good part of Chapter 5 to it.

Unlike the Cartesian model, not only space ceases to be relative to be absolute, i.e. real, but also extension ceases to be the (Cartesian) fundamental attribute of matter, which is now its mass. Although the definition given by Newton is circular. The same happens with the definition of impressed force and centripetal force [260, p. 121-123]:

> Definition I. *The quantity of matter is the measure of the same, arising from its density and bulk conjunctly.*
>
> Definition IV. *An impressed force is an action exerted upon a body, in order to change its state, either of rest, or of moving uniformly forward in a right line.*
>
> Definition V. *A centripetal force is that by which bodies are drawn or impelled, or any way tend, towards a point as to a centre.*
>
> (The innate force (inertia) of Newton's Definition III is discussed in Chapter 19 of this book).

Newton immediately calls the quantity of matter of Definition I *mass* or *body* [260, p. 121]. The primitive nature of the concepts of space and time are discussed beginning in the next section of this chapter and in other chapters of this book.

10.3 Space, time and motion in the Principia

In the scholium to his first eight Initial Definitions, Newton tells us that he thought it appropriate to explain the lesser known terms and the sense in which they are to be taken in the future [260, p. 127]. These are the terms included in his first eight Initial Definitions. As for time, space, place and motion, Newton tells us in that same scholium:

> I do not define time, space, place and motion, as being well known to all.

It is Newton's (not too honest) way of getting around the inevitable problem of primitive concepts. Or perhaps he did not realize that this was indeed the problem he faced. And this was done by, in my opinion, the most important scientist in the history of science, to whom I once again express my deepest admiration.

Although there is no formal definition of space (if there were, it would have to be in terms of at least another more basic concepts that would also have to be defined, or declared as new primitive concepts), in the scholium that follows his first eight Initial Definitions, Newton explains the differences between absolute space, time and motion and their corresponding relative versions, as well as how absolute motion could be detected experimentally [260, p. 127-134]:

> Hitherto I have laid down the definitions of such words as

are less known, and explained the sense in which I would have them to be under stood in the following discourse. I do not define time, space, place and motion, as being well known to all. Only I must observe, that the vulgar conceive those quantities under no other notions but from the relation they bear to sensible objects. And thence arise certain prejudices, for the removing of which, it will be convenient to distinguish them into absolute and relative, true and apparent, mathematical and common.

I. Absolute, true, and mathematical time, of itself, and from its own nature flows equably without regard to anything external, and by another name is called duration: relative, apparent, and common time, is some sensible and external (whether accurate or unequable) measure of duration by the means of motion, which is commonly used instead of true time; such as an hour, a day, a month, a year.

II. Absolute space, in its own nature, without regard to anything external, remains always similar and immovable. Relative space is some movable dimension or measure of the absolute spaces; which our senses determine by its position to bodies; and which is vulgarly taken for immovable space; such is the dimension of a subterraneous, an aereal, or celestial space, determined by its position in respect of the Earth. Absolute and relative space, are the same in figure and magnitude; but they do not remain always numerically the same. For if the Earth, for instance, moves, a space of our air, which relatively and in respect of the Earth remains always the same, will at one time be one part of the absolute space into which the air passes; at another time it will be another part of the same, and so, from the absolute point of view, it will be perpetually mutable.

III. Place is a part of space which a body takes up, and is according to the space, either absolute or relative. I say, a part of space; not the situation, nor the external surface of the body. For the places of equal solids are always equal; but their surfaces, by reason of their dissimilar figures, are often unequal. Positions properly have no quantity, nor are they so much the places themselves, as the properties of places. The motion of the whole is the same thing with the sum of the motions of the parts; that is, the translation of the whole, out of its place, is the same thing with the sum of the translations of the parts out of their places; and therefore the place of the whole is the same thing with the sum of the

10.3 Space, time and motion in the Principia

places of the parts, and for that reason, it is internal, and in the whole body.

IV. Absolute motion is the translation of a body from one absolute place into another; and relative motion, the translation from one relative place into another. Thus in a ship under sail, the relative place of a body is that part of the ship which the body possesses; or that part of its cavity which the body fills, and which therefore moves together with the ship: and relative rest is the continuance of the body in the same part of the ship, or of its cavity. But real, absolute rest, is the continuance of the body in the same part of that immovable space, in which the ship itself, its cavity, and all that it contains, is moved. Wherefore, if the Earth is really at rest, the body, which relatively rests in the ship, will really and absolutely move with the same velocity which the ship has on the Earth. But if the Earth also moves, the true and absolute motion of the body will arise, partly from the true motion of the Earth, in immovable space; partly from the relative motion of the ship on the Earth; and if the body moves also relatively in the ship; its true motion will arise, partly from the true motion of the Earth, in immovable space, and partly from the relative motions as well of the ship on the Earth, as of the body in the ship; and from these relative motions will arise the relative motion of the body on the Earth. As if that part of the Earth, where the ship is, was truly moved toward the east, with a velocity of 10010 parts; while the ship itself, with a fresh gale, and full sails, is carried towards the west, with a velocity expressed by 10 of those parts; but a sailor walks in the ship towards the east, with 1 part of the said velocity; then the sailor will be moved truly in immovable space towards the east, with a velocity of 10001 parts, and relatively on the Earth towards the west, with a velocity of 9 of those parts.

Absolute time, in astronomy, is distinguished from relative, by the equation or correction of the vulgar time. For the natural days are truly unequal, though they are commonly considered as equal, and used for a measure of time; astronomers correct this inequality for their more accurate deducing of the celestial motions. It may be, that there is no such thing as an equable motion, whereby time may be accurately measured. All motions may be accelerated and retarded; but the true, or equable, progress of absolute time is liable to no change. The duration or perseverance of the existence of things remains the same, whether the motions are swift or slow, or none at all: and therefore it ought to be distinguished from what are only sensible measures thereof; and out of which we collect it, by means of the astronomical equation. The necessity of which equation, for determining the times of a phenomenon, is evinced as well from the experiments of the pendulum clock, as by eclipses of the satellites of Jupiter.

As the order of the parts of time is immutable, so also is the order of the parts of space. Suppose those parts to be moved out of their places, and they will be moved (if the expression may be allowed) out of themselves. For times and spaces are, as it were, the places as well of themselves as of all other things. All things are placed in time as to order of succession; and in space as to order of situation. It is from their essence or nature that they are places; and that the primary places of things should be moveable, is absurd. These are therefore the absolute places; and translations out of those places, are the only absolute motions.

But because the parts of space cannot be seen, or distinguished from one another by our senses, therefore in their stead we use sensible measures of them. For from the positions and distances of things from any body considered as immovable, we define all places; and then with respect to such places, we estimate all motions, considering bodies as transferred from some of those places into others. And so, instead of absolute places and motions, we use relative ones; and that without any inconvenience in common affairs; but in philosophical disquisitions, we ought to abstract from our senses, and consider things themselves, distinct from what are only sensible measures of them. For it may be that there is no body really at rest, to which the places and motions of others may be referred.

But we may distinguish rest and motion, absolute and relative, one from the other by their properties, causes and effects. It is a property of rest, that bodies really at rest do rest in respect to one another. And therefore as it is possible, that in the remote regions of the fixed stars, or perhaps far beyond them, there may be some body absolutely at rest; but impossible to know, from the position of bodies to one another in our regions whether any of these do keep the same position to that remote body; it follows that absolute rest cannot be determined from the position of bodies in our regions.

It is a property of motion, that the parts, which retain given positions to their wholes, do partake of the motions of those wholes. For all the parts of revolving bodies endeavour to recede from the axis of motion; and the impetus of bodies moving forward, arises from the joint impetus of all the parts. Therefore, if surrounding bodies are moved, those that are relatively at rest within them, will partake of their motion. Upon which account, the true and absolute motion of a body cannot be determined by the translation of it from those which only seem to rest; for the external bodies ought not only to appear at rest, but to be really at rest. For otherwise, all included bodies, beside their translation from near the surrounding ones, partake likewise of their true motions; and though that translation were not made they would not be really at rest, but only seem to be so. For the surrounding bodies stand in the like relation to the surrounded as the exterior part of a whole does to the interior, or as the shell does to the kernel; but, if the shell moves, the kernel will also move, as being part of the whole, without translation of the shell vicinity.

A property, near akin to the preceding, is this, that if a place is moved, whatever is placed therein moves along with it; and therefore a body,which is moved from a place in motion, partakes also of the motion of its place. Upon which account, all motions, from places in motion, are no other than parts of entire and absolute motions; and every entire motion is composed of the motion of the body out of its first place, and the motion of this place out of its place; and so on, until we come to some immovable place, as in the before-mentioned example of the sailor. Wherefore, entire and absolute motions can be no otherwise determined than by immovable places: and for that reason I did before refer those absolute motions to immovable places, but relative ones to movable places. Now no other places are immovable but those that, from infinity to infinity, do all retain the same given position one to another; and upon this account must ever remain unmoved; and do thereby constitute immovable space.

The causes by which true and relative motions are distinguished, one from the other, are the forces impressed upon bodies to generate motion.True motion is neither generated nor altered, but by some force impressed upon the body moved: but relative motion may be generated or altered without any force impressed upon the body. For it is sufficient only to impress some force on other bodies with which the former is compared, that by their giving way, that relation may be changed, in which the relative rest or motion of this other body did consist. Again, true motion suffers always some change from any force impressed upon the moving body; but relative motion does not necessarily undergo any change by such forces. For if the same forces are likewise impressed on those other bodies, with which the comparison is made, that the relative position may be preserved, then that condition will be preserved in which the relative motion consists. And therefore any relative motion may be changed when the true motion remains unaltered, and the relative may be preserved when the true suffers some change. Upon which accounts; true motion does by no means consist in such relations.

The effects which distinguish absolute from relative motion are, the forces of receding from the axis of circular motion. For there are no such forces in a circular motion purely relative, but in a true and absolute circular motion, they are greater or less, according to the quantity of the motion. If a vessel, hung by a long cord, is so often turned about that the cord is strongly twisted, then filled with water, and held at rest together with the water; after, by the sudden action of another force, it is whirled about the contrary way, and while the cord is untwisting itself, the vessel continues for some time in this motion; the surface of the water will at first be plain, as before the vessel began to move: but the vessel, by gradually communicating its motion to the water, will make it begin sensibly to revolve, and recede by little and little from the middle, and ascend to the sides of the vessel, forming itself into a concave figure (as I have experienced), and the swifter the motion becomes, the higher will the water rise, till at last, performing its revolutions in the same times with the vessel, it becomes relatively at rest in it. This ascent of the water shows its endeavour to recede from the axis of its motion; and the true and absolute circular motion of the water, which is here directly contrary to the relative, discovers itself, and may be measured by this endeavour. At first, when the relative motion of the water in the vessel was greatest, it produced no endeavour to recede from the axis; the water showed no tendency to the circumference, nor any ascent towards the sides of the vessel, but remained of a plain surface, and therefore its true circular motion had not yet begun. But afterwards, when the relative motion of the water had decreased, the ascent thereof towards the sides

10.3 Space, time and motion in the Principia

of the vessel proved its endeavour to recede from the axis; and this endeavour showed the real circular motion of the water perpetually increasing, till it had acquired its greatest quantity, when the water rested relatively in the vessel. And therefore this endeavour does not depend upon any translation of the water in respect of the ambient bodies, nor can true circular motion be defined by such translation. There is only one real circular motion of any one revolving body, corresponding to only one power of endeavouring to recede from its axis of motion, as its proper and adequate effect; but relative motions, in one and the same body, are innumerable, according to the various relations it bears to external bodies, and like other relations, are altogether destitute of any real effect, any otherwise than they may perhaps partake of that one only true motion. And therefore in their system who suppose that our heavens, revolving below the sphere of the fixed stars, carry the planets along with them; the several parts of those heavens, and the planets, which are indeed relatively at rest in their heavens, do yet really move. For they change their position one to another (which never happens to bodies truly at rest), and being carried together with their heavens, partake of their motions, and as parts of revolving wholes, endeavour to recede from the axis of their motions.

Wherefore relative quantities are not the quantities themselves, whose names they bear, but those sensible measures of them (either accurate or inaccurate), which are commonly used instead of the measured quantities themselves. And if the meaning of words is to he determined by their use, then by the names time, space, place and motion, their measures are properly to be understood; and the expression will be unusual, and purely mathematical, if the measured quantities themselves are meant. Upon which account, they do strain the sacred writings, who there interpret those words for the measured quantities. Nor do those less defile the purity of mathematical and philosophical truths, who confound real quantities themselves with their relations and vulgar measures.

It is indeed a matter of great difficulty to discover, and effectually to distinguish, the true motions of particular bodies from the apparent; be cause the parts of that immovable space, in which those motions are per formed, do by no means come under the observation of our senses. Yet the thing is not altogether desperate: for we have some arguments to guide us, partly from the apparent motions, which are the differences of the true motions; partly from the forces, which are the causes and effects of the true motions. For instance, if two globes, kept at a given distance one from the other by means of a cord that connects them, were revolved about their common centre of gravity, we might, from the tension of the cord, discover the endeavour of the globes to recede from the axis of their motion, and from thence we might compute the quantity of their circular motions. And then if any equal forces should be impressed at once on the alternate faces of the globes to augment or diminish their circular motions, from the increase or decrease of the tension of the cord, we might infer the increment or decrement of their motions: and thence would be found on what faces those forces ought to be impressed, that the motions of the globes might be most augmented; that is, we might discover their hinder most faces, or those which, in the circular motion, do follow. But the faces which follow being known, and consequently the opposite ones that precede, we should likewise know the determination of their motions. And thus we might find both the quantity and the determination of this circular motion, even in an immense vacuum, where there was nothing external or sensible with which the globes could be compared. But now, if in that space some remote bodies were placed that kept always a given position one to another, as the fixed stars do in our regions, we could not indeed determine from the relative translation of the globes among those bodies, whether the motion did belong to the globes or to the bodies. But if we observed the cord, and found that its tension was that very tension which the motions of the globes required, we might conclude the motion to be in the globes, and the bodies to be at rest; and then, lastly, from the translation of the globes among the bodies, we should find the determination oi their motions. But how we are to collect the true motions from their causes, effects, and apparent differences; and, vice versa, how from the motions, either true or apparent, we may come to the knowledge of their causes and effects, shall be explained more at large in the following tract. For to this end it was that I composed it.

For Newton, his First Law of Mechanics is a fact of immediate experience that requires absolute space. Thus, space becomes a logical and ontological necessity. But Newton's mechanics is invariant with respect to Galileo's Transformation, therefore his reference frame (the reference frame of the world) is not determined in a single way. And so

10.4 Critique of Newton's absolute space

he admits it in Corollary V, which anticipates the modern Principle of Relativity [260, p. 144]:

> The motions of bodies included in a given space are the same among themselves, whether that space is at rest, or moves uniformly forwards in a right line without any circular motion.

In order to solve this problem, Newton introduces Hypothesis I in Book III of the Principia [260, p. 641]:

> The center of the world system is at rest.

A hypothesis that cannot be experimentally tested, but which he uses to prove Theorem XI of Book III [260, p. 641]:

> The common center of gravity of the Earth, the Sun and all planets is at rest.

Note that Newton does not take into account the stars, whose relative motions were not known to Newton. But space was not only a logical and ontological necessity for Newton, in the general scholium of Book III, and in line with the thought of H. More, Newton identifies space and time with attributes of God (italics mine) [260, p. 783]:

> And from his true dominion it follows that the true God is a living, intelligent, and powerful Being; and, from his other perfections, that he is supreme, or most perfect. He is eternal and infinite, omnipotent and omniscient; that is, his duration reaches from eternity to eternity; his presence from infinity to infinity; he governs all things, and knows all things that are or can be done. He is not eternity or infinity, but eternal and infinite; he is not duration or space, but he endures and is present. He endures for ever, and is every where present; and by existing always and every where, *he constitutes duration and space.*

It is therefore surprising that a 21st century quantum physicist writes in the foreword of a book (about physics and God) [319, p. 7]:

> In the 17th century, Isaac Newton gave birth to a mathematical science that almost eliminated the idea of God's intervention in the material world of physics and chemistry.

The authors of this elimination were Newton's critics, all of them convinced relationists and relativists. As the reader of this book will see in Chapter 5, it is possible to develop a formal demonstration that the observable universe had to have a first cause exterior to the universe itself.

10.4 Critique of Newton's absolute space

The real or illusory nature of Newton's absolute space does not seem an irrelevant scientific matter. It is therefore striking that some relativists deny the real existence of space simply because that existence is not necessary to explain motion. For that same reason they accused Newton of being contradictory, since his First Rule for Philosophizing says [260, p. 615]:

> No more causes of natural things should be admitted than those that are true and sufficient to explain their phenomena.

As if the real or illusory nature of space were not itself a major scientific issue in describing the universe. In any case we would have two ways of explaining motion, which does not detract from the importance of the problem of the physical reality of space. Naturally, one can argue for or against the reality of space, as G. Leibniz (1646-1716) did, for example, in this case against that physical reality and in favor of a purely relational interpretation of space (quoted in [176, p. 117]):

> I will here show, how Men come to form to themselves the Notion of Space. They consider that many things exist at once, and they observe in them a certain Order of Co-existence, according to which the relation of one thing to another is more or less simple. This Order is their Situation or Distance. When it happens that one of those Co-existent Things changes its Relation to a Multitude of others, which do not change their Relation among themselves; and that another Thing, newly come, acquires the same Relation to the others, as the former had; we then say it is come into the Place of the former; and this Change we call Motion in That Body, wherein it is the immediate Cause of Change. And though Many, or even All the Co-existing Things, should change according to certain known Rules of Direction and Swiftness; yet one may always determine the Relation of Situation, which every Co-existent acquires with respect to every other Co-existent; and even That Relation, which any other Co-existent would have to this, or which this would have to any other, if it had not changed or if it had changed any otherwise. And supposing, or feigning, that among those Co-existents, there is a sufficient Number of them, which have undergone no Change; then we may say, that Those which have such a Relation to those fixed Existents, as Others had to them before, have now the same Place which those others had. And That which comprehends all those Places, is called Space.

10.4 Critique of Newton's absolute space

So, for Leibniz, space is relational, a simple order of co-existence; a situation of bodies among themselves. He also claim that all empirical knowledge can be derive from a simple axiom: the Principle of Sufficient Reason, that reads:

> There ought to be some sufficient reason why things should be so, and not otherwise.

from which one could prove the existence of God, but not that God is space, for if such were the case God, like space, would be infinitely divisible, which is absurd. And although Newton did not say that space was an organ of God, Leibniz used this assertion in his criticism of Newtonian space (quoted in [176, p. 114]):

> Sir Isaac Newton says, that Space is an Organ, which God makes use of to perceive Things by. But if God stands in need of any Organ to perceive Things by, it will follow, that they do not depend altogether upon him, nor were produced by him.

Statement answered by S. Clarke with the following words ([176, p. 114]):

> Sir Isaac Newton doth not say, that Space is the Organ which God makes use of to perceive Things by; nor that he has need of any Medium at all, whereby to perceive Things; But on the contrary, that he, being omnipresent, perceives all Things by his immediate Presence to them, in all Space whereever they are, without the Intervention or Assistance of any Organ or Medium whatsoever.

Leibniz developed other arguments against Newton's absolute space, two of the best known being the static shift argument and the kinematic shift argument (taken from [164, p. 162-163]):

> **The static shift**: Imagine a second universe just like ours except that all the matter is located in (i.e. shifted to) another place in absolute space, without any change in the relations of one object to another. Since space is a Euclidean plane, the two places are exactly alike, and so no differences will be seen.

> **The Kinematic shift**: Imagine a second universe just like ours except that the absolute velocity of every piece of matter differs by (i.e. is shifted by) a fixed, constant amount, without any change in the relations of one object to another. Since the two velocities differ only by a constant amount, no differences will be seen.

With respect to Newton's bucket experiment, Leibniz had to admit (quoted in [176, p. 119]):

> However, I grant there is a difference between an absolute true motion of a Body, and a mere relative Change of its Situation with respect to another Body. For when the immediate Cause of the Change is in the body, That Body is truly in Motion; and then the Situation of other Bodies, with respect to it, will be changed consequently, though the Cause of that Change be not in Them.

Leibniz seems to have been trapped in an awkward situation: on the one hand assuming kinematic relativism, and on the other hand the phenomenon of circular motion claiming the existence of an absolute space.

For G. Berkeley (1685-1753), space is a theoretical concept formed by the perception and abstraction of extension. He argues that it is impossible to imagine the motion of a body without imagining it moving with respect to another object (quoted in [164, p. 171]:

> ... no motion can be understood without some determination or direction, which in turn cannot be understood unless besides the body in motion our own body also, or some other body, be understood to exist at the same time.

In addition, Berkeley rejects that Newton's bucket experiment proves the reality of absolute space, and also rejects that the motion of water in the bucket is circular if the rotational and translational motions of the Earth around the Sun are taken into account. The argument of the tension in the cord joining the two rotating globes is not valid for Berkeley either, because without a material reference it is not possible to conceive the motion of the globes and therefore no inertial effect can be attributed to the motion of the two globes. According to Berkeley, the idea of absolute space and motion is a mere fiction without empirical foundation, and he relates all such motions (such as those of water in Newton's bucket) to the reference frame of the fixed stars (quoted in [176, p. 109]):

> If we suppose the other bodies were annihilated and, for example, a globe were to exist alone, no motion could be considered in it; so necessary is it that another body should be given by whose situation the motion should be understood to be determined.

Though different, Berkeley's statement resembles Mach's Principle: The inertia of a body is determined by the masses of the universe and their distribution.

Another critique of Newton's famous bucket experiment is that of C. Huygens (1629-1695), this time from a different relativistic perspective

10.4 Critique of Newton's absolute space

(quoted in [176, p. 125]):

> For a long time I had thought that rotational motion by means of centrifugal forces contains a criterion for true motion. Indeed, with regard to other phenomena it is the same whether a circular disk or a wheel rotates near me, or whether I circle round the stationary disk. However, if a stone is put on the circumference this will be projected only if the disk rotates, and therefore I formerly thought that circular motion is not relative to any other body. Still, this phenomenon showed only than the parts of the wheel, owing to the pressure acting on the circumference, are driven in relative motion among themselves in different directions. Rotational motion is therefore only a relative motion of the parts, which are driven to different sides, but held together by a rope or other connection.

We are then before an argument of pure relativistic dynamics that anticipates a good part of contemporary physics.

In any case, and in spite of its criticisms, Newton's absolute space eventually prevailed in science, philosophy and theology, at least until the first decades of the eighteenth century. As we will see in the following chapters, the eighteenth, nineteenth and twentieth centuries are especially significant for our discussion on the nature of space.

11. Questioning Leibniz's Principle of Sufficient Reason

Abstract.-In the formal framework defined by the Principle of Directional Evolution of the Universe (towards its maximum entropy) and by the Theorem of the Inconsistent Infinity, this chapter proves the incompleteness of Leibniz's Principle of Sufficient Reason, which renders inconclusive Leibniz's critique of Newton's absolute space, and also makes inevitable the existence of first causes that cannot be explained in terms of other causes deduced from our present knowledge of the observable universe.

11.1 Two Leibniz's Principles

As is well known, after the publication of Newton's Principia [260] a famous epistolary debate took place between S. Clarke and G.W. Leibniz, the former defending Newton's absolute space and the latter denying it, and both cases considering its consequences on the very existence of God. Without going into the details of the debate (for which the reader may consult, for instance [233, 234, 176, 291, 164, 352]), Leibniz introduced in it two of his famous principles:

> **Principle of Sufficient Reason**: *There must be a sufficient reason for things to be one way and not another.*

According to Leibniz, the above principle would turn metaphysics into a deductive science.

> **Principle of Identity of the Indiscernibles**: *There cannot exist two different things that are indistinguishable from each other.*

Since (according to Leibniz) two different and indiscernible things cannot exist, and since in Newton's absolute space things could be located in several different and indiscernible ways, Leibniz argued that God would have had to choose one of these indistinguishable ways, without any reason to choose one of them to the detriment of the others, which for Leibniz is not proper to God. Therefore, absolute space cannot exist. Clarke argued in the opposite sense, not defending the possibility of contingent events, but making God's will intervene as the only reason why things were one way and not another.

The Principle of Identity of Indiscernibles is no longer accepted by contemporary science, but the Principle of Sufficient Reason (PSR) is at least partially accepted. So here we respond to this principle. The following answer could be given without the advantages of the knowledge accumulated from Leibniz's time to ours: it would differ very little from the one given in this article. Since the important thing is the PSR answer, whether or not Leibniz is present to answer it, this advantageous knowledge will be used here, including that which has been published but not yet sufficiently accepted in contemporary science, as is the case with the inconsistency of the actual infinity. A key inconsistency for the future of mathematics and especially for the future of physics and the logical understanding of the physical world.

11.2 The formal setting of the discussion

The PSR will be discussed here within a formal scenario whose two fundamental pillars will be the Principle of Directional Evolution of the Universe and the inconsistency of the actual infinity, the latter not as a principle but as a formally demonstrable theorem. From both of them we can deduce the rest of the formal elements that constitute the formal scenario in which Leibniz's famous principle will be contested. All of these formal elements are briefly demonstrated in the appendices to this chapter. As usual, I invite the reader to jump to the proof of the inconsistency of the actual infinity (Theorem 6 of the Axiom of Infinity). I could have chosen any of the more than forty demonstrations contained in [212]. The one included in the first appendix is a very simplified variant of one of those demonstrations, which was also one of the first I was able to develop. It contains less than 300 words that can be read in less than 3 minutes, and if the reader does not find it a correct argument, he/she can stop reading the rest of the article right there.

The formal elements to be used in the PSR discussion, which are formally proved in chapters 3-5, are the following:

1. **Principle of Directional Evolution**: *The observable universe always evolves independently of its rational observers and in the same direction of increasing its global entropy.*

2. **Theorem of the Inconsistent Infinity**: *The actual infinity subsumed in the Axiom of Infinity is inconsistent.*

3. **Theorem of the Consistent Universe**: *The universe evolves under the control of a unique set of invariant and consistent physical laws.*

4. **Theorem of Identicality**: *All particles of the same type have the same properties and behave the same way under the same conditions.*

5. **Theorem of Formal Dependence**: *No concept defines itself; no statement proves itself; no physical object is the cause of itself; and no cause is the cause of itself.*

6. **Theorem of the First Element**: *A consistent sequence in which there is a last element and each element has an immediate predecessor is a complete totality only if it has a first element without predecessors.*

7. **Corollary!of the First Cause**: *No physical object or process can be fully explained without a first cause that cannot be explained in terms of other causes.*

The PSR could also be stated in terms of logical causes: *there is always a logical cause which explains why things are as they are and not otherwise.* In the following, both forms, Leibniz's original and the latter, will be used interchangeably.

11.3 The Principle of Sufficient Reason

The infinite regress of arguments was already considered by Aristotle [13, I.3]:

> We, on the other hand, hold that not every form of knowledge is demonstrative, but that the knowledge of ultimate principles is indemonstrable. The necessity of this fact is obvious, for if one must needs know the antecedent principles and those on which the demonstration rests, and if in this process we at last reach ultimates, these ultimates must necessarily be indemonstrable.

This, of course, is why we have always needed, and will always need, axioms and inductive laws in the foundations of all sciences. In our case, this need is demonstrated by the Theorem of Formal Dependence, a consequence of the Principle of Directional Evolution of the Universe, which is inductively based on overwhelming empirical evidence.

In fact, no one expects the shards of broken glass to spontaneously reassemble into the exact original shape of the broken glass; or that the gas released from a bottle of champagne spontaneously returns to the champagne in the bottle. These typical examples are often used to illustrate the Second Law of Thermodynamics, which is immediately incorporated into the Principle of the Directional Evolution of the Universe. This principle, moreover, permits the formal deduction of results that extend the Aristotelian infinite regress of arguments to definitions, and causes of objects and natural phenomena. As indicated elsewhere in this book, the case of first causes certainly goes far beyond the content of this book. And the reader can easily see why.

As noted above, contemporary science still allows the PSR to be applied, with the exception of contingent events. But both in contemporary science and in Leibniz's arguments, applying the PSR implies applying the principle of infinite regress of causes. And this is the key fact that was initially absent in Leibniz's arguments and is still absent in contemporary physics, although in Leibniz's case he came to admit a first cause of why things are as they are and not otherwise:

Because the universe had to be the best of universes.

Which is obviously an ARBITRARY cause. Indeed, since potential infinity is the only consistent infinite, it would be impossible for humans to fully explain any object or natural phenomenon without recourse to a first cause that cannot be explained in terms of other causes (Corollary 17 of the First Cause, page 49). In the case of God (if there is one), if he is a consistent being, he could not do this either, just as he could not count the last natural number if, as we may suppose, even God cannot count a non-existent number. So the Corollary 17 of the First Cause applies to Him as well, which, as we shall see, has significant consequences for the origin of the universe itself.

One of these consequences is that Leibniz's theological objection to Newton's absolute space is inapplicable: one cannot always give a sufficient reason (a cause) for things being as they are and not otherwise, because in the end we will fall into an inevitable infinite regress of causes, from which it is only possible to get out by means of a first cause that cannot be explained in terms of other causes. Not even God could do that. But Leibniz, perhaps aware of this difficulty, proposed a first cause (the universe had to be the best of the universes) which, as we have just pointed out, is as arbitrary a cause as any other that cannot be explained by other causes. Actually, Leibniz would be quite satisfied with the Corollary 17 of the First Cause: he would only have to think of the universe as the physical object that it is. Since no object can be the cause of itself, every physical object, including the universe, must have a first cause external to the object itself.

What if the universe were eternal? Well, in that case its duration would be infinite, it would have, for example, an infinite number of seconds or any other arbitrary unit of time. That is, it would have an inconsistent duration (Corollary 4). What if the universe had arisen from a fluctuation of nothing? Well, then nothing would not be nothing, but something with the ability to fluctuate, and we would have to apply the Corollary 17 of the First Cause to that something with the ability to fluctuate. What if the present universe were a stage in a cyclic succession of universes being continuously created and destroyed? Well, in this case the number of cycles could only be finite (Corollary 4) and therefore there would be a first universe (Theorem 23

11.3 The Principle of Sufficient Reason

of the First Element) in the cyclic succession of universes to which the Corollary 17 of the First Cause could be applied.

The majority of contemporary physicists, all of them strictly relativistic, deny the existence of physical absolute space. According to them it is only a fiction useful to describe the evolution of the (always) relative positions of natural objects (see the final appendix to Chapter 23). At the same time, and according to these same physicists, space expands, bends, vibrates and transmits its own vibrations. And one wonders how something that does not exist can expand, deform, vibrate and transmit its own vibrations?

Since 2015, we have empirical evidence of gravitational waves, and this changes everything. The vibrations of space are no longer a theoretical matter, they are real, they interact with material objects (by changing the distances between the mirrors of the interferometers that detect them), and the interactions can be detected and measured. Therefore, space is real; it is a real and unique physical object; it is the same for all material objects; it is absolute; it is Newtonian. Chapters 23 and 24 discuss the physical consequences of absolute space and the nature of its substance, respectively.

12. Newton's bucket and absolute rotations

Abstract.-The content of this chapter links the famous Clarke-Leibniz discussion on Newton's bucket experiment with a new and independent argument confirming Newton's idea about the absolute nature of the rotation of the water in his famous bucket. Literally, the existence of billions of objects in the universe animated by absolute motion is demonstrated here: their rotations around their respective internal axes of rotation. Newton's thought experiment of the two balloons connected by a string is also recalled in this chapter. Finally, the problem of the relation between inertial mass, gravitational mass and preinertia is raised.

12.1 A real Newton's experiment

Newton's famous bucket experiment is, in my opinion, one of the most important in the history of physics, both for its results and for the discussions it sparked, discussions that still continue to this day [137, p. 43-108]. The experiment was designed to demonstrate the reality of absolute motion, and therefore of absolute space, as opposed to those who, like Leibniz, defended the relative nature of space. As is well known, Newton's position in favor of absolute motion prevailed for several centuries, until the beginning of the twentieth century, when the theories of relativity finally prevailed in a hegemonic manner that was and still is very hostile to dissent. But as the reader will see throughout this chapter, the last word on Newton's bucket has not yet been said.

First of all, let us recall Newton's own description of his famous experiment in the Definitions prefatory to Book I of his PHILOSOPHIAE NATURALIS PRINCIPIA MATHEMATICA, published in 1687 [260, p. 131-132] [259, p. 80-81]:

> The effects which distinguish absolute from relative motion are, the forces of receding from the axis of circular motion. For there are no such forces in a circular motion purely relative, but in a true and absolute circular motion, they are greater or less, according to the quantity of the motion. If a vessel, hung by a long cord, is so often turned about that

the cord is strongly twisted, then filled with water, and held at rest together with the water; after, by the sudden action of another force, it is whirled about the contrary way, and while the cord is untwisting itself, the vessel continues for some time in this motion; the surface of the water will at first be plain, as before the vessel began to move: but the vessel, by gradually communicating its motion to the water, will make it begin sensibly to revolve, and recede by little and little from the middle, and ascend to the sides of the vessel, forming itself into a concave figure (as I have experienced), and the swifter the motion becomes, the higher will the water rise, till at last, performing its revolutions in the same times with the vessel, it becomes relatively at rest in it. This ascent of the water shows its endeavour to recede from the axis of its motion; and the true and absolute circular motion of the water, which is here directly contrary to the relative, discovers itself, and may be measured by this endeavour. At first, when the relative motion of the water in the vessel was greatest, it produced no endeavour to recede from the axis; the water showed no tendency to the circumference, nor any ascent towards the sides of the vessel, but remained of a plain surface, and therefore its true circular motion had not yet begun. But afterwards, when the relative motion of the water had decreased, the ascent thereof towards the sides of the vessel proved its endeavour to recede from the axis; and this endeavour showed the real circular motion of the water perpetually increasing, till it had acquired its greatest quantity, when the water rested relatively in the vessel. And therefore this endeavour does not depend upon any translation of the water in respect of the ambient bodies, nor can true circular motion be defined by such translation. There is only one real circular motion of any one revolving body, corresponding to only one power of endeavouring to recede from its axis of motion, as its proper and adequate effect; but relative motions, in one and the same body, are innumerable, according to the various relations it bears to external bodies, and like other relations, are altogether destitute of any real effect, any otherwise than they may perhaps partake of that one only true motion. And therefore in their system who suppose that our heavens, revolving below the sphere of the fixed stars, carry the planets along with them; the several parts of those heavens, and the planets, which are indeed relatively at rest in their heavens, do yet really move. For they change their position one to another (which never happens to bodies truly at rest), and being carried together with their heavens, partake of their motions, and as parts of revolving wholes, endeavour to recede from the axis of their motions.

It is clear, then, that for Newton the force responsible for the separation of the water from the axis of rotation is present only in the absolute motion of rotation. The ascent of the water up the walls of the bucket is a proof of absolute motion, which in turn implies changes of position in an absolute space.

12.2 Criticism of Newton's bucket experiment

Chapter 11 recalled the epistolary (and theologically motivated) discussion between G.W. Leibniz and S. Clarke, in which Leibniz rejected Newton's absolute space by making use of his Principle of Sufficient Reason and his Principle of Identity of Indiscernibles (see pag. 119):

> Since (according to Leibniz) two different and indiscernible things cannot exist, and since in Newton's absolute space things could be located in several different and indiscernible ways, Leibniz argued that God would have had to choose one of these indiscernible ways, without any reason to choose one of them in preference to the others, which for Leibniz is not proper to God. Therefore, absolute space cannot exist.

However, the thought experiment of Newton's rotating globes (which is recalled in the next section), made Leibniz change his opinion slightly (quoted in [234, p. 44]:

> I grant there is a difference between an absolute true motion of a body, and a mere relative change of its situation with respect to another body. For when the immediate cause of the change is in the body, that body is truly in motion; and then the situation of other bodies, with respect to it, will be changed consequently, though the cause of that change be not in them.

Although Leibniz never came to admit absolute space nor did he renounce his relational position.

Some 200 years later, Ernst Mach (1838-1916) resumed his critique of Newton's bucket experiment from the same relational perspective as Leibniz. The most prominent and well-known aspect of Mach's critique was his conclusion that the motion of the water in Newton's bucket is a relative motion: the water actually rotates WITH RESPECT TO the background of the fixed[1] stars (BFS from now on). I have highlighted

---

[1] Obviously, the stars are not fixed, as one might have believed in E. Mach's time.

the words "with respect to" because it is from them that my criticism of Mach's criticism and that of all those authors who make use of the same semantic trick will be deduced.

Although the expression *spinning with respect to something* is not entirely wrong, it is not the best way to describe a spin or rotation. Indeed, bodies that rotate, actually rotate AROUND something, usually a straight line (axis of rotation) that can even be materialized for example with a thin wire. Each point of a body rotating AROUND an axis describes a circle AROUND a point of that axis, the axis being internal or external to the object rotating AROUND it. Naturally, in the case of the molecules of water of Newton's bucket, they all describe concentric circles around a vertical axis passing through the geometrical center of the bucket. All of which can be materialized with floating beads and a vertical wire passing through the center of the bucket. The most honest way to describe the motion of these floating balls (or water molecules) is to say that they move AROUND the wire following CIRCULAR TRAJECTORIES AROUND the wire (axis of rotation), and that they are pushed by a force that tends to separate them from the axis of rotation, which is why they will move away from the axis of rotation as far as possible taking into account the walls of the bucket and the complete swarm of molecules subjected to this force. The beads do not describe circles around BFS, in fact it is impossible to describe a circular trajectory around a surface, only around a central point is possible to describe a circular trajectory, simply because a circle is exclusively defined by a point (its center) and a fixed distance to the center (its radius). So to rotate around (with respect to, as Mach would say) a bi-dimensional or three-dimensional surface is meaningless. I think that if he had used the word "AROUND" instead of the expression "WITH RESPECT TO," Mach would have realized his mistake.

As is well known, the circular motion of a material object around an axis imparts a centrifugal force on the rotating body that tends to move it away from the axis of rotation. That force (which we now know is proportional to the distance to the axis of rotation) is responsible for the fact that a fluid such as the water in Newton's bucket, which is rotating around a vertical axis passing through the center of the bucket, tends to move away from that axis and up the walls of the bucket, creating the famous concave absolute surface of the water in Newton's bucket. Therefore, and according to Newton, the concave absolute surface of the water shows the presence of a force caused by the real, absolute motion of the water molecules AROUND the axis of rotation, not WITH RESPECT TO the BFS, a force that is the reaction to the force that must be continuously applied to continuously change the direction of motion of each rotating molecule of water. It is a proof of absolute motion, as the following argument will also prove. But before continuing, let us

12.2 Criticism of Newton's bucket experiment

recall Galileo's words on a subject similar to the one under discussion here [125, p. 183-184]:

> Now, if in order to achieve the same effect in a precise way, it is just as important that the Earth alone should move, stopping all the rest of the Universe, as it is that the whole Universe should move with a single movement, who would want to believe that Nature (which, according to common agreement, does not do by the intervention of many things what it can do by means of a few) has chosen to make an immense number of very large bodies move, with inestimable speed, in order to achieve what can be obtained by the moderate movement of a single body around its own center?

Back to our discussion, consider the daily rotation of the Earth around its north-south geographic axis, without considering the precession and nutation motions of this axis. The trajectory T of any point P of the Earth during this rotation is the complete circle that P describes in about 24 hours, where the center of this circle is a unique point Q of the axis of rotation of our planet, and its unique radius is the distance between the two points P and Q. Because of preinertia [209, Link], observers on the Earth do not notice this rotation; what we do observe is that the Sun and the rest of the celestial bodies rotate around the center of the Earth. Although we have long since discovered that such daily rotational motions of celestial bodies observed from the Earth are only apparent, not real, it is necessary to give a formal, physical demonstration that this is the case, simply to be able to infer from such a case some conclusion about the nature of this motion. As will be seen below, it is possible to demonstrate that such rotational motions are indeed only apparent, not even relative, because, as we shall see, they are physically and logically impossible.

Indeed, in addition to the Earth, there are other planets in the solar system that also rotate around an internal axis under similar conditions to the Earth. From these planets, and for the same reasons as for the Earth, the Sun and the other celestial bodies appear to rotate around each of these planets. Consequently, each point of the Sun and the other celestial bodies would simultaneously describe different circular orbits with different centers of rotation, which would mean that each of the points of each of the celestial bodies would have to be in different places at the same time, describing different orbits. Furthermore, a star located, say, a billion light-years from Earth and on Earth's equatorial plane would have to move at a speed 3.3×10^{13} times greater than the speed of light, which we assume to be physically impossible.

Consequently, and since it is impossible for the same point to be in different places at the same time, describing different orbits, and since we also assume that it is impossible to exceed the speed of light, we must conclude that the rotations of the Sun and of all the celestial bodies observed from the Earth and from the rest of the planets of the solar system are logically and physically impossible. Therefore, they are not real, they do not exist, they are only apparent motions. The only things that exist are the planetary rotations around their respective internal axes of rotation. The same argument applies to all the planets of any other planetary system (star-planets) in the universe. It must be concluded that all the rotations around their respective internal axes of rotation of all the planets of all the planetary systems in the universe are absolute rotations, as Newton argued for the case of the rotation of the water in the bucket in his experiment. There are literally billions of objects in absolute motion (rotation) in the universe. A conclusion that should have been universally accepted for centuries, but still is not: for relativistic officialism, absolute motion does not exist.

12.3 A thought experiment: Newton's rotating globes

In the same text in which Newton presented his famous bucket experiment, he also presented another experiment, this time a thought experiment, which would also confirm the absolute nature of physical space. [260, p. 133] [259, p. 82]:

> It is indeed a matter of great difficulty to discover, and effectually to distinguish, the true motions of particular bodies from the apparent; be cause the parts of that immovable space, in which those motions are performed, do by no means come under the observation of our senses. Yet the thing is not altogether desperate: for we have some arguments to guide us, partly from the apparent motions, which are the differences of the true motions; partly from the forces, which are the causes and effects of the true motions. For instance, if two globes, kept at a given distance one from the other by means of a cord that connects them, were revolved about their common centre of gravity, we might, from the tension of the cord, discover the endeavour of the globes to recede from the axis of their motion, and from thence we might compute the quantity of their circular motions. And then if any equal forces should be impressed at once on the alternate faces of the globes to augment or diminish their circular motions, from the increase or decrease of the tension of the cord, we might infer the increment or decrement of their motions: and thence would be found on what faces those forces ought to be impressed, that the motions of the

12.4 Mass and Mach's Principle

> globes might be most augmented; that is, we might discover their hinder most faces, or those which, in the circular motion, do follow. But the faces which follow being known, and consequently the opposite ones that precede, we should likewise know the determination of their motions. And thus we might find both the quantity and the determination of this circular motion, even in an immense vacuum, where there was nothing external or sensible with which the globes could be compared. But now, if in that space some remote bodies were placed that kept always a given position one to another, as the fixed stars do in our regions, we could not indeed determine from the relative translation of the globes among those bodies, whether the motion did belong to the globes or to the bodies. But if we observed the cord, and found that its tension was that very tension which the motions of the globes required, we might conclude the motion to be in the globes, and the bodies to be at rest; and then, lastly, from the translation of the globes among the bodies, we should find the determination of their motions. But how we are to collect the true motions from their causes, effects, and apparent differences; and, vice versa, how from the motions, either true or apparent, we may come to the knowledge of their causes and effects, shall be explained more at large in the following tract. For to this end it was that I composed it.

In this case, Leibniz partially agreed with Newton. [233, p. 82-83]:

> I agree, however, that there is a difference between a true absolute motion of a body and a mere relative change of its situation by reference to another body. For when the immediate cause of the change is in the body, it is truly in motion and then the situation of the others in relation to it will consequently be changed even though the cause of this change is not in them.

Two hundred years later, E. Mach took up the problem of the spinning balloons to declare it outside our possibilities of analysis because it is based on an unreal situation of which we have no experience to serve as a guide, and then we cannot analyze the situation. But, we should answer him, this eliminates the possibility of thought experiments and the pure exercise of logic in the practice of science. Indeed, even if we have no experience with the two-balloon universe, we have logical tools to analyze the problem and deduce formal consequences that can be applied to the real universe if the real universe is formally consistent, which is possible to demonstrate on the basis of its directional evolution (Chapter 5).

12.4 Mass and Mach's Principle

Mach's Principle states that the inertial mass of material objects is produced by all the masses present in the universe. It would then be what we could call a collective property, so that if we were to leave a single body in the universe, it would have no inertial mass. Among contemporary physicists there is a division of opinion regarding Mach's Principle. In any case, even that collective property involves the basic and general concept of mass, which is probably a primitive concept, not definable in terms of other more basic concepts. Gravitational mass, on the other hand, does not seem to be a collective but an individual property of each of the material bodies, this mass is what modifies the properties of physical space making possible the gravitational interaction, one of the four fundamental interactions responsible for the evolution of the universe.

Preinertia is another particular property of every object in the universe, including photons (see Chapter 19), by virtue of which the object inherits the velocity vector of the material object(s) that sets it in motion. This property, despite making continuous unconsciously (implicit) use of it, contemporary physicists have not yet discovered it and, therefore, they cannot use it in its arguments and experiments. We have, then, three physical facts in search of explanation:

1. Inertial mass: The resistance of all material objects to change their state of motion.
2. Gravitational mass: The ability of all material objects to modify the physical properties of the real physical space[2] by creating gravitational fields around them.
3. Preinertia: The ability of any physical object to inherit the velocity vector of the proper reference frame where it is set in motion.

Mach's Principle proposes an explanation of the first fact: it is produced by the mass of the whole universe, it would be a sort of collective property. The other two facts can only be thought of as individual properties of each object. Thus, although much remains to be discussed on mass, the simplest explanation of the above three facts would be that all material objects have a property capable of modifying the properties of space, of offering a certain resistance to their own changes of motion, and of inheriting the velocity vector of the proper reference frame from which they start its own independent motion. This latter property manifests itself even in supposedly massless objects such as photons. Although there is a possibility that photons have a mass of the order of

---

[2]Space is in fact a real physical object as the empirical detection and measurement of its vibrations (gravitational waves) proves.

12.4 Mass and Mach's Principle

10^{-64} Kg, which could be called quantum mass m_q [208, Link p. 235]:

$$m_q = \sqrt{\frac{G\hbar^3 R_\infty^4}{c^5}} = \hbar t_p R_\infty^2 = 6.845023 \times 10^{-64} Kg \qquad (1)$$

where t_p is the Planck time and R_∞ is the universal Rydberg constant, which is specific to each chemical element and varies slightly with its mass. In any case, I repeat here some of the mass-related questions that should be asked and the answers sought:

- Of an object in uniform motion, what determines and controls its linear trajectory, its successive positions along the successive instants?
- How does a body remember that it was pushed? Where lies the imprint of that action?
- What changed, if any, in its internal structure as a consequence of being set in motion?
- What distinguishes a ball that has been pushed from another that was not?
- Are space and time somehow affected by a ball set in motion?
- Knowing that a body A was pushed and other body B was not pushed, being initially A and B at relative rest, is it the same to say that A moves with respect to B as to say that B moves with respect to A?
- To put A in motion is the same as to put the rest of the universe in motion?
- Is there any absolute describable reality?
- If there is no reality describable in absolute terms, are there as many realities as there are relative forms of observing it? To observe what?
- Could the universe be described, as such an object, from outside the universe?
- Are we living beings with the capacity to reason but not to observe reality?
- Is the theory of special relativity the ultimate theory?
- What relation, if any, does exist between inertia and preinertia?
- Can be preinertial a massless particle?
- Are inertial and preinertial objects affected in the same way by gravitational fields?

- Is preinertia sufficient to explain light deviation by massive objects?
- Is preinertia a fundamental attribute, as mass, charge or spin, of elementary particles?
- If not, which fundamental attribute of elementary particle could account for preinertia?
- Does preinertia result from the interaction between matter and physical space?
- Are all waves preinertial?
- etc.

13. Space in the XVIII and XIX centuries

13.1 Introduction

The history of space in the two centuries covered by this chapter is a busy one. Although theological discussions about absolute space and divine omnipresence continued, from a purely physical perspective, Newton's mechanics prevailed in a resounding way (it is still the mechanics we use today to explain the physics of everyday events). Not so resounding and general was the acceptance of Newton's absolute and real space as opposed to the unreal relational space, i.e. without physical reality, of his critics. In this sense, I. Kant and E. Mach were two of the most relevant figures of the time. As we will see here, Kant did not always defend the same ideas about space. The relationalist Mach finally imposed his famous "Principle of the Fixed Stars".

In the second half of the eighteenth century, and in the face of the persistence of the Euclidean enigma of parallels, some non-Euclidean alternatives of geometry emerged, one of which will be used later, already in the twentieth century, to explain the geometry of gravitational-relativistic deformations of space. Modern physics assume that space can extend, deform, vibrate and be the transmitting medium of its own vibrations, although for the majority of physicists space does not exist, it is only a relational illusion! In this we have not advanced much, in our days things exist according to convenience.

The old Aristotelian quintessence, the ether, reappears at the end of the 19th century as the transmitting medium of electromagnetic waves, and with it one of the most famous series of experiments in the history of science: the Michelson-Morley experiments, whose aim was finally to determine the absolute motion of the Earth. In the name of rigor, it is obligatory to include the word "practically" in the statement: the practically negative results of the Michelson-Morley experiment practically put an end to the existential expectations of absolute space.

The Michelson-Morley experiments have been repeated at least 30 times, the last one already in the 21st century, always with the same practically negative results. Since 1887, the Michelson-Morley experiments have received a great deal of scientific and even media attention (in the media of the time). They were crucial to the first theory of special relativity: Lorentz's theory. Although published in 1905, Einstein's special theory of relativity would not become famous until 1916, when Einstein published his general theory of relativity. Incidentally, Einstein did not consider the Michelson-Morley experiments relevant, at least he does not cite them in his famous 1905 paper [87], which I find difficult to justify.

13.2 The initial success of Newtonian absolute space

Despite the persistent criticism of Newton's absolute space by Leibniz, Berkeley and Huygens, this Newtonian concept eventually prevailed in physics, at least during the first third of the 18th century. And the same happened with the rest of Newton's Principia, in this case the success will last until our days, at least for speeds far from the speed of light. As a result, the Cartesian model was replaced by the Newtonian model. And the Principia are still considered to be the foundation of mechanics for speeds far from the speed of light.

Newton's initial success is well reflected in the words written in 1745 by the mathematician and astronomer John Keill (1671-1721), which also describe very well the characteristics that should be proper to absolute space (quoted in [176, p.128]):

> We conceive Space to be that, wherein all Bodies are placed, or, to speak with the Schools, have their Ubi; that it is altogether penetrable, receiving all Bodies into itself, and refusing Ingress to nothing whatsoever; that is immovably fixed, capable of no Action, Form or Quality; whose Parts it is impossible to separate from each other, by any Force however great; but the Space itself remaining immovable, receives the Successions of things in motion, determines the Velocities of their Motions, and measures the Distances of the things themselves.

Another advocate of the Newtonian idea of absolute space was Leonard Euler (1707-1783), who defined the concept of place as a part of the infinite space that constitutes the world. The place of a physical body would therefore be its absolute, not relative, place. In his Reflections on Space and Time he further states that (quoted in French [176, p. 130]):

> Place is a part of the immense or infinite space of which the

whole world consists. Place accepted in this sense is usually called absolute, to distinguish it from relative place, of which mention will soon be made.

Euler even tries to prove the existence of absolute space by resorting to the Principle of Sufficient Reason, since (quoted in Latin in [176, p. 131]):

> A body that is absolutely at rest, if it has not been subjected to any external action, will remain perpetually at rest.
> There is no inherent reason why a body should begin to move in one direction instead of in all others, and if every cause of motion were removed from the exterior, it would not be able to conceive of motion according to any direction. Therefore, this truth certainly rests on the Principle of Sufficient Reason.

It was common in the 18th century to use the principle of sufficient reason to prove the real existence of absolute space. The argument is similar to that of Newton. The famous mathematician C. MacLaurin (1898-1746) recalls this Newtonian argument in the following words (quoted in [176, p. 130])

> This perseverance of a body in a state of rest or uniform motion, can only take place with relation to absolute space, and can only be intelligible by admitting it.

13.3 The nature of space according to Kant

Space was one of the recurring themes in Immanuel Kant's (1724-1804) philosophy, but he did not always have the same view of the concept. In 1755 he defended relational positions, although not exactly the same as those of Leibniz. For Kant, the spatial relations between bodies went beyond the mere ordering of things in space: these spatial relations were the consequences of mutual causal interactions between physical objects. And since this causal interdependence was not present in matter itself, but was added in the divine creation, space must have an existence independent of matter.

A little later, in 1758, Kant defends purer relational positions, in the style of Leibniz [178, p. 13]:

> Now I can consider a body in relation to certain external objects that surround it at first, and then, if it does not change this relation, I will say that it is at rest. But as soon as I look at it in relation to a sphere of further extent, then it is possible that the body together with its near objects changes its position in relation to those, and I will inform it

of a movement from this point of view. Now I am at liberty to extend my circle of vision as much as I want and to observe my body in relation to ever more distant circles, and I understand that my judgment of the movement and the rest of this body is never constant, but can always change with new perspectives... Now I begin to realize that I am missing something in the expression of movement and rest. I should never use it in an absolute sense, but always in a relative one. I should never say that a body is at rest without also stating in what respect it is at rest, and I should never say that it moves without at the same time stating the objects in respect of which it changes its relation. Even if I wanted to imagine a mathematical space empty of all creatures as a container of bodies, this would still not help me. For how am I to distinguish the parts of it and the various places that are not occupied by anything corporeal?

Around 1763, Kant changes his point of view and begins to defend Newton's absolute space. He goes so far as to give a proof of the existence of this absolute space based on the existence of what we now call enantiomorphic forms, such as the right and left hand, which cannot be superimposed with the same up-down and left-right orientation. The different arrangement of their parts, according to Kant, could only be explained by the different arrangement of their parts with respect to absolute space. Absolute space and time, according to Kant, would turn out to be a priori intuitions, knowledge that precedes experience. And he tries to prove this with the following argumentation [181, p.157-158]:

Space is not an empirical concept that has been drawn from outer experiences. For in order for certain sensations to be related to something outside me (i.e., to something in another place in space from that in which I find myself), thus in order for me to represent them as outside one another, thus not merely as different but as in different places, the representation of space must already be their ground) Thus the representation of space cannot be obtained from the relations of outer appearance through experience, but this outer experience is itself first possible only through this representation.

In 1786 Kant published *Metaphysische Anfangsgrünce der Naturwissenschaft* (Metaphysical Principles of the Science of Nature), a treatise in his hypothetical deductive style in which Kant makes clear his ideas about physical space and time. Some of the formal elements of that treatise are [180, p. 19-73]:

DEFINITION. Matter is that which is mobile in space. Space, which is itself mobile, is called material space or also relative space. Finally, that space in which all motion must be thought of (and, consequently, is absolutely immobile) is called pure space and even absolute space.

DEFINITION. The motion of an object is the modification of its external relations with reference to a given space.

DEFINITION. The rest is the permanent presence in the same place; but the permanent thing is that which exists in certain time, that is to say, that lasts.

AXIOM. Any movement, object of a possible experience, can be considered at will as movement of a body in a space at rest, or the body being at rest and the space in movement (moving) in the opposite direction and with the same velocity.

DEFINITION. Matter is the mobile as it fills a space. To fill a space is to resist any mobile that strives to penetrate such space, due to its motion. A space that is not filled is an empty space.

THEOREM. Matter occupies a space, not by its mere existence, but by virtue of a particular motive force.

DEFINITION. A matter in its movement penetrates another when it suppresses, by compression, the space of its expansion.

THEOREM. Matter can be compressed to infinity, but it can never be penetrated by another matter, whatever the compression force of the latter.

THEOREM. Matter is divisible in infinitum in parts which, in turn, each one is matter.

THEOREM. The essential attraction to every material body is an immediate action of some on others through the empty space.

THEOREM. The original force of attraction on which the possibility of matter as such is founded, extends immediately in the space of the universe, from one part of it to another, to infinity.

13.4 Mach's Principle

Despite Newton's success, many scientists between the 18th and 19th centuries, such as J. L. Lagrange (1736-1813) or P. S. Laplace (1749-1827), had little interest in absolute space. And by the 19th century, the idea of its practical uselessness had become widespread even among its adherents. A situation that is well reflected in the following words of J. C. Maxwell (1831-1879) [235, p. 110-111]:

Absolute space is conceived as remaining always similar to itself and immovable. The arrangement of the parts of space cannot be altered any more than we can do with the order of the parts of time. To imagine that they move from their places is to imagine that a place moves away from itself.

But just as there is nothing to distinguish one part of time from another, except the different events that take place in them, so there is nothing to distinguish one part of space from another, except its relation to the place in which material bodies are found. We cannot describe the time of one event except by reference to another event, or the position of a body except by reference to another body. All our knowledge of both time and place is essentially relative.

E. Mach (1838-1916) went a little further by calling absolute space a conceptual aberration, which must be eliminated from mechanics (quoted in [262, p. 12]):

> No one is competent to predicate things about absolute space and absolute motion; they are pure things of thought, pure mental construct, that cannot be produced in experience.

His words about Newton's bucket experiment are also very revealing:

> Newton's experiment with the rotating vessel of water simply informs us, that the relative rotation of the water with respect to the sides of the vessel produces no noticeable centrifugal forces, but that such forces are produced by its relative rotation with respect to the mass of the earth and the other celestial bodies. No one is competent to say how the experiment would turn out if the sides of the vessel increased in thickness and mass till they were ultimately several leagues thick. The one experiment only lies before us, and our business is, to bring it into accord with the other facts known to us, and not with the arbitrary fictions of our imagination. (quoted in [262, p. 13]).

> Concerning the conceptual monstrosities of absolute space and absolute time, I could not retract anything. I have only shown here more clearly than before that Newton talks a lot about these things, but has made absolutely no serious application of the same. His Coroll. V. (Principia, 1687, p. 19) contains the only practically useful (probably approximated) inertial system. (quoted in German in [176, p. 142]).

> For me there is only a relative movement at all and I cannot make a difference between rotation and translation in it. If a body turns relatively against the fixed star sky, then

centrifugal forces appear, it turns relatively against another body, but not against the fixed star sky, then the centrifugal forces are missing. I have nothing against it if one calls the former rotation an absolute one, if one only does not forget that this means nothing else than a relative rotation against the fixed star sky. Can we perhaps hold Newton's water glass, rotate the fixed star sky against it, and prove the absence of the centrifugal forces now? The attempt is not executable, the thought at all senseless, since both cases are not to be distinguished sensually from each other so. Accordingly, I consider both cases to be the same case and the Newtonian distinction to be an illusion. (quoted in German in [176, p. 143])

All of which leads to Mach's Principle (named after Einstein):

Principle 2 (Mach's) *The inertial forces experienced by a body in non-uniform motion are determined by the quantity and distribution of matter in the universe.*

13.5 The birth of non-Euclidean geometries

By the second half of the 18th century, despite considerable efforts, no progress had been made on the problem of parallels, called "the scandal of elementary geometry" by Lagrange (quoted in [311]); or "the shameful part of mathematics" by Gauss (quoted in [264, p. 9]). Some mathematicians, such as Gauss, began to explore the possibility of non-Euclidean geometries in the late eighteenth and early nineteenth centuries. Geometries without the Fifth Postulate, or even geometries based on postulates claiming that parts of Euclid's postulates are false.

The speculations on non-Euclidean geometries were initially developed as a way to prove the Fifth Postulate by *reductio ad absurdum* (this was the case, for instance, of the attempts by Giovanni Girolamo Saccheri). The idea of true non-Euclidean geometries came into the scene in the first half of the nineteenth century, with some precedents as 'the geometry of visibles' discussed by the philosopher Thomas Reid [292, Chp. 6 §9], [373]:

> The shape of visible figures are geometrically equivalent to their projection onto the surfaces of spheres.

It is not unusual for many authors to have the same new idea at the same time. This was the case with the discovery of non-Euclidean geometries in the nineteenth century. In fact, Karl F. Gauss, Ferdinand Schweikart, Nicolái Lobachevsky, János Bolyai and Bernhard Riemann were contemporary mathematicians of the 19th century, and it was they who laid the foundations of the new non-euclidean geometries. However, Gauss, one of the most important mathematicians of all time, never published his ideas on non-euclidean geometry (a term he coined himself in 1824). Bolyai could only publish his work in an appendix to his father's book *Tentament* (a compendium of mathematics). And Lobachevsky's first publication (1829) on his *'Imaginary Geometry'* could only be made in a rather unknown journal. The situation was somewhat different for Schweikart, who managed to publish his *'astral geometry'* in 1818, perhaps the first serious publication aimed at exploring the new geometries.

B. Riemann (1826-1866), one of the forefathers of non-Euclidean geometry, had the most profound insight into the new Copernican revolution in geometry. He proposed abstract geometric surfaces, independent of Euclidean space, with precisely determined curvatures. All geometries exist on such surfaces: elliptic and spherical geometries when the surface curvature is positive, hyperbolic geometry (the geometry of Bolyai and Lobachevsky) when the surface curvature is negative, and Euclidean geometry when the surface curvature is zero (Figure 13.1).

In 1868, Eugenio Beltrami (1835-1900) proved the consistency of

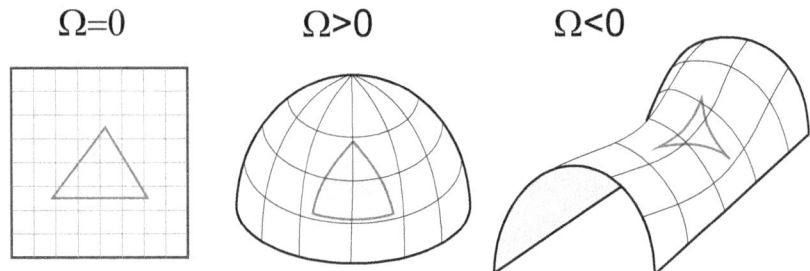

Figure 13.1 – A triangle on three surfaces: Euclidean (curvature (Ω) zero); elliptic (curvature greater than zero); and hyperbolic (curvature less than zero)

non-Euclidean geometries [26], and then the impossibility of deriving the Fifth Postulate from the other four Euclid's postulates. The proof was to develop an Euclidean model of non-Euclidean geometries. Non-Euclidean geometries are thus legitimized. And not only legitimized, but Albert Einstein popularized Riemann geometry in his general theory of relativity. The contrast between Euclidean and non-Euclidean geometries is quite clear.

The Hyperbolic Axiom reads:

> There exists a line l and a point P not in l such that at least two distinct coplanar lines parallel to l pass through P.

The Elliptic Axiom states:

> Through a point exterior to a given line, there is no line parallel to the given line.

While Playfair's Axiom (a variant of Euclid's Fifth Postulate) reads:

> Through a given point one, and only one, parallel can be drawn to a given straight line.

Apart from the non-existence of parallels, another notable difference between Euclidean geometry and Riemann elliptic geometry is that in the latter there are infinitely many different straight lines passing through the same two points, which contradicts the strong version of Euclid's First Postulate, according to which there is only one straight line between any two points. Euclid's original statement (the weak version of the first postulate) establishes the existence of (at least) one straight line between two points. His statement is therefore compatible with the existence of more than one straight line between two points. Although it seems unlikely that this was Euclid's belief, nor that of the majority of subsequent Euclidean authors.

In 2020 the author proposed and published [196, 200] a new foundational basis for Euclidean geometry consisting of 29 functional[1] definitions and 10 axioms that, among other things, legitimize the definitions. The foundational basis includes 45 corollaries formally deduced, and in an immediate way, from the definitions and axioms. One of these corollaries is the strong form of Euclid's first postulate: through two points passes one, and only one, straight line. The new basis also makes it possible to prove the famous postulate of parallels (see Appendix**??**).

13.6 Michelson-Morley experiment

(This section is taken from a chapter of [208])

One of the best-known and most famous experiments in the history of science is the Michelson-Morley experiment, which was first carried out in 1887 and has since been repeated in various ways about thirty times, the last times (known to me) in 2009 [101, 155] and 2015 [258]. In addition, their corresponding data are continually analyzed and reinterpreted from different perspectives [51, 313, 50, 75].

The aim of the original experiment was to detect the absolute motion of the Earth through the hypothetical luminiferous ether (an aim often expressed in confusing terms as "detecting the ether wind" or "ether drift" or even "the speed of the Earth with respect to light"). Basically, the apparatus used by Michelson and Morley was an interferometer that analyzed the interference of two rays of light (both resulting from the division of an initial beam of light) propagating in two orthogonal

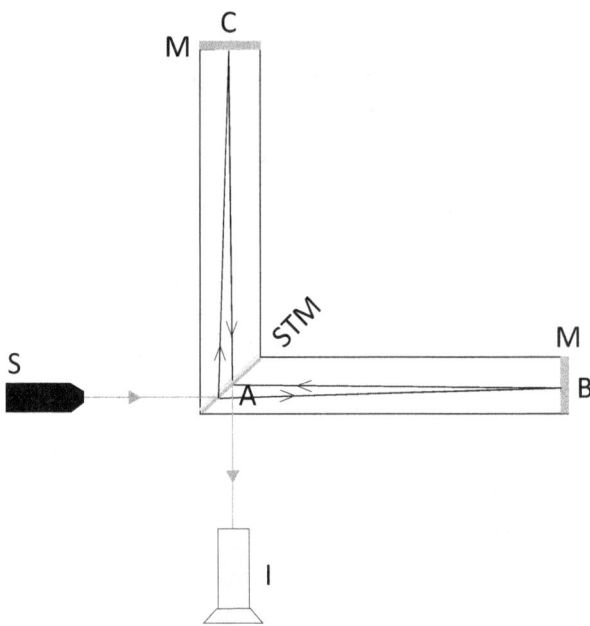

Figure 13.2 – Schematic diagram of the Michelson-Morley interferometer. S: Source of light; M: Mirror; STM: Semitransparent mirror; I: Interferometer; AB: Horizontal arm; AC: Vertical arm.

directions through the supposed ether (Figure 13.2).

Obviously, experiments a la Michelson-Morley are based on the belief that light behaves as a non-preinertial object (an object that does not inherit the relative velocity vector of the frame in which it is set in motion), because it is surely impossible to detect the velocity of the Earth by means of objects that inherit its velocity vector, i.e. the velocity we are trying to detect (except perhaps in the experimental conditions of the Santiago del Collado experiment [208, p. 371-378]). Think, for example, of the impossibility of measuring the speed of a train by dropping a coin in the train: the coin will fall in the same place on the floor of the train, whether the train is at rest or moving with any velocity relative to the platform.

The Earth moves around the Sun at nearly 30 km/s, but it also moves with the Solar System around the center of our Galaxy, and our Galaxy moves around the center of its group of galaxies (Local Group), and the Local Group moves with the Virgo (or Local) Cluster, and this with the Virgo (or Local) Supercluster, superclusters being surely the largest structures in the Universe. We do not know exactly how the Earth moves, although the speed of the Earth with respect to the CMB (Cosmic Microwave Background) could be in the order of 367 Km/s in the direction of the galactic coordinates (264.4, 48.4). Without

---

[1] Usable in demonstrations.

13.6 Michelson-Morley experiment

going into details, let us analyze the schematic argument behind the Michelson-Morley experiment.

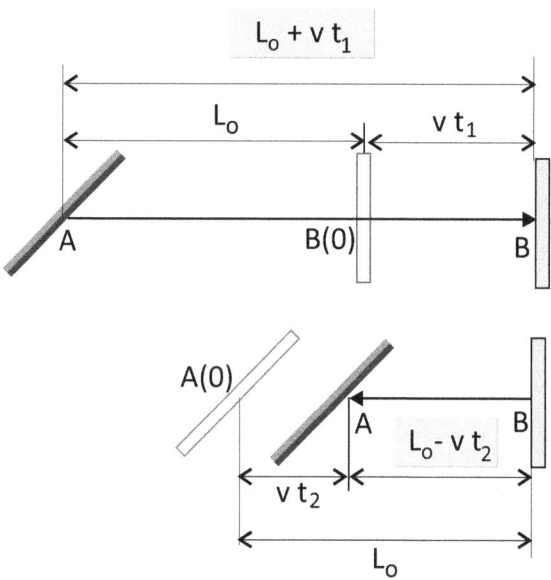

Figure 13.3 – The travel of light along the arm AB. $B(0)$ is the position of the mirror B when light leaves the semi-transparent mirror A; $A(0)$ is the position of the mirror A when light leaves B.

Assume one of the arms of the apparatus (for instance the arm AB in Figure 13.2), whose proper length is L_o, moves in the same direction as the velocity v of the Earth through the luminiferous ether (which in the original experiments it was assumed to be the orbital velocity of the Earth around the sun, i.e. $v \approx 30$ Km/s). Assume the ray of light that moves parallel to this arm take a time t_1 to go from A to B (Figure 13.3) and a time of t_2 to go from B to A. We will have:

$$ct_1 = L_o + vt_1 \tag{1}$$

$$ct_2 = L_o - vt_2 \tag{2}$$

Therefore:

$$t_1 = \frac{L_o}{c - v} \tag{3}$$

$$t_2 = \frac{L_o}{c + v} \tag{4}$$

$$t_1 + t_2 = \frac{2cL_o}{c^2 - v^2} \tag{5}$$

$$= \frac{2L_o/c}{1 - v^2/c^2} \tag{6}$$

where $t_1 + t_2$ is the time it takes light to complete its trip from A to B and from B to A.

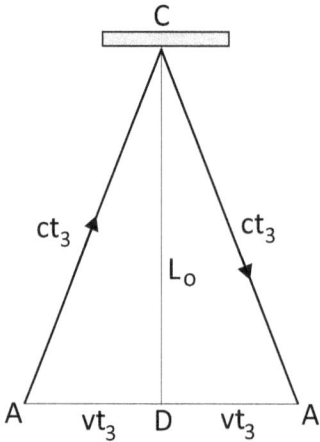

Figure 13.4 – The travel of light along the arm AC.

In the arm perpendicular to v (arm AC), whose proper length is also L_o, light travels from A to C and then from C to A, being AC and CA the hypotenuses of two equal right triangles whose legs are L_o and vt_3, where t_3 is the time light lasts to go from A to C as well as from C to A (Figure 13.4). The orthogonality between DC and AD and DA makes it inevitable classic Pythagorean theorem, and we will have:

$$(ct_3)^2 = L_o^2 + (vt_3)^2 \tag{7}$$

$$t_3^2(c^2 - v^2) = L_o^2 \tag{8}$$

$$2t_3 = \frac{2L_o}{\sqrt{c^2 - v^2}} \tag{9}$$

$$= \frac{2L_o/c}{\sqrt{1 - v^2/c^2}} \tag{10}$$

where $2t_3$ is the time it takes light to complete its round trip along this arm AC.

By comparing (6) with (10) we immediately conclude that the ray moving along the arm AB lasts more time in completing its trip than the ray moving along the arm AC:

$$\Delta t = (t_1 + t_2) - 2t_3 \tag{11}$$

$$= \frac{2L_o/c}{1 - v^2/c^2} - \frac{2L_o/c}{\sqrt{1 - v^2/c^2}} \tag{12}$$

13.6 Michelson-Morley experiment

$$= \frac{2L_o}{c}\left(\frac{1}{1-v^2/c^2} - \frac{1}{\sqrt{1-v^2/c^2}}\right) > 0 \tag{13}$$

where Δt is detectable from the interference between both rays. But although Δt is experimentally detectable it was, and continue to be, undetected in significant terms.

The negative results of Michelson's experiment were well known by 1905 and, as expected, they had a major impact on relativity research before Einstein, although Einstein does not cite any of them, not even the popular Michelson-Morley experiment, which was carried out 18 years before the publication of Einstein's paper *On the electrodynamics of moving objects*.

14. Infinity, language, and non-Euclidean geometries

Abstract. This chapter reminds us that contemporary mathematics, geometry and mathematical physics are infinitist disciplines that assume the Axiom of Infinity. The second part of the chapter recalls the infinite (in this case potential) regress of arguments, definitions and causes that, among other things, makes inevitable the use of primitive concepts in all sciences. As we will see here, from this necessity derive certain abuses of language that, in the case of non-Euclidean geometries, have ended up giving a very distorted and unreal image of their relationship with Euclidean geometry. It is also recalled that since 2021 there is a new foundational basis of this Euclidean geometry in which the Parallel Postulate can be proved.

Keywords: actual infinity, potential infinity, infinite regress, straight lines, parallelism, equidistance, non-Euclidean geometries, Euclidean geometries.

14.1 Introduction

Geometries, Euclidean and non-Euclidean, are infinitist: they assume the Hypothesis of the Actual Infinite and its metric consequences. And not only geometries, contemporary physics is also unequivocally infinitist: its mathematical language includes among its foundations the Hypothesis of the Actual Infinity subsumed in the Axiom 1 of Infinity. An axiom that could be inconsistent, although this possibility has not even been contemplated for more than a century. In Chapter 3 of this book, and after a brief explanation of the differences between the actual infinity and the potential infinity, a short basic demonstration of such inconsistency is provided. The last two appendices of the book include two other very simple proofs of this inconsistency. The reader is also invited to analyze other different more advanced proofs in [212]. Obviously, the inconsistency of the actual infinity would change everything in mathematics, geometry, and mathematical physics.

Related to infinity is the infinite (in this case, potential) regress of arguments, definitions, and causes, which imposes certain insurmountable limits on human knowledge, and which explains, at least in part, the abuses of language that are common in all scientific disciplines, including formal sciences such as geometry. In the case of geometry, the abuses are committed especially in non-Euclidean geometries and in relation to the concepts of straight line and parallelism. Abuses that have never been considered in the two centuries of its history and that give a very distorted, even false, image of its relations with Euclidean geometries. The last section of this chapter deals with these abuses and distortions.

14.2 The inevitable incompleteness of human knowledge

Everything seems to indicate that our observable universe evolves in the direction marked by the Principle 1 of Directional Evolution (page 41):

> The observable universe always evolves independently of its rational observers and in the same direction of increasing its global entropy.

It is, on the other hand, significant that from this inductive principle it immediately follows the Theorem 20 of Formal Dependence (page 43):

> No concept defines itself; no statement proves itself; no physical object is the cause of itself; and no cause is the cause of itself.

in which one can recognize the Aristotelian infinite regress of arguments, extended also to definitions and causes. Naturally, such restrictions inevitably limit human knowledge, and in a much more general and severe way than Gödel's incompleteness theorems, to which, however, much more attention is paid [201, 199].

Here we are interested in the potentially infinite regress of definitions, which makes the use of primitive (undefinable) concepts inevitable in all sciences, including formal sciences such as geometry. Obviously, since concepts do not define themselves (Theorem 20 of the Formal Dependence), if one were to succeed in defining a primitive concept, the definition would have to include at least another concept that would become the new primitive concept replacing the newly defined one. There is no way to get rid of primitive concepts.

But instead of explicitly admitting the need for indefinable basic concepts, attempts are often made to define them more or less ambiguously, or circularly, or invalid for other reasons. This fact has given rise to a certain chaos in the formal use of the most basic concepts in geometry (and in the rest of sciences), and of some not so basic ones such as the concept of straight line, for which it is possible to give a formal definition, although based on two primitive concepts: the concept of point and the concept of line. In the last section of this chapter we will have the opportunity to prove that the definition of straight line poses a serious conflict in the commonly accepted relations of non-Euclidean geometries with Euclidean geometries, particularly the one developed by Playfair in 1813 [279, 280] and by the author in 2021 [200].

14.3 Straight lines and parallelism

In Euclid's Elements the following definitions appear [150, p. 153-155]:

a) A point is that of which there is no part.
b) A line is a length without breadth.
c) The extremities of a line are points.
d) A straight-line is a line which lies evenly with the points on itself.

All of them are unsatisfactory: Euclid himself did not use them explicitly in his demonstrations. However, admitting that the straight line is a primitive concept is equivalent to admitting a primitive concept (straight line) that includes another primitive concept (line) that includes another primitive concept (point). Perhaps too many primitive concepts involved in the same concept. As will be seen below, straight lines can indeed be explicitly and uniquely defined, although such a definition is unusual in classical and modern geometries, and could be the reason for the formal conflicts between Euclidean geometry and non-Euclidean geometries. In J. Playfair's Elements of Geometry [279, 280] we can read [279, p. 8]:

a) A Point is that which has position, but not magnitude.

b) A line is length without breadth.
   Corollary. The extremities of a line are points; and the intersections of one line with another are also points.

c) If two lines are such that they cannot coincide in any two points, without coinciding altogether, each of them is called a straight line.
   Corollary. Hence two straight lines cannot enclose a space. Neither can two straight lines have a common segment; for they cannot coincide in part, without coinciding altogether.

In Playfair's definition of a straight line, an essential (topological) characteristic of straight lines, of straightness, already appears: two straight lines cannot have a common segment if they are not part of the same straight line, nor can they enclose a space. And in the foundational basis (29 definitions, 10 axioms and 45 corollaries) of the author's New Elements of Euclidean Geometry[1] we find [200, p. 28-66]:

1. Points and segments that do not belong to the same line are said non-collinear. Non-collinear lines with at least one common segment are said locally collinear.

2. Lines whose segments have all of them the same definition as the whole line are said uniform. Two or more uniform lines are said mutually uniform iff any segment of any of them has the same definition as any segment of any of the others.

3. To extend a given line by a given length is to define a line, said extension of the given line, that is adjacent to the given line, has the given length, and the extension and the extended line are lines of the same class as the given line. Lines that can be extended from each endpoint and by any given length are called extensible lines.

4. **Definition 17 (of Straight Lines 2)** *Straight lines: Extensible and mutually uniform lines that can neither be locally collinear nor have non-common points between common points.*

Therefore, it is possible to give an exclusive definition of straight line not based on metric concepts alien to the nature of lines[2] but on concepts proper to the topological nature of lines. It is also a functional, productive, definition. That is, a definition that is explicitly used in the demonstrations. Note how "being the line of least length joining two points" does not appear in any of the above definitions of straight line. That is a metrical result that can be deduced, at least in some particular cases [200].

For a general demonstration the reader may consider the demonstrable fact that the straight line AB joining two points A and B has a length less than the sum of the lengths of the other two sides of any triangle ABC constructed on AB; and that the same conclusion applies to the sides AC and CB of any of the triangles constructed on AC and CB; etc.

14.4 Language abuses in non-Euclidean geometries

In spherical (Riemannian) geometry, the straight lines (or lines of maximum straightness [291, p. 8]) are the great circles (also called geodesics). And parallel straight lines are those that do not intersect each

---

[1] In which the Parallels Postulate is proved.
[2] As is the case of the non-Euclidean definition of straight line as the line that minimizes the distance between any two given points.

14.4 Language abuses in non-Euclidean geometries

other. Since all the great circles (straight lines in spherical geometry) intersect at two points, it is concluded that in this geometry there are no parallel "straights" lines, and therefore Euclid's Postulate of Parallels is not necessary. The reason why the maximal circles are called straight lines is because they are the lines of least length that join any two points of the sphere that defines the geometry under consideration. In fact, any two points on the sphere together withe the center of the sphere determine a plane that intersects the sphere in a great circle (straight line).

The problem is that the great circles of spherical geometry, their supposedly "straight" lines, do not fulfill the basic properties of Euclidean straight lines, and therefore should not be considered straight lines in any way:

1. Two great circles intersect at two points, and all points between those two common points are non-common points, which is impossible in Euclidean straight lines (see Definition 17 of straight lines given above).

2. Two great circles enclose non-zero surfaces, which is impossible with Euclidean lines.

3. Great circles are not extensible, they all have the same length. Euclidean lines are extensible and can exist with any length and extend (produced) from each endpoint to any given length.

4. There are points outside a Euclidean straight line that are in a straight line with that straight line, and can be joined to it by a straight line forming a new straight line of greater length. This is also impossible with the great circles of spherical geometry.

5. If P is a point between the endpoints A and B of a Euclidean straight line AB, in the direction from P to B the point A is never reached, which is the case if AB is a great circle.

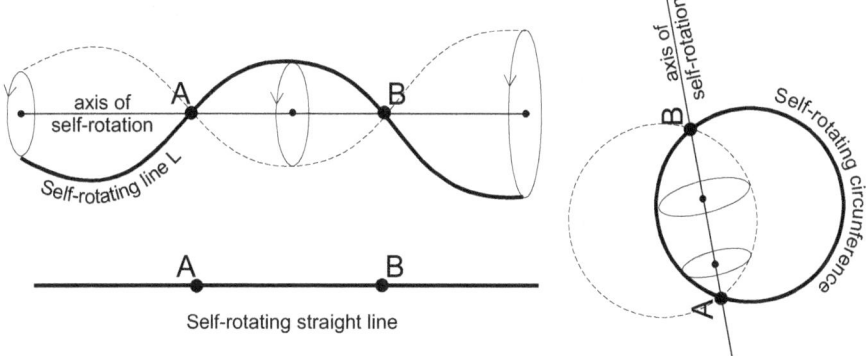

Figure 14.1 – Self-rotating line L. The points A and B of L define an axis of self-rotation.

6. Euclidean straight lines remain invariant in shape and position while self-rotating (a line self-rotates if each of its points describes a circle whose center is a point of the same straight line, the axis of self-rotation, defined by any two points of the self-rotating line, see Figure 14.1), which is not the case with great circles.

Comment: this condition could be used to define in geometrical terms lines of zero curvature, and include in Definition 17 that straight lines have zero curvature, since it seems reasonable to require that straight lines are not curved lines. I will do so in the next issue of [200]. In such a case the non-Euclidean lines could never be straight since they all have non-zero curvature.

On the other hand, parallel lines (not necessarily straight lines) should not be defined in terms of whether they intersect (non-parallel) or do not intersect (parallel), since there are non-parallel lines that do not intersect each other: any curve with asymptotes and one of its asymptotes. And since parallelism is a metrical relation between different lines rather than a topological property of a line, it makes sense to define parallelism in terms of equidistance. If this is done, and there is no reason not to do so, in spherical geometry there would be a potentially infinite number of parallel lines: all minor circles equidistant from each other.

In the case of hyperbolic geometries the problem is the opposite (apart from calling here also straight lines to curved lines): for a given point there exists an infinite number of parallels to a given "straight" line. But here again the problem is solved with the same solution: defining parallelism in terms of equidistance. In this case, from that infinitude of parallels to a given line (so called because they do not cut the given line) all those that are not equidistant would have to be eliminated, and only one would remain.

Non-Euclidean geometries have always been presented as alternatives to Euclidean geometry in which the Postulate of Parallels does not exist. But in reality what do not exist are the Euclidean straight lines, the lines of zero curvature. And there are not problems with parallel lines either if parallelism is defined in terms of equidistance. Furthermore, and taking into account that in the New Elements of Euclidean Geometry the Postulate of Parallels is proved as a theorem, to continue insisting on the non-existence of parallels or on the existence of more than one parallel through a given point to a given line will imply rejecting some of the 10 axioms of those New Elements, axioms which are the following:

Axiom 1.-Point, line and surface are primitive concepts of which any number, and in any arrangement, can be considered and drawn.

Axiom 2.-A line has at least two points, at least one point between any two of its points, and at most two endpoints, whether or not in the line.

Axiom 3.-Two adjacent lines make a line, and a point of a line can be common to any number of any other different lines, either collinear, or non-collinear, or locally collinear.

Axiom 4.-Being not a figure, each point of a line, except endpoints, has just two sides in that line, whose lengths are greater than zero and sum the length of the whole line.

Axiom 5.-Any two points can be the endpoints of a straight line, and only both points are necessary to draw the straight line.

Axiom 6.-Any three points lie in a plane, in which any straight line has two, and only two, sides. Any other line is in one of such sides iff its endpoints are in that side.

Axiom 7.-The distances from the points of a line to a fixed point or to another line vary in a continuous way. The distances from a point to itself and to a line to which it belongs are zero.

Axiom 8.-Any point in a plane can be the center of a circle of any radius, and its complementary arcs are each on a different side of its chord.

Axiom 9.-It is possible for two adjacent straight lines to make any angle at their common endpoint. The angle is zero iff both straight lines belong to the same straight line.

Axiom 10.-The area of a polygon is greater than zero, and is the sum of the areas of the two adjacent polygons defined by any of its divisors. Equal polygons have equal areas.

15. Theories of inertial relativity

15.1 Physics and mathematics

It is a truism to say that we use ordinary and formal languages to describe the world, even to understand it. But all ordinary and formal languages (such as mathematics) have the same two limitations, both consequences of the Aristotelian (potentially) infinite regress:

1. The need for concepts that cannot be defined.
2. The need for statements that cannot be proved.

Since concepts do not define themselves, in order to define a concept we have to define it in terms of other different concepts, which in turn have to be defined in terms of other different concepts, which in turn have to be defined in terms of other different concepts... Therefore, we have to accept the use of undefined basic concepts, concepts that in science we call primitive concepts (set, point, space, mass, etc.). But the main problem is not the limitation of human knowledge by the need to use primitive concepts, the main problem is the existence of circular definitions that are closed in different ways, all of them useless, but which end up creating false certainties. As a science, physics should explicitly declare its primitive concepts and be consistent in their subsequent use, which is not (and never has been) the case in contemporary physics. Other linguistic problems are the imprecise use of polysemantic terms and the imprecise use of different terms to denote the same object or process and the same term to denote different objects or processes.

Statements do not prove themselves either, so in order to prove them, and for the same reason as in the case of definitions, we have to make use of unproved statements. In modern science there are two types of these unproved (primitive) statements:

1. Inductive principles of the experimental sciences: Conclusions from observation of nature and experimentation. We know them as the

fundamental laws (principles or postulates) of experimental sciences: biology, physics, geology and chemistry.

2. Axioms of the formal sciences (and of some experimental sciences): statements that are arbitrarily constructed, either:

   a) trying to make them self-evident; or:

   b) ignoring their level of evidence.

   Some axioms of contemporary mathematics, such as the Axiom of Infinity (discussed in Chapter 3), correspond to the type 2b.

Since Galileo, it has been relentlessly repeated that the language of physics is mathematics. And it is repeated proudly, as if that in itself gives truth to physics, the science of the inorganic (and part of the organic) physical world. Iconic are the images of large blackboards full of abstract symbols used as signs of intelligence and wisdom, when in fact they could be signs of stupidity or nonsense if the corresponding mathematics were inconsistent.

So it makes a lot of sense to ask: is the infinitist mathematics of contemporary physics consistent? But contemporary physicists does not even ask this question. Nor does mathematicians. Chapters 3 and appendixes A and B prove the inconsistency of the Hypothesis of the Actual Infinity, a hypothesis subsumed in the Axiom of Infinity. And it is not an irrelevant axiom for physics: it is the axiom that formally justifies, among many other things, the infinite division of space and time. In other words, it formally justifies the spacetime continuum. But if the axiom of infinity is inconsistent, then the omnipresent spacetime continuum is also inconsistent. I find the lack of attention given by contemporary theoretical physicists to these essential issues of their mathematical formalism irresponsible.

15.2 The Newton-Maxwell relativistic conflict

It is well known that M. Faraday's empirical electromagnetism eventually found its mathematical expression in J. C. Maxwell's theory of electromagnetism, summarized and expressed in his famous equations. It is also well known that these equations posed a serious relativistic problem: according to the Galileo-Newton theory of relativity, it was not possible to distinguish experimentally between two reference frames in uniform relative motion (inertial reference frames). According to Faraday-Maxwell electromagnetism, however, such a distinction is possible:

> An electric charge at rest produces only an electric field around it, while in uniform rectilinear motion it also produces a magnetic field.

15.2 The Newton-Maxwell relativistic conflict

It would then be possible to distinguish in empirical terms an inertial reference frame at rest from another in uniform motion. This is naturally incompatible with Galileo-Newton relativity. The conflict will give rise to different solutions, including the theories of relativity of Lorentz (1904) and Einstein (1905).

Indeed, Maxwell's equations of electromagnetism lead almost immediately to the wave equation for electric fields ($\vec{E}$) and magnetic fields ($\vec{B}$):

$$\nabla^2 \vec{E} = \mu_o \epsilon_o \frac{\partial^2 \vec{E}}{\partial^2 t} \tag{1}$$

$$\nabla^2 \vec{B} = \mu_o \epsilon_o \frac{\partial^2 \vec{B}}{\partial^2 t} \tag{2}$$

where:

- ∇^2 is a second order differential operator.
- μ_o is the magnetic permeability of a vacuum, which is a universal constant.
- ϵ_o is the electric permittivity of a vacuum, which is also a universal constant.

Comparing (1) and (2) with the standard form of a wave equation:

$$\nabla^2 \vec{Y} = \frac{1}{v^2} \frac{\partial^2 \vec{Y}}{\partial^2 t} \tag{3}$$

one immediately infers, as Maxwell did, that:

$$v = (\mu_o \epsilon_o)^{-1/2} \tag{4}$$

$$= (4\pi \times 10^{-7} mKgC^{-2} \times 8.8541878 \times \tag{5}$$

$$\times 10^{-12} C^2 s^2 Kg^{-1} m^{-3})^{-1/2} = 299792.458 \, Km/s \tag{6}$$

is the speed of the electromagnetic waves, that coincides with the speed c of light in a vacuum. As Maxwell proposed, and as we now know very well, light is a set of electromagnetic waves. Obviously, and since c is the arithmetic inverse of the arithmetic product of two universal constants, it is also a universal constant.

The universality of c led Maxwell to propose, contrary to Faraday, that electromagnetic waves propagate in a physical medium with certain physical properties: the *ether*. And as we already know from Chapter 13, the Michelson-Morley experiment used the universality of c to try to detect and measure the absolute speed of the Earth through this ether, the famous "ether wind", which should be perceived as interference fringes between the two orthogonal rays of light. But the expected interferences were not detected, at least not with the expected minimum evidence.

The negative results of the Michelson-Morley experiment could be explained in two different ways. The first way was proposed independently by H. A. Lorentz and G. F. FitzGerald. In the case of Lorentz based on a contraction of material objects in the direction of their motion through the ether; in the case of FitzGerald based on an expansion of material objects in the direction perpendicular to the direction of motion through the ether. The second way to face the negative results of Michelson-Morley experiment was the assumption of a new principle that assumed the constancy of the speed of light in all inertial reference frames, making the existence of the ether unnecessary. The first solution led to Lorentz's theory of relativity in 1904, and the second, published in 1905, led to Einstein's special theory of relativity, which was almost ignored until 1916, the year of the publication of Einstein's general theory of relativity (which includes all reference frames, not just inertial ones). It seems correct to refer to Lorentz's and Einstein's theories of relativity as inertial theories, since they refer exclusively to inertial reference frames (with uniform relative velocity vectors). In the following sections we will briefly examine the formal foundations of both theories.

15.3 Lorentz theory of inertial relativity

As mentioned above, both Lorentz and FitzGerald proposed a contraction (expansion) of all material objects in the parallel (orthogonal) direction of their motion through the ether to explain the negative results of the Michelson-Morley experiments. If such a contraction (expansion) were real, the distances traveled by the two orthogonal light beams could be equalized, thus explaining the absence of the expected interference fringes between these two beams when they arrive at their common detector.

It appears that FitzGerald left no written documentation of his ideas on this subject. They only appear in O. J. Lodge's book *The Ether of Space* [218] in which the exchange of ideas between both authors is recorded. But according to this text there is no contraction of the interferometer arm parallel to its direction of motion but an expansion of its orthogonal arm [218, p. 39-41]:

> That fact has now come clearly to light. It was first suggested by the late Professor G. F. FitzGerald, of Trinity Col-

15.3 Lorentz theory of inertial relativity

lege Dublin, while sitting in my study at Liverpool and discussing the matter with me. The suggestion bore the impress of truth from the first. It independently occurred also to Professor H. A. Lorentz, of Leiden, into whose theory it completely fits, and who has brilliantly worked it into his system. It may be explained briefly thus:

Electric charges in motion constitute an electric current. Similar charges repel each other, but currents in the same direction attract. Consequently two similar charges moving in parallel lines will repel each other less than if stationary -less also than if moving one after the other in the same line. Likewise two opposite charges, a fixed distance apart, attract each other less when moving side by side, than when chasing each other. The modification of the static force, thus caused, depends on the squared ratio of their joint speed to the velocity of light.

Atoms of matter are charged; and cohesion is a residual electric attraction (see end of Appendix 1). So when a block of matter is moving through the ether of space its cohesive forces across the line of motion are diminished, and consequently in that direction it expands, by an amount proportioned to the square of aberration magnitude.

A light journey, to and fro, across the path of a relatively moving medium is slightly quicker than the same journey, to and fro, along (see p. 64). But if the journeys are planned or set out on a block of matter, they do not remain quite the same when it is conveyed through space: the journey across the direction of motion becomes longer than the other journey, as we have just seen. And the extra distance compensates or neutralizes the extra speed; so that light takes the same time for both.

Independently, Lorentz proposed a more complete theory based on a contraction of the interferometer arm parallel to its velocity v. In addition to the contraction of the interferometer arm, Lorentz also proposed a dilation of all rhythmic processes by a factor:

$$\left(1 - \frac{v^2}{c^2}\right)^{1/2} \tag{7}$$

where v is the velocity of the proper reference frame of the objects producing the rhythmic processes, and c is the speed of light. From these two propositions, Lorentz deduced the mathematical operator that transforms measurements made in two inertial reference frames into each other. Thus, for a uniform relative motion parallel to the X

axes of both systems, the spatial and temporal coordinates (x, t) and (x', t') of both systems would be related by:

Lorentz Transformation (X axis)
$$\begin{cases} x' = \dfrac{x - vt}{\sqrt{1 - \dfrac{v^2}{c^2}}} \\ t' = \dfrac{t - vx/c^2}{\sqrt{1 - \dfrac{v^2}{c^2}}} \end{cases} \qquad (8)$$

The above equations have come to be known as the Lorentz Transformation, and they make Maxwell's equations of electromagnetism invariant, ensuring that all measurements made in either system are related to each other in the same way.

Consequently, Lorentz's theory, published in 1904 [219] and known as Lorentz's theory of relativity, not only explained the negative result of the Michelson-Morley experiment, but also resolved the Newton-Maxwell conflict: it was once again impossible to determine the state of rest or uniform rectilinear motion of a material object by experiment with that object. Until 1916, Lorentz's theory of relativity was practically the only theory of relativity considered by most physicists and mathematicians. It was supported, for example, by H. Poincaré, one of the most important scientists in the field.

15.4 Einstein's theory of special relativity

(This section is partially taken from [208])

The two papers containing Einstein's theories of relativity were published respectively in 1905 [87] (the part known as the special theory of relativity, which includes only inertial reference frames) and in 1916 [90] (the part known as general relativity, which includes non-inertial reference frames). The 1905 publication was virtually unknown to the scientific community until 1916. As mentioned above, until that year the only theory of relativity considered by the scientific community was the Lorentz theory just recalled in the previous section.

The special theory of relativity is a theoretical construct which, as is well known, is based on two fundamental principles: [87, p. 895]:

> Principle of Relativity: The laws by which the states of physical systems undergo change are not affected, whether these changes of state be referred to the one or the other of two systems of coordinates in uniform translatory motion.

> Principle of the Constancy of the Speed of Light: Any ray

15.4 Einstein's theory of special relativity

of light moves in the 'stationary' system of coordinates with the determined velocity c, whether the ray be emitted by a stationary or by a moving body.

Or in a modern, more compact, form:

Principle of Relativity: The laws of physics have the same form in all inertial reference frames.

Principle of the Constancy of the Speed of Light: The speed of light is the same in all inertial reference frames.

The last statement of the principle of the constancy of the speed of light (very common in modern texts) is misleading. It could be interpreted to mean ONLY that the speed of light is a universal constant. But Maxwell already said this in his equations of electromagnetism: in these equations, the speed of light is defined as an algebraic combination of two universal constants (and then as a universal constant): the electric permittivity of the vacuum and the magnetic permeability of the vacuum.

The Principle of the Constancy of the speed of light is usually expressed in modern texts in terms of the independence of the speed of light from the relative speed of its emitting source:

The speed of light is independent of the relative motion of its emitting source.

This is a confusing statement, and I would say it is also a tricky statement. It is confusing because velocity is a vector magnitude, and the velocity of light depends (in vector terms!) on the velocity of its emitting source: In fact, photons, like any other physical object, are preinertial (see Chapter 19): they always inherit the relative velocity vector of their emitting source as a component of their velocity vector. For example (Figure 15.1), a photon emitted in the vertical direction, i.e. parallel to the Y_o axis of the proper reference frame RF_o of its emitting source, will be observed in the reference frame RF_v which coincides at a given moment with RF_o and from whose perspective RF_o moves with a velocity v parallel to its X_v axis, along an inclined trajectory with a velocity vector $\vec{c}_v$ whose horizontal component is exactly v. *What is universal is the modulus of the velocity vector of light (its speed).* It is also a tricky statement because it hides its most polemical facet, which becomes evident with the following statement:

The speed of light is independent of the speed of the instrument that measures it.

Indeed, the great novelty introduced by the second relativistic principle is that the speed of light emitted by a body does not change as the instrument measuring it, approaches or moves away from the emitting

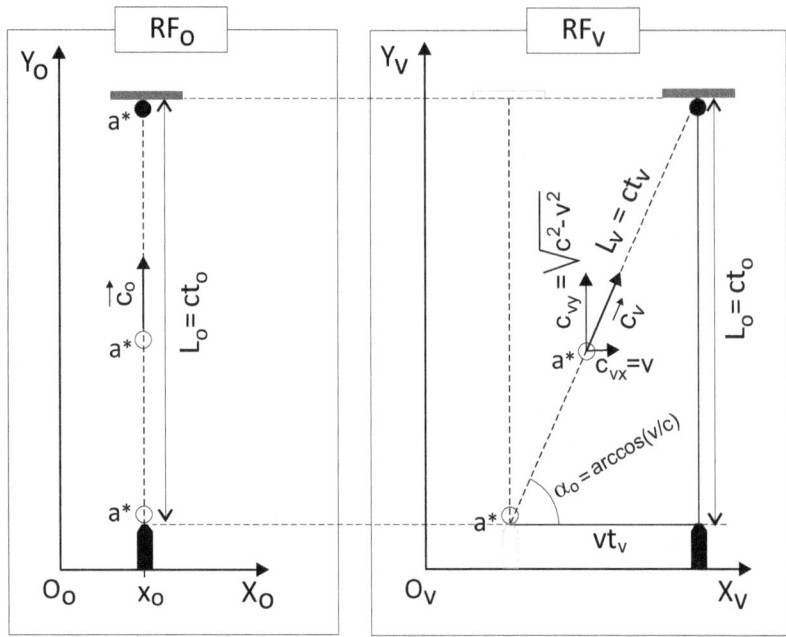

Figure 15.1 – Left: Velocity vector $\vec{c}_o$ of light in RF_o. Right: Velocity vector $\vec{c}_v$ of light in RF_v.

body, regardless of the speed at which that instrument approaches or moves away from the emitting body, even if the speed of that instrument is close to the speed of the light it is measuring, even (if possible) the speed of light. Of course, this is the source of all the relativistic weirdness like length contraction, time dilation, non-existence of a "universal now" (local simultaneity instead of universal simultaneity).

From his two fundamental principles, Einstein deduces the mathematical operator that converts between measurements made in two inertial reference frames, which turns out to be the same Lorentz Transformation that appears in Lorentz's theory of relativity, although Einstein does not name it as such, nor does he cite Lorentz's article (Einstein's 1905 article contains no bibliographic reference). A fundamental difference between the two theories, Lorentz's and Einstein's, is that the former requires the ether, while the latter does not:

> The introduction of a "luminiferous ether" will prove to be superfluous inasmuch as the view here to be developed will not require an "absolutely stationary space" provided with special properties, nor assign a velocity-vector to a point of the empty space in which electromagnetic processes take place.

Einstein's change of opinion with respect to the ether, or rather with respect to physical space, is discussed in the last section of this chapter.

15.5 Relativistic consequences on space and time.

Einstein's theory of special relativity has at least three consequences incompatible with classical mechanics. We will call them relativistic inertial deformations of spacetime (RDST), they are the following:

1. Contraction of all material objects in the direction of their relative velocities.

2. Time dilation in the direction of their relative velocities.

3. Lack of universal simultaneity between events in the direction of their relative velocities, except if both directions are orthogonal.

With regard to the contraction of material bodies in the direction of the relative velocity with which they are observed, there is a certain division of opinion among contemporary relativists: for some it is real, for others it is only apparent. The latter defend the apparent nature of the contraction because, in their view, it is not possible for a material object to have an infinite number of different sizes at the same time. This seems to be a reasonable position, but not everyone agrees with it, such as Einstein himself [183, p. 43] and some modern authors of our time [330, 129, 21]. Other authors, such as A. P. French, proposed a way out of this notorious controversy, as M. Born called it [43]. In fact, A. P. French wrote in 1968 [123, pp. 113-114]:

> This discussion should make it clear that the question "Does the FitzGerald-Lorentz contraction really take place?" has no single, unequivocal answer from a relativistic point of view. The whole emphasis is on defining what actual observations we must make if we want to measure the length of some object that may be in motion relative to us. And the prescription is simply that we measure the positions of its ends at the same instant as judged by us. What else could we possibly do? Thus the contraction, when we observe it, is not a property of matter but something inherent in the measuring process.

The following argument (taken from [208, p. 86-87]) points to the alternative of apparent deformations, which in this case are, moreover, apparent deformations incompatible with the behavior of elastic objects. In the reference frame RF_o an elastic and flexible cord rests free of forces on the plane X_oY_o. The elastic cord is scaled with yellow and black marks of equal length L_o, some of which are parallel to the X_o axis and some of which are parallel to the Y_o axis. Since the cord is at rest and no force is acting on it, all yellow and black marks have the same length, and this is indeed what is observed in RF_o (Figure 15.2, left). Things are quite different when this elastic cord is observed from the reference frame RF_v that, as always in this book, coincides with

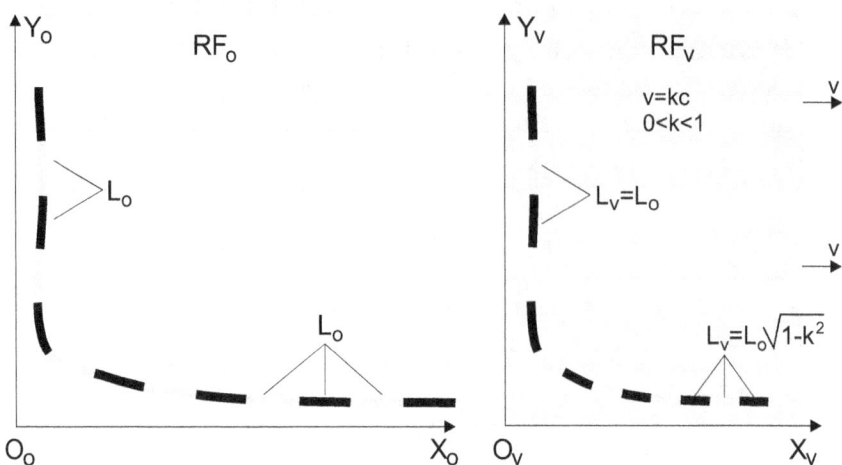

Figure 15.2 – The elastic cord at rest on the plane X_oY_o of its proper reference frame RF_o (left), and from the reference frame RF_v (right).

RF_o at a certain instant, and from whose perspective RF_o moves according to our conventions: with a uniform velocity v parallel to the increasing X_v of RF_v'. As Figure 15.2 (right) illustrates, all marks parallel to the X_v axis are observed with a length L_v, that according to the Lorentz Transformation will be:

$$L_v = \gamma^{-1} L_o < L_o \tag{9}$$

In consequence, while all marks parallel to its Y_v axis are observed with the same length L_o, all marks parallel to X_v have a length $L_v < L_o$. The observers in RF_v will, therefore, observe an elastic cord free of forces with some marks more stretched than others, which is impossible for an elastic cord completely free of forces. In consequence, for all observers, except those in RF_o and those moving parallel to Z_o, the elastic cord is observed with some parts more stretched than others, without any force acting on it. Obviously, this goes against the laws of mechanics governing the behaviour of elastic materials. The conclusion can only be that FitzGerald-Lorentz contraction is apparent, as apparent as the deformation of a rod partially submerged in water.

Other arguments point in the same direction of the mechanical inconsistency of the FitzGerald-Lorentz contraction, whether it is real or apparent [208]. Now, if the FitzGerald-Lorentz contraction is only apparent, are the inertial time dilation and the lack of universal simultaneity (phase inertial difference in synchronization) also apparent? And if some of these deformations are real and some are apparent, should not SR indicate which are real and which are apparent, and why? In [208] other arguments are developed that point very clearly to the apparent nature of time inertial dilation and phase inertial differ-

15.6 Experimental confirmations of special relativity

ence in synchronization.

15.6 Experimental confirmations of special relativity

For several decades now, it has been tirelessly repeated that special relativity is duly confirmed by experience. What is practically never remembered is that a confirmation of the special relativity must be (and almost never is):

1. SYMMETRIC.-If from an inertial reference frame A it is observed that in all physical objects of another inertial reference frame B there is a contraction of length in the direction of the relative motion between the two frames (FitzGerald-Lorentz contraction), then at the same time from the reference frame B it must also be observed that in all objects of A there is exactly the same contraction of length in the same direction of the relative motion between the two frames. The same is true for the inertial dilation of time: if from the reference frame A it is observed that all clocks of the reference frame B run slower than the proper clocks of A, then at the same time from the reference frame B it must be observed that all clocks of A run slower than the proper clocks of B. And whatever the events are, two simultaneous events in A that occur at a non-zero distance in the direction of relative motion will be observed as non-simultaneous in B. And two simultaneous events in B separated by a non-zero distance in the direction of relative motion will be observed as non-simultaneous in A.

2. UNIVERSAL.-Provided they are observed with the same relative velocity $v = kc$, $(0 < k < 1)$, all objects will be contracted in the direction of relative motion by the same factor $\sqrt{1-k^2}$, irrespective of their size, composition and internal structure. A steel cube and a foam-rubber cube will undergo the same degree of contraction in the direction of relative motion. Time dilation and phase difference in synchronization are also universal: the same for all conceivable kinds of clocks: mechanical, electrical, electronic, atomic, chemical, biological, etc.

3. ACAUSAL.-The only and exclusive cause of all relativistic inertial spacetime deformations is the relative velocity at which the involved physical objects and events are observed. No physical agent is involved in these deformations. Only the relative velocity at which they are observed.

On the other hand, the observed confirmations could be confirmations of apparent deformations, as is the case with all refractive deformations: Snell's Law explains them, but the observed and measured deformations are not real. Moreover, the observed and measured relativistic deformations could be the consequence of a discrete world (including space and time) interpreted in terms of an indiscrete continuum-based mathematics. In fact, it turns out that the factor that converts between the discrete and continuous versions of the Pythagorean Theorem is precisely the relativistic Lorentz factor [203], and Pythagoras Theorem is a key piece in the calculation of distances.

Considering that the spacetime continuum is modeled by the set $\mathbb{R}^4 = \mathbb{R} \times \mathbb{R} \times \mathbb{R} \times \mathbb{R}$ of all real 4-tuples (x, y, z, t) of real numbers, where the first three represent a point in space and the last one a time instant, the following two questions are inescapable (although, as far as I know, physics has never asked them):

1. How can contract a line in space if its points cannot contract (they have zero extension and cannot contract), nor can it change its number of points: that number of points will always be the same before and after the contraction: exactly $2^{\aleph_0}$ points?

2. How can dilate a time interval if its instants cannot dilate (they would cease to be instants of null duration), nor can it change its number of instants: that number of instants will always be the same before and after the dilation: exactly $2^{\aleph_0}$ instants?

15.7 Space in the 20th century

(This section is partially taken from [213])

The theories of relativity (special and general) constitute an absolutely hegemonic current of thought in contemporary physics. For the vast majority of contemporary physicists, neither space nor time have a real existence, they are purely relational instruments useful to describe the evolution of relations between the coexisting objects of the universe. For example, L. Smolin wrote [325, p. 266]:

> ... space and time, like society, are in the end also empty conceptions. They have meaning only to the extent that they stand for the complexity of the relationships between the things that happen in the world.

But naturally, there are also some (few) physicists who do not think the same way about the reality of space and time. This is the case, for example, of F. Wilczek [371, p. 180]:

> Spacetime is also a form of matter.
>
> Spacetime has a life of its own.

Or the case of N. A. Tambakis: [342, p. 146]:

> It seems to me that in this way we can confirm the well-known epistemological assumption that space and time are

15.7 Space in the 20th century

not fictions but rather modes of the dynamic existence of matter.

The case of A. Einstein is a bit more complex. Let's remember some of his words through the years about space and time:

> 1905: The introduction of a "luminiferous ether" will prove to be superfluous inasmuch as the view here to be developed does no longer need an absolute space, at absolute rest, with physical properties [87, p. 891].

> 1913: For me it is absurd to attribute physical properties to "space" (Letter to E. Mach cited in [185, p. 135]).

> 1914: As much I am not disposed to believe in ghosts so I do not believe un the enormous thing about which you are talking and which you call space [89, p. 345].

> 1915: Thereby (through the general covariance of the field equations) space and time lose the last remnant of physical reality (Letter to M. Schlick cited in [185, p. 134]).

However, in 1916 Einstein changed his mind about the physical nature of space and the existence of the ether

> 1916: I agree with you that the general theory of relativity is closer to the ether hypothesis than the special relativity (Letter to H. A. Lorentz cited in [185, p. 135]).

> 1919: Thus, once again "empty" space appears as endowed with physical properties, i.e. no longer as physical empty, as seemed to be the case according to special relativity. One can thus say that the ether is resurrected in the general theory of relativity, though in a more sublimated form (Morgan Manuscript, cited in [185, p. 137]).

> 1938: Our only way out seems to be to take for granted that space has the physical property of transmitting electromagnetic waves... We may still use the world ether, but only to express some physical property of space... At the moment it no longer stands for a medium built up of particles [97, pp. 159-160] [98, p. 115]

In later writings he defended that the physical notion of space is linked to the existence of rigid bodies, but he rejects the idea that space is an *a priori form of intuition* [37], as Kant defended [179]. Einstein *"always supported an objective description of physical reality, without interference of the observer"* [177, p. 128].

But as indicated above, for a good part of contemporary physicists, space is not real, it is only a useful theoretical fiction. At the same time, they all affirm without the slightest doubt that space expands, deforms, vibrates, and is the transmitting medium of its own vibrations. And one wonders, how can something that does not exist expand, deform and vibrate? A deformation of something that does not exist does not exist (at least that is what logic suggests).

In the other direction, and considering that preinertia would prevent the detection of absolute motion, if space were real, would it not be the receptacle of all physical objects in the universe? Would not these physical objects move THROUGH space? Would this motion not be absolute motion? Would not the different absolute motions of different objects through the same real physical space be the cause of all observed relative motions? Would not everything be much simpler?

16. Discrete versus continuous

16.1 Introduction

This chapter continues with the proofs of some formal results that will be used in the analysis of the nature of physical space. Among them are the theorems of decimal expansions, finite divisions, inconsistent divisions, and finite lengths. I repeat, more or less partially, the content of some sections already included in previous chapters because it is also convenient to include it here in order to give a complete view of the main subject of the chapter: the discrete versus continuous nature of space (and time).

16.2 The problem of the continuous

Although related to the modern spacetime continuum (see next Section 16.3), the problem of the continuous has a Pythagorean origin [231]. In my opinion, its importance in the history of science has not been sufficiently appreciated. The firsts Pythagorean believe in the existence of indivisible geometrical points with an extent δ greater than zero, consequently they believed that all lengths would have to be commensurable: the ratio between any two of these lengths, say L_1 and L_2, would be a ratio between two natural numbers [231, pp. 11-16]:

$$L_1 = n_1\delta; \; L_2 = n_2\delta \tag{1}$$

$$\frac{L_1}{L_2} = \frac{n_1\delta}{n_2\delta} = \frac{n_1}{n_2} \tag{2}$$

Somewhat later, the Pythagorean discovered the existence of non-commensurable lengths: the length of the diagonal L_d of a square with the length of its side. For example, if the length of the side is 9δ, we would have:

$$L_d = \sqrt{9^2\delta^2 + 9^2\delta^2} \tag{3}$$

$$= 9\delta\sqrt{2} \tag{4}$$

$$\frac{L_d}{L_s} = \frac{9\delta\sqrt{2}}{9\delta} = \sqrt{2} \tag{5}$$

Unfortunately, they did not consider the possibility of a discrete arith-

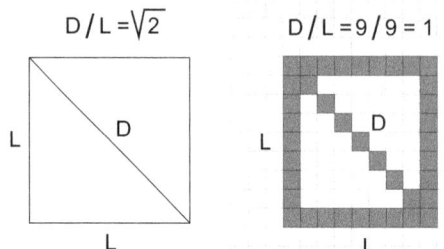

Figure 16.1 – Left: In continuous geometry the diagonal D and the side L of a square are not commensurable. Right: In discrete geometry the diagonal D and the side L of a square are commensurable.

metic, for instance:

$$L_d = \lfloor \sqrt{9^2\delta^2 + 9^2\delta^2} \rfloor \tag{6}$$

$$L_d = \delta \lfloor \sqrt{9^2 + 9^2} \rfloor \tag{7}$$

$$= 9\delta \lfloor \sqrt{2} \rfloor \tag{8}$$

$$= 9\delta \tag{9}$$

$$\frac{L_d}{L_s} = \frac{9\delta}{9\delta} = 1 \tag{10}$$

where $\lfloor x \rfloor$ stands for the integer part of x.

Perhaps due to the enormous influence of our sensory perception of the physical world as a continuous scenario of space and time (see Section 4.6), this type of discontinuous arithmetic is yet to be developed in formal and universal terms. In any case, one of the consequences of the Pythagorean discovery of incommensurable lengths was the abandonment of the extensive points in favor of the non-extensive points, which are the same points that we still use today in all continuous geometries. But the discussions about the continuous (and in general about infinity) lasted until the beginning of the 20th century, when the existence of an actual infinite set was axiomatically proposed and accepted in the nascent set theory. Since then, the hegemony of infinitist mathematics (and geometry) has been as absolute as the submission of physics to infinitist mathematics. As is often the case with hegemonic positions (even of thought), in addition to being hegemonic they have been, and continue to be, quite hostile to dissidence, which has been ostracized for more than a century. The power of the academy is not different from other more or less absolutist human powers.

16.3 The spacetime continuum

The word "*continuum*" is a rather polysemous term in mathematics and physics. It can be used, for example, as:

- The set of the real numbers (Harper Collins Dictionary of Mathematics [44, p. 118]).
- A continuous distribution of matter (Harper Collins Dictionary of Mathematics [44, p. 118]).
- A system of axes that form a frame of reference (Oxford Dictionary of Physics [74, p. 94]).
- The set of points on a line (linear continuum) (Oxford Dictionary of Philosophy [39, p. t8]).
- A compact connected metric space (Wolfram MathWorld [368]).
- Said of a magnitude: that takes values that are not separate from each other (DRAE).

The expression "power of the continuum" is less polysemous: it is the cardinal $2^{\aleph_o}$ of the set $\mathbb{R}$ of the real numbers, or of any of its non-denumerable subsets. Although the word "power" in "power set" also means: the set of all subsets of a given set. And the power of the continuum is also the cardinal of the power set of the set $\mathbb{N}$ of the natural numbers, i.e. the cardinal of the set of all subsets of the set $\mathbb{N}$ of the natural numbers when considered as a complete totality, which is the natural consideration in infinitist mathematics.

For any given set A, its cardinal number is usually written $|A|$, and its power set $P(A)$. So, according to the above definitions, we can write the well known equalities:

$$|\mathbb{N}| = |\mathbb{Q}| = \aleph_o \tag{11}$$

$$|\mathbb{R}| = |P(\mathbb{N})| = |P(\mathbb{Q})| = 2^{\aleph_o} \tag{12}$$

where $\mathbb{Q}$ and $\mathbb{R}$ are respectively the set of (all) rational numbers and the set of (all) real numbers. In physics, the continuum is almost always associated with spacetime, the relativistic assumed four-dimensional manifold of space and time. In fact, relativistic spacetime is a four-dimensional continuum of spatial points and temporal instants modeled by the set $\mathbb{R}^4$ of all real 4-tuples (x, y, z, t).

One of the most picturesque properties of the infinitist spacetime continuum is that any linear segment, for instance of a Planck's length:

$$l_p =\approx 0.000000000000000000000000000000016 \text{ millimeters}$$

has the same number of points as the entire observable three-dimensional universe. Or that any interval of time, for example the duration of Planck time:

$$t_p =\approx 0.0000000000000000000000000000000000000053 \text{ seconds.}$$

has the same number of instants as the entire history of the observable universe (over 13800 million years). As is well known, this is the so called Dimension Problem proved by Cantor[1] [19, 78, 324, 363, 133, 76, 59, 66]. It is only necessary to define the corresponding one-to-one correspondence between both sets and make an appropriate use of ellipsis (the magic wands of infinitism) to prove it. This has little discussed consequences in physics: for example, infinitist physics assumes that a segment of Planck length contains as many point entities (point charges, point masses, virtual point particles etc) as the whole three dimensional observable universe.

The concept of point, on the other hand, is a primitive concept; an undefined concept for which we do not even have an intuitive notion (as we do, for example, for the primitive concept of set). The same is true for the concept of instant, *"the points of time"*. The mark we make with the tip of a pencil on a sheet of paper, or with chalk on a blackboard, is not a point, but a conventional graphic mark drawn for representational purposes, a graphic mark which, by the way, contains the same number of points as the entire three-dimensional universe. I have the impression that physicists are not interested in these logical problems related to the actual infinity, although they are anything but irrelevant to physical models and theories. In Chapter 18 it will be proved *formally* that points can have neither size nor shape, which, as will be discussed there, has very important physical consequences.

The most significant use of the continuum in physics is in the spacetime continuum, a tetra-dimensional continuum of spatial points and temporal instants modeled, as noted above, by the set $\mathbb{R}^4$, the set of all 4-tuples (x, y, z, t) of real numbers. The spacetime continuum constitutes the fundamental core of special relativity, which at the end is a theory of the spacetime continuum [343, 243]. The interest in the geometry of space and in the nature of time, including the debates on their infinite divisibility, has interested philosophers at least since the

---

[1] It is also well known Cantor's reaction of not publishing any more in the magazine The Journal of Crelle that rejected the referred proof [78], the same reaction of Einstein with the magazine The Physical Review that in 1936 dared to review one of his articles [184, 267]

16.3 The spacetime continuum

14th century [222], and continues to be in the center of a few number of modern philosophical debates [291, 256, 233, 234, 153], now including the (still irrelevant) alternative of a discrete and finite space and time.

Some pioneering authors were interested in discrete spaces in the first half of the 20th century [36], [70]. W. Heisenberg, for instance, conceived the idea of space as a kind of crystal lattice made up of tiny cells of the size of elementary particles, although he did not develop the idea in the end. Things have started to change, especially in the last two decades [118, 241, 187, 122, 28, 302, 29, 20, 296]. A growing number of physicists now suspect that Planck length and Planck time define a kind of granularity of space and time that could be an effective alternative to the infinitist continuum; an alternative that could be experimentally tested [254, 73, 119, 216, 64]. The problem is that, although the discrete nature of spacetime has been proposed in different areas of physics [176, 138, 354, 112, 326, 327, 339, 24, 220, 228, 9, etc.], the proposals have been invariably developed in the framework of infinitist mathematics. There can be no greater blunder.

To briefly examine the involvement of points and instants in physical phenomena, we will end this section by recalling two well-known physical phenomena involving the points and instants that supposedly form the spacetime continuum: the diffraction of light through a narrow slit (Fraunhofer diffraction), and the spacetime propagation of a field perturbation. In relation to the first phenomenon, whose current explanation is also well know, you can read things like:

- Each *point* of the wavefront become a secondary source of waves, emitting new waves, called diffracted waves ...

- Each *point* of the slit become a secondary source of waves ...

- Each photon at a *point* ...

- etc

But there is a problem with this classical explanation (a problem, by the way, representative of the lack of rigor in the use of language in physics literature): any slit has the same number of points as any other, which is also the number of points in the whole observable universe: $2^{\aleph_0}$ points. Therefore, for a given source of light of a given wavelength and a given distance from the slit to the viewing screen, the corresponding diffraction pattern cannot be explained in terms of points: since all slits have the same geometry and the same number of points, they would all produce the same diffraction pattern, whatever the slit width. And this is not the case: the pattern also depends on the width of the slit.

We could consider the alternative that some points are involved in the formation of the pattern and others are not. But which ones are involved and which ones are not? We must conclude that the diffraction of light through a single Fraunhofer slit cannot be satisfactorily explained by infinitist mathematics, in this case by the geometry of the infinitist continuum. And the reason for this is precisely the continuum of points that space is supposed to consist of. In fact, this problem does not arise in the context of the CALMs introduced in Chapter 18 of this book.

Even more dramatic is the situation with the propagation of a field disturbance (e.g., electromagnetic) through the spacetime continuum. Here we will inevitably find erroneous expressions such as: adjacent points, contiguous points, successive points, next instants, successive instant etc. all of them impossible: in the spacetime continuum no point (instant) has an immediate successor, a next, adjacent, contiguous... point (instant) as, for example, 4 has its immediate successor 5 in the case of natural numbers. The points of a continuum do not touch each other. It is only possible to jump from one point (instant) to another through a non-numerable infinity of points (instants): again $2^{\aleph_0}$ points (instants). So, in the end, only discontinuous and incomplete descriptions of this propagation through the spacetime continuum can be made, incomplete because it is not possible to know what will be the next point (instant) affected; and it will not be possible because there is no point (instant) immediately succeeding (adjacent to) a given point (instant), whatever the given point. A real drama ignored by the theoretical infinitist physics (although it is not aware of its infinitism).

16.4 Finite but non-computable natural numbers

This section proposes to the reader an exercise of imagination and at the same time of humility. It proposes an objective reflection on the incommensurable size of most of the natural numbers included in the list of natural numbers in their natural order of precedence, when that list is considered as a complete totality (Definition 3). All these natural numbers are finite, and each of them is just one unit greater than its immediate predecessor, except the first of them: the number 1 which has no predecessor and is defined just by one of such unities. It will be worth trying to imagine the greatness of the vast majority of these finite numbers. I say try to imagine because, as we will see here, it is not even possible to imagine them.

Well, at the end of the 19th and beginning of the 20th century we ended up assuming that all these numbers (all!) exist in the act, as a complete totality (Hypothesis of the Actual Infinity subsumed in the

16.4 Finite but non-computable natural numbers

Axiom of Infinity). And that there is another number greater than all of them: $\aleph_o$, the smallest infinite (cardinal) number greater than all finite natural numbers, which is also the total number of natural numbers (the cardinal, among others, of the set $\mathbb{N}$ of natural numbers). Or in other words, although there is not a last natural number completing the ordered list of natural numbers, we <u>assume</u> there is a precise number of numbers in that list, and that this number is the least (infinite) number greater than all numbers in that list. This is the foundational core of infinitist mathematics.

Indeed, modern infinitist mathematics assumes the existence of the set $\mathbb{N}$ of the natural numbers in their natural order of precedence $\{1, 2, 3,\dots\}$ as a complete totality (Definition 3) with just $\aleph_o$ elements, each one unit greater than its immediate predecessor, except the first, and all of them finite. Modern infinitist mathematics assumes, therefore, that once defined the least natural number, the number 1, as one unit, infinitely many other natural numbers (all of them finite) can be successively defined simply by adding one unit to the last defined natural number (Peano's Axioms [272, p. 1]). And, above all, that the list of natural numbers so defined exists as a complete totality (Definition 3) that contains in the act all! natural numbers (Axiom of Infinity).

So, after adding infinitely many successive units to one initial unit we do not reach a number with infinitely many units, but infinitely many finite natural numbers; each with finitely many units and with one unit more than its immediate predecessor. This is quite conflicting, and the reason for the conflict is the assumption that the ordered list of the natural numbers exists as a complete totality despite the fact that no last natural number completes the list (see [198] or [212, Chapter 31]). But not only conflicting, this assumption is also inconsistent as is proved in Chapters 3 of this book, and in other more than forty different ways in [212].

Two well known attributes of the natural numbers are their immediate successiveness (adjacency) and their discreteness: each natural number n has an immediate successor $n + 1$ and (except the first one) an immediate predecessor, so that there is no other natural number between $n - 1$ and n, nor between n and $n + 1$. In addition, and being all of them finite, between any two natural numbers m and n (being $m < n$) only a finite number $n - m$ of natural numbers do exist. That said, it is worth considering the gigantic size that successive finite natural numbers can reach. It is an unusual but very convenient exercise to have a comparative reference on the size of certain natural numbers and on the number of decimals of the real numbers that are discussed in the next section.

Indeed, it is possible to define in precise arithmetic terms natural numbers that written in standard text (e.g., 5 millimetre per digit) would occupy a length millions of times greater than the diameter of the visible universe. And they are not anodyne sequences of zeros, but precise sequences of different digits (for example, of the decimal numbering system). This is the case of the expo-factorial numbers (in symbols $n^{!}$, note the factorial symbol ! is here an exponent, a power) and, specially, the case of n-expo-factorial numbers (in symbols $n^{!n}$) [212, Chapter 14]:

> The grandeur of, for example, $9^{!9}$ (9-expofactorial of 9) is far beyond human imagination. Three standard arithmetic symbols, just 9, !, 9, is all we need to define a finite number so large that the standard writing of its precise sequence of figures would surely be a string of numerals of a length millions of times greater than the diameter of the observable universe. If we use the hexadecimal numeral system, $F^{!F}$ would be inconceivable greater.

It is a good exercise for our imagination (and humility) to imagine a finite natural number that would take us millions of years to go through all its digits, moving at 300000 kilometers per second. Not to say the impossibility to represent all those digits in material terms taking into account that the estimated number of elementary particles of ordinary matter in the observable universe is only in the order of 10^{80} [356, 255]. Obviously, numbers as $9^{!9}$ are finite but uncomputable. They can only be defined theoretically.

Imagine now we place a zero and a decimal point before the first digit of $9^{!9}$. We would have a precise rational number $q_{!9}$ with a finite number $9^{!9}$ of decimals, being its sequence of decimals the same as the sequence of digits of $9^{!9}$, which, uncomputable as it may be, will be a precise sequence of digits, for example (the sequence of digits is invented):

$$q_{!9} = 0.3412983247520983\ldots 908734 \qquad (13)$$

Now consider any physical magnitude M whose mathematical definition includes an irrational number as π, e, $\sqrt{2}$ etc. To measure M with a total precision would require to know the precise ω-ordered sequence of all its decimals, which contains $\aleph_o$ decimals, and then infinitely many decimals more than $q_{!9}$. To measure a physical magnitude as M with a precision of, say 30 decimals, would qualify as an extraordinary success (and in my opinion it would be an extraordinary success). But an extraordinary success only within the finitist scenario, because from the infinitist point of view, 30 decimals are absolutely insignificant compared with its total number of decimals, which exist as a complete totality of $\aleph_o$ decimals. Is it appropriate to talk about the extremely high degree of precision of a measurement with 30 decimal places if the actual number of decimal places is, for instance, $9^{!9}$, which is infinitely less than $\aleph_o$?

16.5 Numbers with infinitely many decimals

From the infinitist perspective, the infinitely many decimals of a real number with an infinite decimal expansion, as for example π, do exist as a complete and ω-ordered totality, which means that it is a mind-independent entity because our mind cannot embrace the actual infinity (we cannot even imagine a number with a finite number of $9^{!9}$ decimals). But as will be shown below, numbers with an infinite number of decimals are inconsistent.

Let a be the decimal expansion of any real number with an infinite decimal expansion:

$$a = .d_1 d_2 d_3 \cdots = d_1 \times 10^{-1} + d_2 \times 10^{-2} + d_3 \times 10^{-3} + \ldots \tag{14}$$

and let A be the set of all its successive decimals:

$$A = \{d_1, d_2, d_3, \ldots\} \tag{15}$$

It is immediate to define a one to one correspondence f between A and the set of natural numbers $\mathbb{N}$:

$$f(d_i) = i, \ \forall d_i \in A \tag{16}$$

Among dozens of others [212], Theorem 5 (page 19) proves the set $\mathbb{N}$ is inconsistent when considered as a complete totality. In consequence, the same must apply to the set A and to the infinite decimal expansion of the above infinite decimal expansion a when considered as a complete totality (Definition 3, Corollary 1). Thus, and being a any infinite decimal expansion, the above argument proves the following:

Theorem 27 (of Decimal Expansions) *The infinite decimal expansions are inconsistent.*

From the point of view of the potential infinity, things are very different: from this perspective an irrational number is not a mind independent entity formed by a complete ω-ordered sequence of decimals that exist all at once and by themselves. From this perspective, the irrational numbers result from endless process of calculation that cannot be replaced with a division between two integers, although at each stage of the calculation the number coincides with a rational number with a finite number of decimals. In this sense, the irrational numbers are also definable as (potentially infinite) sequences of rational numbers, and therefore as sequences of ratios between two integer numbers.

In the case of rational numbers, they are defined by a division between two integers, a division that may or may not have an end. In turn, integers would result from the endless process of counting, i.e. from the endless process of adding (or subtracting) successive units to one initial zero. Obviously, the existence of endless processes of counting and calculation does not necessarily mean the existence of their corresponding finished results as complete totalities, as is assumed from the infinitist perspective. It is time to recall the famous quote by Leopold Kronecker (collected in numerous texts, for example in [328, p. 117]):

> God has created the natural numbers, the rest is man's work.

And the commentary appearing on a web page of a certain University:

> He [Kronecker] opposed the work of his student Georg Cantor on infinity, considering that it lacked rigor. How wrong he was!

The author of the above commentary should be reminded that Cantor did not prove the existence of infinities (see Chapter 3 of this book). The existence of infinities had finally to be established by an arbitrary law (Axiom of Infinity). We shall see who is wrong. It has gone so far in mathematical infinitism, and physics has relied so much on infinitist mathematics, that the consequences of its inconsistency could be the greatest in the history of science. So, check out Theorem 5 and its subsequent corollaries; or the arguments on Hilbert machine (Appendix A) and on the last disk supertask (Appendix B), or the more than 40 arguments contained in [212].

Finally, a curiosity: if we remove the decimal point from (14) we get:

$$a' = d_1 d_2 d_3 \ldots \tag{17}$$

But which kind of number is a'? It cannot be a natural number because natural numbers are finite and each of them (except the number 1) is greater than the number of its digits (in the decimal numbering system), and since a' has an infinite number of digits it cannot be a natural number (otherwise it would be an infinite natural number). And then, is it a member of the sequence of powers (18), or of the sequence of alephs (19)?

$$\aleph_o, \; 2^{\aleph_o}, \; 2^{2^{\aleph_o}} \ldots \tag{18}$$

$$\aleph_o, \; \aleph_1, \; \aleph_2, \ldots \tag{19}$$

or a new kind of infinite number? In any case it would be as inconsistent as any other infinite number.

16.6 Discrete and continuous magnitudes

The main difference between a discrete magnitude and a continuous magnitude is that in a discrete magnitude there is always an indivisible minimum value; and in a continuous magnitude there is not. A significant consequence is that the set V of all possible values of a discrete magnitude is completely determined by its particular indivisible minimum μ and the successive natural numbers:

$$V = \{\mu, 2\mu, 3\mu, 4\mu, \dots\} \tag{20}$$

Therefore, each value $n\mu$ of the sequence of successive values of a discrete magnitude has an immediate successor $(n+1)\mu$, and an immediate predecessor $((n-1)\mu$ if $n > 1$; and no other value exists between any value and its immediate successor, or its immediate predecessor, if any. For this reason, the sequence of all possible values of a discrete magnitude is said to have immediate successiveness, adjacency (as the sequence of natural numbers in their natural order of precedence). Furthermore, since the indivisible minimum must be finite (Corollary 4), and can only have a finite number of decimals (Theorem 27), we can conclude that the sequence of the possible values of any discrete magnitude has immediate successiveness, adjacency (each value has an immediate successor, except the last one) and each of its members is finite and have a finite number of decimals, if any.

Things are quite different with continuous magnitudes, because they are densely ordered: between any two values of a continuous magnitude, there are infinitely many other different values; and between any two of these infinitely many values, there are other infinitely many values; and between any two of these infinitely many values, there are other infinitely many values; and so on and so forth. Therefore, a sequence of continuous values does not have immediate successiveness (adjacency): between any two values of this continuous sequence of values, there are always an infinite number of different values.

The definition of some physical constants involves irrational numbers, as is the case with π in the Planck constants: Planck length, Planck time, Planck mass, Planck energy, and Planck charge. So they are all irrational numbers with an ω-ordered sequence of decimals, i.e. with infinitely many decimals. From the infinitist point of view, all these constants are irrational numbers with an infinite number of decimals. And according to the Theorem 27 on numbers with infinitely many decimals, all of them would be inconsistent if the infinite sequence of their decimals were considered as complete totalities (Definition 3). So each of these constants should have an exact finite sequence of decimals if they are consistent constants. Finally, and with respect to the variable magnitudes, the following is verified:

Corollary 18 (of the Discrete Values) *The number of all possible values of a variable magnitude is finite, and all of them can be arranged in a discrete set with a minimum and a maximum value.*

Proof: It is an immediate consequence of Corollary 4 of the Inconsistent Infinite Sets, Theorem 10 of the Strictly Ordered Sets and Theorem 11 of Discrete Sets. □ □

16.7 Inconsistency of the actual infinite divisions

In chapter 3 of this book, the inconsistency of ω-ordered sets was proved, and then the inconsistency of any other infinite collection containing ω-ordered sub-collections, such as the continuum, was proved (Corollary 1 of the Inconsistent Sets). This result is sufficient to prove the inconsistency of the infinite partition of space (or time), since any such partition is defined by an infinite ordinal $\alpha \geq \omega$ and an α-ordered sequence of points (instants) [212]. In this section, other independent proofs will be given that confirm the inconsistency of infinite division of an interval of space or time (in fact, of any finite interval or real numbers). First, the inconsistency of dividing a finite interval of real numbers into an infinite number of equal parts will be demonstrated. And second, the same inconsistency is proved for an infinite number of parts of decreasing size.

Theorem 28 (of the Finite Divisions) *A finite interval of real numbers divided into parts of equal length, except at most the last part if any, can only have a finite number of parts.*

Proof: Let (a, b) be any open finite real interval (the argument also applies to closed and semi-closed real intervals). To divide (a, b) into a certain number of parts of equal length (except the last one, if any) means to define a number of adjacent and disjoints sub-intervals:

$$(a, x_1)[x_1, x_2)[x_2, x_3)[x_3, x_4)\ldots \qquad (21)$$

such that:

$$D \equiv (a, x_1)[x_1, x_2)[x_2, x_3)[x_3, x_4)\cdots = (a, b) \qquad (22)$$

So, the division (partition or segmentation) D is defined by a finite, or infinite, sequence of points $\langle x_i \rangle$ within the interval (a, b), having all parts of D, except at most the last part if any, the same length δ, being δ any real number such that $0 < \delta < b - a$. Let x be any point within (a, b) such that $b - x < \delta$. Evidently, the point x can only be a point of

16.7 Inconsistency of the actual infinite divisions

the last, or second last, part of D. Therefore, D must have a last part $[x_\varphi, b)$. So, and being the successive parts adjacent, disjoint and of the same length, except at most the last part $[x_\varphi, b)$, D satisfies:

1. D has a first element (a, x_1).
2. D has a last element $[x_\varphi, b)$.
3. Each element of D of the form $[x_i, x_{i+1})$ has an immediate predecessor $[x_{i-1}, x_i)$, or (a, x_1).
4. Each element of D of the form $[x_i, x_{i+1})$ of D has an immediate successor $[x_{i+1}, x_{i+2})$, or (x_φ, b).

Assume now that a part of D of the form $[x_\nu, x_{\nu+1})$ has a finite number ν of predecessors. Being different from the last part $[x_\varphi, b)$, it will have an immediate successor $[x_{\nu+1}, x_{\nu+2})$ in D which will have just one more predecessor than $[x_\nu, x_{\nu+1})$. So, the immediate successor of a part with an immediate successor and a finite number ν of predecessors has also a finite number $\nu + 1$ of predecessors (Peano's Axiom of the successor [272, p. 1]). Since there is a part of the form $[x_\nu, x_{\nu+1})$ with a finite number of predecessors, for instance the part $[x_1, x_2)$, and each part of D (except the last one $[x_\varphi, b)$) has an immediate successor, we can inductively conclude that all parts of D with an immediate successor, i.e. all parts of D including its second last part, has a finite number of predecessors. Therefore, D has also a finite number of elements (Peano's Axiom of the successor). □

Other more formalized proof based on ordinal numbers can be found in [212]. The impossibility to divide a line of a finite length into an infinite number of parts of the same length is firmly suspected since the XVIII century, at least for some empiricist as G. Berkeley and D. Hume [32, 166, 121].

We will now analyze the possibility of dividing the above interval (a, b) into an infinite number of parts of decreasing length, which is the usual way a finite length is assumed that can be infinitely divided. For this let $S_1 = \langle x_i \rangle$ be an ω-ordered and strictly increasing and convergent sequence of points within (a, b) whose mathematical limit is just b. The sequence of points S_1 defines in (a, b) an ω-Division $D_{1,\omega}$ that can be expressed in the same way as the above case, though now it will be infinite:

$$S_1 = x_1, x_2, x_3, \ldots \tag{23}$$

$$D_{1,\omega} \equiv (a, x_1)[x_1, x_2)[x_2, x_3)[x_3, x_4)\ldots \tag{24}$$

$$(a, x_1) \bigcup_i [x_i, x_{i+1}) = (a, b) \tag{25}$$

$$\lim_i x_i = b \tag{26}$$

An unavoidable feature of ω-divisions is their enormous asymmetry: in our case, any point x arbitrarily close to b will belong to a part $[x_v, x_{v+1})$, so that only a finite number v of parts precedes $[x_v, x_{v+1})$, and an infinite number of them succeed it. In any case, if we remove x_1 from S_1, the remaining sequence S_2 of points will define in (a, b) a new ω-division $D_{2,\omega}$:

$$S_2 = x_2, x_3, x_4, \ldots \tag{27}$$

$$D_{2,\omega} \equiv (a, x_2)[x_2, x_3)[x_3, x_4)[x_4, x_5) \ldots \tag{28}$$

If we remove x_2 from S_2, the remaining sequence S_3 of points will define in (a, b) a new ω-division $D_{3,\omega}$:

$$S_3 = x_3, x_4, x_5, \ldots \tag{29}$$

$$D_{3,\omega} \equiv (a, x_3)[x_3, x_4)[x_4, x_5)[x_5, x_6) \ldots \tag{30}$$

If we remove x_3 from S_3, the remaining sequence S_4 of points will define in (a, b) a new ω-division $D_{4,\omega}$:

$$S_4 = x_4, x_5, x_6, \ldots \tag{31}$$

$$D_{4,\omega} \equiv (a, x_4)[x_4, x_5)[x_5, x_6)[x_6, x_7) \ldots \tag{32}$$

This suggests the following Procedure P:

> Remove from the successive sequences S_1, S_2, S_3... its first element if, and only if, the remaining elements define in (a, b) an ω-division.

We will prove now by Modus Tollens that all elements of the initial sequence S_1 can be removed (an inductive proof is also possible [212, pp. 185-186]). Assume that not all elements of S_1 can be removed by the Procedure P. In these conditions, at least one point, say x_v, of S_1 could not be removed. But the sequence:

$$S_{v+1} = x_{v+1}, x_{v+2}, x_{v+3}, \ldots \tag{33}$$

Define in (a, b) an ω-division $D_{v+1,\omega}$:

$$D_{v+1,\omega} \equiv (a, x_{v+1})[x_{v+1}, x_{v+2})[x_{v+2}, x_{v+3})[x_{v+3}, x_{v+4}) \ldots \tag{34}$$

And the same applies to any other point $x_i < x_v$. Therefore, it is impossible that the Procedure P does not remove form S_1 all of its elements, while the remaining ones still define in (a, b) and ω-division.

Obviously, the problem is that if we remove all elements of S_1 we get an empty set of points and it is impossible to divide (a,b) in parts without points to define the division. Note that what has just been demonstrated is not an indeterminacy but an impossibility: the set of points of S_1 that cannot be successively removed without the remaining points defining a ω-division in (a,b) is the empty set; while if we had proved an indeterminacy the set of points of S_1 that cannot be removed without the remaining points defining a ω-division in (a,b) would be an indeterminable non-empty set. The above argument, then, proves the following:

Theorem 29 (of the Inconsistent Divisions) *The actual infinite division of any finite real interval is inconsistent.*

What confirms Corollary 5 of Infinite Divisibility. We must, therefore, conclude that ω-divisions of finite intervals are inconsistent. And taking into account that ω is the least infinite ordinal, if α is any other infinite ordinal greater than ω, an α-partition of (a,b) will contain an inconsistent ω-subdivision of a subinterval (a,b') of (a,b) [212, Theorem 19]. So, we can end by recalling David Hume's words [166, p. 32]:

> I conclude [...] that no finite extension is capable of containing an infinite number of parts; and consequently that no finite extension is infinitely divisible.

Although Theorem 29 applies to both space and time, other independent proof of the inconsistency of the infinite division of space and time intimately related to Zeno Dichotomy are available in [212].

16.8 Finite lengths and distances

As noted in Chapter 3, current infinitism assumes that the ordered list of natural numbers in their natural order exists as a complete totality (Definition 3), even though no last number can complete the list. The following argument is related to this supposedly infinite list of natural numbers, each one one unit greater than the previous one, but all of them finite. It proves that in the Euclidean space $\mathbb{R}^3$ every line (whether straight or not) with two endpoints, either open or closed at those endpoints, has a finite length. An immediate consequence is that even in a space of infinite extent it is impossible to find two points that are separated by an infinite distance. Let us then prove the following important theorem:

Theorem 30 (of the Finite Lengths) *In the Euclidean space $\mathbb{R}^3$ every line with two endpoints has a finite length.*

Proof: Let AB be any line in the Euclidean space $\mathbb{R}^3$, and $\lambda > 0$ any finite length. Let $\mathbf{P} = AP_1, P_1P_2, P_2P_3 \ldots$ be a partition of AB all of whose parts have the same finite length $\lambda > 0$, except at most the last one, if any, whose length can be less than λ. A point X such that $XB < \lambda$ can only be in the last, or second last, part of **P**. So, **P** has a last part $P_\phi B$. For every P_i in **P** it holds: any point Y of the segment AP_i such that $YP_i < \lambda$ a can only belong to $P_{i-1}P_i$, or to AP_1. And any point Z of the segment P_iB such that $P_iZ < \lambda$ can only belong to P_iP_{i+1} or to $P_\phi B$. In consequence, each part of **P** of the form $P_{i-1}P_i$ has an immediate predecessor $P_{i-2}P_{i-1}$ (or AP_1) and an immediate successor P_iP_{i+1} (or $P_\phi B$). In additions, AP_1 has an immediate successor P_1P_2, and $P_\phi B$ has an immediate predecessor $P_{\phi-1}P_\phi$. So, **P** has a first element AP_1, a last element $P_\phi B$, and each element has an immediate predecessor (except AP_1), and an immediate successor (except $P_\phi B$). Let us suppose there exists a part in **P** with an immediate successor and a finite number n of predecessors. The immediate successor of this part will also have a finite number $n+1$ of predecessors (Peano's Axiom of the successor [272, p. 1]). Since P_1P_2 has an immediate successor P_2P_3 and a finite number of predecessors, just one predecessor AP_1, we can inductively conclude that all parts of **P** with an immediate successor, i.e. all parts of **P** until its second last part have a finite number of predecessors. So, **P** has a finite number of elements (Peano's Axiom of the successor). And being finite the sum of any finite number of finite lengths, AB has finite length. □

Several immediate corollaries of Theorem 30 can now be easily proved [212, pp. 178-180]. Some of then are the following ones:

Corollary 19 (of Closed Lines) *In the Euclidean space $\mathbb{R}^3$, any closed line has a finite length.*

Corollary 20 (of the Finite Distances) *In the Euclidean space $\mathbb{R}^3$ the distance between any two of its points is always finite.*

Corollary 21 (of Lines Joining Points) *In the Euclidean space $\mathbb{R}^3$ it is impossible to join any two of its points by a line of infinite length.*

To which we can immediately add the following:

Corollary 22 (of Infinite Lengths) *In the Euclidean space $\mathbb{R}^3$ all lines of infinite length are inconsistent.*

Proof: In a line of infinite length there would have to be at least two points separated by an infinite length, which is impossible according to Corollary 21. □

An immediate consequence of the above results is that if we remove from the supposedly infinite $\mathbb{R}^3$ all pairs of points separated by a finite distance, the result can only be either a single point or the empty set. Under these conditions it is not at all clear what meaning an infinite universe could have. In any case, according to the results proved above, and taking into account that the spacetime continuum is modeled by the set $\mathbb{R}^4$, we can state the following:

Theorem 31 (of Finite Distances and Durations) *In the spacetime continuum, the distance between any two points and the time elapsed between any two instants are always finite.*

This agree with Huby's opinion [163, p. 121]:

> That the universe, if real, must be finite in both space and time.

In addition to being finite, the intervals of space and time cannot be divided into an actually infinite number of parts (Theorem 29 of the Inconsistent Divisions). It can only be divided into a finite number of parts. And we must consider the following two alternatives, which are exhaustive and exclusive:

1. There is an indivisible unit defining length intervals and time intervals.

2. There is not indivisible units for at least one of the magnitudes length and time.

If there were no indivisible units, the corresponding space and time intervals, always of finite length (Corollary20 of the Finite Distance), would be compatible with their infinite divisibility, and therefore compatible with inconsistent intervals (Theorem 29 of the Inconsistent Divisions).

16.9 Pythagoras Discrete Theorem

(This section is taken from [208, p. 421-423])

If the continuum is inconsistent, as the Theorem 8 page 23 of the Inconsistent Continuum proves, the only alternative to the spacetime continuum would be a discontinuous, i.e. discrete, space and time made of indivisible units (atoms of space and time in L. Smolin words [327]). The interest in discrete space -times began in the first half of the twentieth century [36, 70], although only in a minority of authors. W. Heisenberg, for instance, considered the idea of space as a sort of crystal lattice composed on tiny cells of the size of an elementary particle [111].

Things have begin to change, especially in the last two decades [118, 241, 187, 122, 28, 302, 29, 20, 296, ...]: An increasing number of physicists suspect now that, in fact, Planck length and Planck time define a sort of spacetime 'granularity' that could be an efficient alternative to the infinitist spacetime continuum. An alternative that could even be tested experimentally [254, 73, 119, 216, 64]. The discrete nature of spacetime has been proposed in different areas of physics [176, 138, 354, 112, 326, 327, 339, 24, 220, 228, 9, etc.], although the proposed models continue to be developed within the framework of infinitist mathematics.

Although discrete geometries already exist, they exist for particular purposes, for example the combinatorial analysis of the relationships between geometric elements [34], or the development of computational algorithms for the representation of geometric objects [80, 63]. There are even general discrete geometries, whether or not applied to quantum gravity, but not independent of infinitist mathematics. The discrete geometry suggested here would be a geometry with indivisible units of space instead of points, a geometry that could only be developed on the basis of a discrete and finitist mathematics. For this geometry, everything remains to be done, starting with the establishment of its foundational base (axioms and definitions). Even so, some non-detailed arguments, as the next one, can be made.

Though the fine structure of (a possible) quantum space is unknown, let us consider the right angled triangles depicted in Figure 16.2.

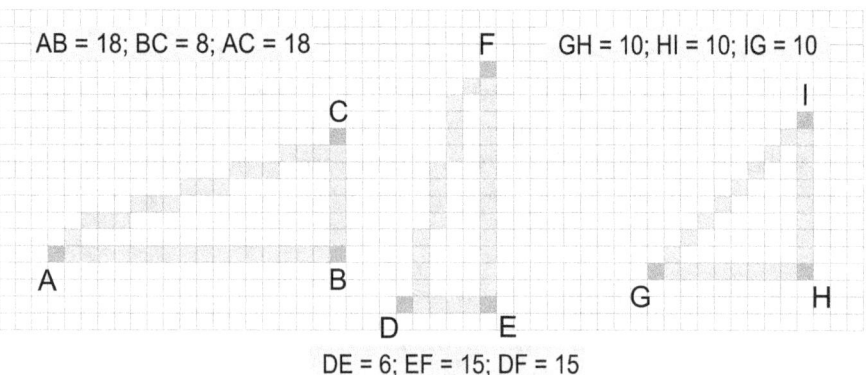

Figure 16.2 – A crude representation of three right angled triangles to test Pythagoras Discrete Theorem. The size of bi-dimensional space absolute quanta has been exaggerated to makes it possible their visual counting.

It can easily be tested the number of space quanta (qusits) of their corresponding hypotenuses is in each case equal to the number of qusits of the greater of their corresponding legs. This discrete version of Pythagoras Theorem could surely be proved once the foundation of discrete geometry had been formally established. If that were the case, the factor for converting between continuous and discontinuous hypotenuses would have the algebraic form of the relativistic factor γ.

Indeed, let h, x and y be the respective number of qusits of the hypotenuse and legs of a right triangle in a discrete space time, and let λ be the length of a qusit in both the discrete and the continuous geometry. Assume $y > x$. In discrete geometry we will have: $h = y$, because

16.9 Pythagoras Discrete Theorem

$y > x$. In the continuous geometry the length of the hypotenuse would no longer be $h\lambda$ but $h'\lambda$, being $h' > h$, because it is greater than the length $y\lambda$ of the greatest leg (note that while h, x and y are natural numbers, λ and h' are real numbers). According to classical Pythagoras Theorem, it can be written:

$$\text{Hypotenuse: } h'\lambda = \sqrt{(y\lambda)^2 + (x\lambda)^2} \tag{35}$$

$$\text{leg: } y\lambda = \sqrt{(h'\lambda)^2 - (x\lambda)^2} \tag{36}$$

$$y = \sqrt{h'^2 - x^2} \tag{37}$$

The ratio between the continuous and the discrete hypotenuse is given by:

$$\frac{h'\lambda}{h\lambda} = \frac{h'}{h} \tag{38}$$

$$= \frac{h'}{y} \tag{39}$$

$$= \frac{h'}{\sqrt{h'^2 - x^2}} \tag{40}$$

$$= \frac{1}{\sqrt{1 - (x/h')^2}} \tag{41}$$

where the last term on the right side of (41) as the algebraic form of the relativistic Lorentz factor γ. It can ve rewritten as:

$$\frac{h'\lambda}{h\lambda} = \frac{1}{\sqrt{1 - (x\lambda/h'\lambda)^2}} \tag{42}$$

Let a^* be a photon that moves through a vertical distance $y\lambda$ in the rest frame RF_o of its source. Assume a^* moves the same vertical distance $y\lambda$ from the perspective of another inertial frame RF_v while RF_o moves with respect to RF_v the horizontal distance $x\lambda$ at a uniform velocity v parallel to X_v for a time t_v. So, a^* moves with respect to RF_v along the hypotenuse of a right triangle whose legs are $y\lambda$ and $x\lambda = vt_v$, i.e. along $h'\lambda$ (35). And it will hold $h'\lambda = ct_v$. Therefore, (42) can be rewritten:

$$\frac{h'\lambda}{h\lambda} = \frac{1}{\sqrt{1 - (vt_v/ct_v)^2}} \tag{43}$$

$$= \frac{1}{\sqrt{1-(v/c)^2}} \qquad (44)$$

$$= \gamma \qquad (45)$$

which proves the ratio between the continuous hypotenuse and its corresponding discrete alternative is the relativistic Lorentz factor γ.

17. Physical versus geometrical space

Abstract.-This chapter proves in formal terms a quality of all points of the continuum geometrical space that is often explicitly or implicitly assumed: points have neither size nor shape. It also highlights the contradiction of, on the one hand, denying the reality of a physical space, which continues to be (together with time and spacetime) just an illusion for most contemporary physicists, and, on the other hand, accepting that this unreal space is continuously undergoing intrinsic deformations, accelerated intrinsic expansions, and propagation of its own deformations (gravitational waves), as if something that is not real, that is just an illusion, had the actual physical ability to warp, expand, vibrate and transmit its own vibrations. If, on the contrary, the physical reality of space (together with time and spacetime) is assumed, almost insoluble problems arise related to the modeling of spacetime through a four-dimensional infinitist continuum of points without size and without shape and instants without duration. As a counterpoint, this work considers the structure and functioning of CALMs, where, apart from being free of the inconsistencies related to infinity proved in Chapters 3, 4 and 5, none of these problems arise, and where it is possible to envisage a theory of gravity in which spacetime warps are unnecessary.

Keywords: sizeless points, physical points, spacetime continuum, physical space, intrinsic deformation of space, gravitational waves, expansion of space, Theorem of Formal Dependence, gravity in CALMs.

17.1 The mathematical language of physics

According to the Theorem of the Consistent Universe proved in Chapter 5 of this book, the universe evolves under the control of a single set of invariant and formally consistent laws. The empirical evidence for this theorem is so enormous that until now we have not considered it necessary its formal demonstration. This book demonstrates it because it is used in several arguments. Naturally, it is the formal consistency of the universe that allows its analysis and description with mathematical and computational languages.

Some theories of modern physics are actually mathematical theories, albeit scarcely concerned with the foundations of mathematics. Frequently, and with a certain arrogance, it is said and written that the universe cannot be explained with ordinary language, that it is only possible to do so with mathematical language. The problem (which physicists do not address) is what is the mathematics that explains the universe? There seems to be only one mathematics, and this is not so. Although in our days there is a mathematical current absolutely dominant over all others (infinitist mathematics), the finitist alternatives also exist.

It is true that for more than a century the mathematics based on infinitist set theory has been overwhelmingly dominant. So dominant that it has ended up having a role in physics similar to that of catechisms in religions. But that does not mean that it is the ultimate and definitive mathematics. This mathematics includes a very dangerous axiom because its inconsistency would have devastating consequences on the whole mathematical edifice built since the end of the 19th century. And naturally on the physical theories most committed to this infinitist mathematics.

To put it briefly, though intentionally provocative, the Axiom of Infinity legitimizes the existence of the complete list of natural numbers in their natural ordering, though there is no last natural number that completes the list. Or said in Aristotelian terms, the Axiom of Infinity legitimizes that the incompletable exists as completed. That statement triggered me. And thirty years and more than forty proofs later, here I am trying to convince the reader of the inconsistency of the actual infinity. So far, not very successfully, but I was warned about that.

The reader can find one such proof, the Corollary 4 in Chapter 3 of this book, and two additional proofs in appendixes A and B. I chose them for their simplicity and brevity, and, in the case of Hilbert machine, as a counterpoint to the iconic Hilbert Hotel, which almost everyone will have heard of, even if they have never stayed in any of its infinitely many rooms. The rest of the proofs can be found in [212]. There are many types of such proofs: based on set theory, on supertasks, on transfinite arithmetic, on geometry, etc.

Returning to the initial dictum that the universe could not be described with ordinary language but with the language of mathematics, I must express my disagreement. If physics is not expressed in the physical terms of ordinary language, then we will not have a physical description of the universe. As B. Russell would surely say, we would not know of which we are talking about [306, p. 959] [309]. But the universe is physical, not mathematical. In the same sense that our nature is physical, not mathematical. We are objects of a physical universe, not a system of mathematical equations.

In any case, prudence recommends the analysis of the foundations of infinitist mathematics and the critique of supremacist infinitism, with its extraordinary and excluding language, because in the end such extraordinary language could be inconsistent. Take a look at Chapter 3 of this book, or at [212]. And let us not forget the arrogance of some theories of modern physics, for example special relativity, which are used as if they were new fundamental laws of logic: if something does not agree with them, that something is automatically declared false. This in the same physics that has been unable to discover something as physically evident as preinertia, surely the most basic property of all physical objects (!) The shame of physics [215, pdf].

17.2 Geometrical points and physical points

As already indicated, in some of the previous chapters the set $\mathbb{R}^4$ (the set of all 4-tuples of real numbers) models the geometrical spacetime continuum that in turn models space and time in physics. The elements of that set represent points in space and instants in time, two primitive concepts (point and instant) that we make intuitive use of in most arguments about space and time in physics. As indicated in Chapter 30, our intuition about them is flawed by our usual way of representing them graphically.

As almost everyone knows, a simple contradiction would allow us to prove anything we wanted to prove. To the extent that one way to analyze the consistency of a theory is to prove the existence of at least one proposition that cannot be proved within that theory [201]. But, unfortunately, physicists are not very fond of questioning the foundations of their mathematical language (neither are most mathematicians, who are more religious than critical with respect to the foundations of their mathematics). Arguments such as those in this section are unheard of in physics textbooks, including the best books on physical space, such as [128]. Indeed, practically in all physics books we can find expressions such as:

1. "the space around a point"
2. "a tiny ball or point"
3. "point-like charge"
4. "mass point"
5. "mass concentrated in a point"

6. "infinitesimal point"
7. "the motion of points with mass and charge"
8. "propagation proceeds point by point"
9. "propagate through the contiguous points"
10. "creating changes at adjacent points"
11. "what happens to the field at adjacent points
12. etc.

that show the use of qualities that points do not have: extension and adjacency (immediate successiveness). The consideration that they do, leads immediately to a contradiction, as we will see here. Of course, the extension of points has never been proposed, let alone measured (otherwise we would all know it), which is almost a proof that points are assumed to have no extension. But one might think that they do, although for unknown reasons it has not yet been proposed, not even the limits that extension could have.

On the other hand, the lack of immediate successiveness (adjacency, contiguity) between points in space (or instants in time) is not even considered in contemporary physics, as if they actually exist. But they do not exist: between any two points (instants) in space (time) there are always a non-numerable infinitude of different points (instants), none of which is contiguous (adjacent) to any other, as it happens, for example, with natural numbers: between n and $n+1$ there is no other natural number; $n+1$ is contiguous, adjacent to n ($\forall n \in \mathbb{N}$). The points do not touch each other; neither do the instants.

To simplify the discussion that follows, without losing an iota of formal rigor, we will consider only the points of a line r in the Euclidean $\mathbb{R}^4$ where a metric and arbitrary units of measurement have been defined, for example those of the IS system based on the multiples and submultiples of the meter. In the IS we have units of length so greater as the Yottameter (Ym) = 10^{24} m, and so small as the yoctometer (ym) = 10^{-24} m.

But we could define other submultiples of the meter inconceivably smaller than a ym using as the negative exponent any expofactorial number (see Chapter 16 in this book or [212] for more details), whose expression written in normal (non-exponential) text would be millions of times longer than the length of the visible universe (and they would still be numbers with a finite number of zeroes after the decimal point and before the first non-zero decimal). Since the unit chosen to express the assumed length of a point is irrelevant, we will assume that a point has length λ defined by a real number r of yoctometers (ym):

$$\lambda = r \text{ ym}, \ r > 0, \ r \in \mathbb{R} \tag{1}$$

17.2 Geometrical points and physical points

Let us consider now three segments AB, CD and EF in a straight line L whose respective lengths are:

$$AB = 1 \text{ Ym} \tag{2}$$
$$CD = 1 \text{ m} \tag{3}$$
$$EF = 1 \text{ ym} \tag{4}$$

Obviously:
$$EF < CD < AB \tag{5}$$

We must also consider the set C of cardinals:
$$C = \{1, 2, 3, \ldots \aleph_o, 2^{\aleph_o}\} \tag{6}$$

in which multiplication is defined [56]. Cantor himself proved that any segment of a line has the same number of points as the entire line, exactly $2^{\aleph_o}$ points (Dimension Problem, see Chapter 16). Therefore, if we assign any length to points, for example the one defined in (1), we will have:

$$AB = 2^{\aleph_o} \text{ points} \times r\frac{\text{ym}}{\text{point}} = r \times 2^{\aleph_o} \text{ ym} = 2^{\aleph_o} \text{ ym} \tag{7}$$

$$CD = 2^{\aleph_o} \text{ points} \times r\frac{\text{ym}}{\text{point}} = r \times 2^{\aleph_o} \text{ ym} = 2^{\aleph_o} \text{ ym} \tag{8}$$

$$EF = 2^{\aleph_o} \text{ points} \times r\frac{\text{ym}}{\text{point}} = r \times 2^{\aleph_o} \text{ ym} = 2^{\aleph_o} \text{ ym} \tag{9}$$

Consequently:
$$EF = CD = AB \tag{10}$$

which contradicts (5). Therefore, points have not extension (size), and then they cannot have a contour, a shape. A similar argument proves that the instants of the spacetime continuum have not duration. We have then proved the following:

Theorem 32 (of abstract points) *In the spacetime continuum, points have neither size nor shape, and instants have not duration.*

But then, what qualities do have the points of the space continuum? It is often said that points only have position and that two points cannot occupy the same position. So we have to admit that positions in space are defined by objects that have no extension. Consequently, the positions would only be defined for objects, or parts of objects, that have no extension. The situation is more untenable from other physical point of view: it does not seem physically reasonable to admit that something without extension can have, for example, mass or electric charge as would be the case of mass points or charge points.

The other non-quality of points that is controversial, especially in physics, is their lack of immediate successiveness (adjacency): in the spacetime continuum no point has an immediate successor as occurs, for example, in the natural numbers, in which each number n has an immediate successor $n+1$, without there being other natural numbers between them. In the case of points, between any two of them, whatever they are, there is always the same uncountable infinite number of points ($2^{\aleph_0}$ points). If you were a point on a straight line, and you looked in either direction of that straight line, you would see no point of the line: if you were seeing a point, that point would be your impossible immediate successor.

Or to put it another way, if a point of that straight line starts to move through that line in one of its two directions, it could travel the entire corresponding semi-line without passing through any of its points: the first one to pass through would be its impossible immediate successor in that direction (this is the nuclear argument in Zeno Dichotomy [212]). It could be said that since the time of Zeno we have been warned of the discrete nature of space and time. But we have never heeded the warning. As we will see later, this situation complicates the supposed curvature and extension of the physical spacetime, when the physical spacetime is modeled by the continuum $\mathbb{R}^4$.

But the most serious problem with the continuum is its formal inconsistency (Theorem 8). Recall that any line of that continuum contains an infinite number of ω-ordered sequences of points, all of them inconsistent (Theorem 2). It is not necessary to go into semiotic and abstract excesses to make a serious critique of the unacceptable inadequacies of the infinitism that has been dominant in mathematics for more than a century now.

17.3 Physical space

The dominant idea in contemporary physics is that neither space nor time are real physical objects. Answers like the next one can be found on some physics well-known FAQ websites (obviously answered by 'expert' physicists):

> Spacetime is not a fabric, it is not material. Space is just an illusion, time is just an illusion therefore spacetime is just an illusion and a good way of simplifying the concept of general relativity to the public.

This has also been the opinion of many relevant authors in the history of science and thought (particularly empiricists): G. Leibniz, D. Hume, C. Huygens, E. Mach, H. Poincaré, E. Borel, L. Wittgenstein etc. And

17.3 Physical space

of the vast majority of contemporary physicists. For example [325, p. 266]:

> ... space and time, like society, are in the end also empty conceptions. They have meaning only to the extent that they stand for the complexity of the relationships between the things that happen in the world.

Although, on the other hand, we can also read the contrary opinion. For instance, according to:

1. A. Einstein
   - I agree with you that the general theory of relativity is closer to the ether hypothesis than the special theory [186, p. 68].
   - According to the general theory of relativity, space is endowed with physical qualities... [186, p. 98].
2. F. Wilczek:
   - Spacetime is also a form of matter [371, p. 180].
   - Spacetime has a life of its own [371, p. 180].
   - According to general relativity, spacetime is extremely rigid [371, p. 181].
   - Dark energy could be a universal density of space itself [371, p. 194].
   - What appears to our eyes as empty space is revealed to our minds as a complex medium full of spontaneous activity [370, p. 1].
3. N. A. Tambakis:
   - It seems to me that in this way we can confirm the well-known epistemological assumption that space and time are not fictions but rather modes of the dynamic existence of matter [342, p. 146].
4. M. Kaku:
   - In a sense, gravity does not exist; it is the distortion of space and time that moves the planets and stars (cited in [40, p. 63]).

The case of A. Einstein is a bit more complex. Let's remember some of his words through the years about space and time:

> 1905: The introduction of a "luminiferous ether" will prove to be superfluous inasmuch as the view here to be developed does no longer need an absolute space, at absolute rest, with physical properties [87, p. 891].

> 1913: For me it is absurd to attribute physical properties to

"space" (Letter to E. Mach cited in [185, p. 135]).

1914: As much I am not disposed to believe in ghosts so I do not believe un the enormous thing about which you are talking and which you call space [89, p. 345].

1915: Thereby (through the general covariance of the field equations) space and time lose the last remnant of physical reality (Letter to M. Schlick cited in [185, p. 134]).

However, in 1916 Einstein changed his mind about the physical nature of space and the existence of the ether.

1916: I agree with you that the general theory of relativity is closer to the ether hypothesis than the special relativity (Letter to H. A. Lorentz cited in [185, p. 135]).

1919: Thus, once again "empty" space appears as endowed with physical properties, i.e. no longer as physical empty, as seemed to be the case according to special relativity. One can thus say that the ether is resurrected in the general theory of relativity, though in a more sublimated form (Morgan Manuscript, cited in [185, p. 137]).

1938: Our only way out seems to be to take for granted that space has the physical property of transmitting electromagnetic waves... We may still use the world ether, but only to express some physical property of space... At the moment it no longer stands for a medium built up of particles [97, pp. 159-160] [98, p. 115]

In later writings he defended that the physical notion of space is linked to the existence of rigid bodies, but he rejects the idea that space is an *a priori form of intuition* [37], as Kant defended [179]. Einstein *"always supported an objective description of physical reality, without interference of the observer"* [177, p. 128].

So, who is right? The next section proves that the attempts to solve this question leads to a serious relativistic conflict related to the real or unreal nature of (physical) space. Maybe this conflict can only be solved within a CALM perspective of space and time.

17.4 A relativistic conflict on the reality of space

Let us first consider the alternative that space is not real. The first and simplest problem would be to explain how something that can be intrinsically deformed can be unreal. One could argue that there is actually nothing that deforms, but that massive objects behave as if there is something between them that is deformed by their presence; but in

17.4 A relativistic conflict on the reality of space

reality there is nothing between them that is deformed by their presence. So, this argument is anything but a physical explanation. That is to say, we would continue ignoring the physical cause of the gravitational behavior of massive objects: the explanation given includes an *'as if'* that can be followed by any other unreal thing: as if X, as if Y, as if Z... For example, massive objects behave as if there were a God that arbitrarily determines their behavior. That doesn't look like physics. In these conditions we would have to admit again that gravitational interactions are ghostly actions at a distance, or mediated by a phantom medium.

More difficult to solve is the second problem posed by the unreal space. In the observable universe, space, and only space, expands in an accelerated way, and it did it in a super-accelerated way in the first instants of its history (inflation stage [144, 145]). Moreover, the expansion is not general, it affects certain areas of space and does not affect others, among the latter the intragalactic space. Here we are already faced with an actual contradiction: it is not possible for space, and only for space, to expand and to test in experimental terms its expansion if space is unreal. Unreal objects have no real properties. It is not logically acceptable to say that something unreal, and only something that is unreal, expands, and that this expansion is detectable in experimental terms. It seems reasonable to conclude that if X expands, and only X expands, and that expansion of X has been actually detected and measured, then X must be real. Whatever X be.

The third problem posed by the physical unreality of space is its vibration by gravitational waves (on gravitational waves there is an abundant and interesting literature: [299, 333, 332, 68, 325, 40, 290, 245, 65, 343, 49] etc.). These waves, which have already been experimentally detected, are produced by very violent gravitational interactions (for example, the rotation of two black holes with respect to each other). Well, gravitational waves would have to be elastic vibrations of something that does not exist (the unreal space), which propagate exclusively through a medium that does not exist (the same unreal space). Moreover, they propagate through this unreal medium at the speed of light, which is a key fact, as we will see below, to explain the physical nature of space. For this explanation, we must also take into account the logical results (principles and theorems) assumed and proved in this book.

Let us now consider the second alternative: space is real; it is a physical entity endowed with certain physical properties. These properties should include the following two:

1. It must permeate all physical objects, offering zero resistance to their motion (as was the case with the primitive pre-relativistic

ether).

2. It can deform reversibly, and allow the transmission of gravitational waves, which consist of the propagation of two mutually perpendicular series of successive dilations and contractions of space, both series in turn perpendicular to the direction of wave propagation (quadrupole radiation). This waves propagates through space at the speed of light, which requires the physical space to be a very rigid medium [371, p. 181].

Of course, both properties raise the same problem that the ancient ether posed when it was believed to be the medium through which electromagnetic waves propagated. But now it is the deformation of space itself that propagates through space itself. It is not possible then to consider that gravitational waves propagate in the vacuum (like electromagnetic waves): they are deformations of space that propagate through space itself.

It seems reasonable to prescribe that we need a new paradigm of physical space that satisfactorily explains the above drawbacks. The structure and functioning of CALMs could serve as a guideline for the construction of that new paradigm of physical space and time in which this and other fundamental problems, could be solved. The next section is an introduction to that finitist and discrete alternative.

According to the relativistic orthodoxy, the new physical space could not serve as a reference frame since it contains no perceptible elements that can be used to refer motion. Chapter 17 of this book addresses this issue. Here it will suffice to recall that due to the preinertia of all physical objects (including massless objects as photons), it is impossible to detect absolute motion (except, perhaps, under the conditions indicated in [208, p. 371-378]).

To complete the problems raised in this section on the physical reality of space, we must also consider all the formal results obtained in the preceding chapters of this book, which conclude in the existence of a real, physical, space and time made of contiguous (successive) indivisible units: qusits and qutits. The steps of that argument are now summarized in the following demonstrated conclusions and proposed fundamental principles:

1. The infinity in the Axiom of Infinity can only be the actual infinity (Theorem 6 of the Axiom of Infinity, page 21).

2. All ω-ordered sets are inconsistent (Corollary 2 of the ω-Ordered Sets, 21).

3. The Axiom of Infinity is inconsistent (Theorem 6 of the Inconsistent Axiom of Infinity, page 21).

4. The actual infinity is inconsistent (Theorem 1 of the Actual Infinity, page 17).

5. The spacetime continuum is inconsistent (Theorem 8 of the Inconsistent Continuum, page 23).

6. A consistent universe cannot contains an actual infinite number of physical objects. (Corollary 11 of the Finite Universe, page 25).

7. The laws of physics do not apply in spaces smaller than the minimum unit of space nor in times smaller than the minimum unit of time, both being of non-zero extension (duration) (Theorem 16 of the Discrete Threshold, page 34).

8. No space exists between any two successive space minimum units, and no time elapses between two successive time minimum units. (Theorem 17 of Adjacency, page 35).

9. The actual infinite division of any finite real interval is inconsistent (Theorem 29 of the Inconsistent Divisions, page 185).

10. In the Euclidean space $\mathbb{R}^3$ every line with two endpoints has a finite length (Theorem 30 of the Finite Length, page 185).

11. In the spacetime continuum, the distance between any two points and the time elapsed between any two instants are always finite (Theorem 31 of Finite Distances and Durations, page 187).

12. Every space interval (or time interval) is finite and can only be divided into an integer number of adjacent qusits (qutits). (Theorem 24 of Finite Space and Time, page 210).

It is the latter result that establishes the real, physical nature of space and time. Note that this is a sub-theorem, a result whose proof uses formal elements and empirical inductive evidence. In this case the empirical inductive evidence is provided by natural objects, in none of which we have ever seen the existence of internal clocks to measure qutits, nor the existence of internal yardsticks to measure qusits (not to mention the very measurement procedures that all those objects would have to be continuously performing). Finally, note also the incompatibility of all these results with the spacetime continuum, and their compatibility with cellular automata like models (CALM)s.

17.5 Gravity from the CALM perspective

As noted above, for most physicists space does not exist, nor does time. They are not real, they are fictitious, mere illusions. But then it turns out that the presence/absence of a massive body intrinsically deforms space, reversibly transforms it from Euclidean to non-Euclidean. And certain space deformations travel through space itself at the speed of light (gravitational waves). But this is only possible if space is a real deformable medium. It is not possible to deform what does not exist, because what does not exist has no properties, not even the property of being deformed. There are no deformations of objects that do not exist. One gets the impression that some theoretical physicists are *lost in abstraction*. The actual infinity is not a good guide.

From now on, and to avoid excessive redundancies, we will only discuss on space, though the discussion can be immediately extended to time and spacetime. As will be seen, the dialectical tension (real space versus unreal space) can only be resolved by admitting the reality of physical space, which is in accordance with all formal results listed in the above section and assumed/proved in the precedent chapters of this book.

We will now analyze in physical (not in mathematical) terms how a physical space modeled by the continuum $\mathbb{R}^3$ could be deformed. It has been proved above that points have no extension and no shape (Theorem 32). Therefore, the model $\mathbb{R}^3$ cannot be deformed by deforming its points: the points of space have no extension or shape to deform. And if the deformation of $\mathbb{R}^3$ is not possible by deforming its points, the only way to deform space would be by either by differential movements of its points, or by removing/creating points as needed.

In the first case we would have to face the problem of explain the way something that has no size can be moved. Moreover, the boundary between moving and non-moving points would also be impossible (there is no adjacency between the points of the space-time continuum). And if that were not enough, the differential motion of the points would create impossible gaps in the continuous space without this space ceasing to be continuous. In the second case, we would have to admit either the violation of the Theorem of Formal Dependence 20, or that the modeled universe is an open system capable of exchanging points with another unknown external reality.

The consequences of Theorem 32 are disastrous for the role of $\mathbb{R}^3$ as a model of a real space that can vibrate and can be continuously expanded and deformed in the Euclidean and non-Euclidean directions. It is also important to recall at this point the extraordinary difficulties that appear when applying the equations of gravity (both classical and relativistic) to cases of three or more bodies. Difficulties so far insurmountable that can only be solved with the technique of approximations and the use of powerful computational resources. More than beautiful or elegant, those equations should be qualified for what they surely are: an approximate (and diabolically complex) way of describing real gravitational interactions in real systems of three or more real objects.

17.5 Gravity from the CALM perspective

Things are quite different from the perspective of CALMs. Evidently, there does not exist (yet) a CALM alternative to relativistic gravity and its corresponding intrinsic space deformations and vibrations. We are not even at the entrance of that alternative. We are pointing out that there could be such an alternative. And from that position simply indicative of a new path, we can consider some of its peculiarities:

1. Every object in a CALM would be defined by the state of a certain set of qusits, being that state defined in terms of a determined set of variables.

2. Each object could modify the state of other CALM's qusits (even of all CALM's qusits), the more the closer they are to the object. In this sense, each object defines its own field of interactions (forces) with any other CALM's object.

3. Thus, what the dynamic of an object of a CALM could modify would not be the Euclidean/non-Euclidean geometry of space, but the values of the state variables of each qusit of the CALM, i.e. its field of interactions.

4. As discussed in Chapter 18, the state of each qusit is updated at each successive qutit thanks to the two modes of existence of each qusit: the permanence mode (perceptible) and the interactive mode (executed in the background, not perceptible). These modifications would be similar to the dynamic of a field of forces (including gravitational force).

5. In a CALM it would be possible to reinterpret gravity in terms of interactions (forces), rather than in terms of intrinsic deformations of the physical space defined by qusits.

6. Being gravity exclusively additive, if a body A of mass m_a gravitationally accelerates two other objects B and C of masses respectively m_b and m_c such that $m_a > m_b > m_c$, the gravitational accelerations of B and C produced by A will be the same because only the additive gravitational pull of A counts, from which it is not possible to subtract the gravitational effects of B and C on A because gravity is only additive. This is in accord with the classical:

$$m_b a_b = G \frac{m_b m_a}{d^2} \tag{11}$$

$$m_c a_c = G \frac{m_c m_a}{d^2} \tag{12}$$

$$a_b = a_c = G \frac{m_a}{d^2} \tag{13}$$

7. Since all objects are preinertial and inertial, there would exist in

all of them a fundamental mass (rest mass?) responsible for preinertia, inertia and gravitational interactions.

8. Gravitational interactions would determine the trajectories of CALM objects (including photons) through the fabric of the CALM's qusits.

Chapter 19 of this book dealt with preinertia, a universal attribute of all physical objects, including photons, whereby they all inherit (in vector terms) the relative velocity of the reference frame in which they are set in motion. As one might say to a classical Greek, it's the reason we land in the same place where we jumped vertically, and not 37 km further (367 Km in modern terms). The reason for preinertia could be the rest mass, or some other universal property not yet determined. Although the most reasonable and simple thing is that it be the rest mass. If that were the case, photons would have to have some rest mass, however minuscule.

The problem with the rest mass of photons is that it makes infinity appear in the Standard Model (by breaking its gauge symmetry), but the problem is not the rest mass of photons, the problem is to make use of an infinitist mathematics language founded on an inconsistent hypothesis: the Hypothesis of the Actual Infinity subsumed in the Axiom of Infinity (see Chapter 3). In Chapter 16 a universal constant m_q was defined with the dimensions of a mass, although much smaller than Planck mass m_p:

$$m_q = \sqrt{\frac{G\hbar^3 R_\infty^4}{c^5}} = \hbar t_p R_\infty^2 = \approx 6.238883052 \times 10^{-64} Kg \qquad (14)$$

where R_∞ is Rydberg universal constant. Being defined in terms of universal constants, m_q is also a universal constant, be it or not the rest mass of the photon. This mass m_q can also be written as a fraction of m_p:

$$m_q = \frac{G m_e^2 e^8}{8^4 \pi^6 \epsilon_o^4 \hbar^5 c^5} m_p = 3.146 \times 10^{-57} m_p \qquad (15)$$

where m_e is the rest mass of an electron, e the unit of electrical charge, and ϵ_o the electric permittivity. Be it or not the rest mass of a photon, m_q is in the order of magnitude of other estimations, most of which range from $< 10^{-51}$g to $< 10^{-64}$g [349, 322, 355, 120].

17.6 Expanding geometrical space and physical space

Apart from the initial inflationary stage, the physical space of the universe, and only it, is expanding since its formation, and it is expanding faster and faster. Furthermore, there are areas of space that are expanding and areas that are not expanding, such as intra-galactic space. Here we encounter the same problem of space deformation discussed above. How can something that has no real existence, something that is only a fiction, an illusion, be expanding for more than 13.8 billion years? As in the case of geometric deformation, we will have to admit that if space expands it is because it is something with the physical capacity to expand. And only real objects have physical capabilities. Therefore, space must be real. Even if it is modeled by the continuum of real numbers.

If physical space is modeled by the set $\mathbb{R}^3$, then the physical version of the points of $\mathbb{R}^3$ must be real physical elements that have neither size nor shape. Consequently, the expansion of the universe cannot be caused by expanding its points, because these would cease to be points. Nor can space be expanded by creating gaps between its points, in this case the continuum of points would cease to be a continuum to become a discontinuum with gaps that over time would grow in number and/or size.

The only solution is that new physical points appear (whatever these physical points modeled by the geometric points of $\mathbb{R}^3$ are). And here problems also appear because that continuous creation of physical points would imply one of the following two alternatives.

1. The universe violates the Theorem of Formal Dependence 20 according to which no formal object is self-defining, self-proving, or self-causing.
2. The universe is not an isolated system, and there would have to be another unobserved reality from which comes the new space that makes expansion possible.

By contrast, from the discrete and finitist perspective of a CALM, it does not seem necessary any expansion of space, it would be enough to analyze the possibilities of motion of CALM's objects through the space defined, once and for all, by the CALM's fabric of qusits.

The physical reality of space deduced from its ability to deform, vibrate and expand raises the question of absolute motion, which is anathema to modern physics. But if physical space is real, then why shouldn't it be possible to move THROUGH it?

17.7 Fields and CALMs

One of the most fruitful and relevant concepts in the history of physics is the concept of field. Although the basic idea of a field can already be found in Leibniz, it was explicitly introduced by Faraday (an experimental physicist with little mathematical training). Shortly after, Maxwell expressed the electromagnetic field in mathematical terms with his famous equations. Since then, the use of the concept of field has been generalized in almost all areas of physics (theoretical and experimental) with remarkable success. The Oxford Physics Dictionary and The Oxford Philosophy Dictionary give the following definitions:

1. **field**: A region in which a body experiences a force as a result of the presence of some other body or bodies. A field is thus a method of representing the way in which bodies are able to influence each other [74, p. 184].

2. **field**: A central concept of physical theory. A field is defined by the distribution of a physical quantity, such as temperature, mass density, or potential energy, at different points in space [39, p. 134]

There are two basic ways of looking at the concept of field:

1. A physical medium from whose variations result the interactions of the objects contained in that physical medium.

2. A way of describing the way in which different physical objects interact with each other, without there being an actual physical medium from whose variations these interactions might result.

Faraday was in favor of the first alternative. For him, the similarity between different fields was a proof of the physical reality of the corresponding media. For instance [108, 3284][109, p. 20]:

> All these effects and expedients accord with the view that the space or medium external to the magnet is as important to its existence as the body of the magnet itself.

Faraday's view invites to consider the possibility of a reinterpretation of physical fields from the point of view of the structure and functioning of CALMs. An interesting possibility would be the development of a quantum field theory within a CALM.

18. Cellular Automata Like Models

18.1 Introduction

This chapter uses the logic of cellular automata as a formal tool to initiate a new discrete analysis of the basic structure and functioning of the universe. It is only a change of perspective, though it may be extravagant. But extravagance is preferable to inconsistency, and the present infinitist alternative could be inconsistent if any of the more than forty proofs of the inconsistency the of the actual infinity given in [212] were valid.

18.2 Indivisible units of space and time

The speed c of light in the vacuum (used here as the carrier medium for the physical fields) is one of the universal constants of physics of which we have the greatest empirical evidence. It is the speed of an object (as a photon) that takes a Planck time to traverse a Planck length:

$$c = \frac{l_p}{t_p} = \frac{1.616255 \times 10^{-35}\,m}{5.391247 \times 10^{-44}\,s} = 299792423\,ms^{-1} \qquad (1)$$

The speed c of light in a vacuum can also be defined in terms of other pair of universal constants, the electric permittivity ϵ_o and the magnetic permeability μ_o of the vacuum:

$$c = \frac{1}{\sqrt{\epsilon_o \mu_o}} = 299792423\,ms^{-1} \qquad (2)$$

(In SI, c is defined as 299792458 m/s because a meter is defined in the SI as the distance light travels in 1/299792458 s). According to (1) and (2), it is clear that the speed c of light in the vacuum is a universal constant. And it is not only the speed of electromagnetic waves through the vacuum, it is also the speed of the propagation through the same vacuum of other perturbations of other physical fields, for example the propagation of gravitational waves. On the other hand, from (1) and (2) it immediately follows:

$$t_p = l_p/c = l_p\sqrt{\epsilon_o \mu_o} \tag{3}$$

which is a rather enigmatic relation between the possible unit of discrete time (qutit) t_p and the possible unit of discrete space (qusit) l_p defined through two universal constant: the electric permittivity and the magnetic permeability of the vacuum.

The electromagnetic spectrum is considered in contemporary physics as continuous and (virtually) infinite, for instance [362, p. 891]:

> The wavelengths of electromagnetic waves have no inherent upper or lower bound.

But since the wavelength of any wave is defined as the distance between two successive points in the same state of vibration, then, and according to the Theorem 30 of the Finite Lengths, all wavelengths will be finite. And if the set of all possible wavelengths exist, then, and according to Theorems 10 and 11 that set can be discretely ordered, with a minimum and a maximum.

We will now demonstrate four important results from the perspective of a finite and discrete universe:

Theorem 33 (of non-extensive points) *The points (instants) of the spacetime continuum have not extension (duration).*

Proof: Suppose that the points of the spacetime continuum have an extension of δ meters in a given metric, being δ a real number greater than zero. Let AB and CD be the lengths, in the same metric, of any two lineal intervals of that continuum such that:

$$AB < CD \tag{4}$$

Since the number of points of AB and CD is the same, just $2^{\aleph_o}$, we would have:

$$AB = 2^{\aleph_o} \times \delta \text{ m} = 2^{\aleph_o} \text{ m} \tag{5}$$

$$CD = 2^{\aleph_o} \times \delta \text{ m} = 2^{\aleph_o} \text{ m} \tag{6}$$

$$\therefore AB = CD \tag{7}$$

which contradicts (4). Therefore, points cannot have an extension greater than zero. For the same reasons, instants cannot have a duration greater than zero. □

Theorem 34 (of the Physical Laws) *In a consistent universe the laws* of physics cannot be applied to a point, nor during an instant. Neither can they be applied to an interval of points nor during an interval of instants.

Proof: Since, according to Theorem 33, points (instants) have no extension (duration), to apply a physical law to a point during an instant is to apply that law to no space during no time. Nor can it be applied to an interval of space during an interval of time because if the interval has an infinite number of points (instants) then it is inconsistent (Corollary 4); and if it has a finite number of points (instants) then it has no extension (duration) (Theorem 33), and it would be the same as applying it to a point during an instant, i.e. applying it to no space during no time. □

Theorem 35 (of Indivisible Units) *There is an indivisible minimum of space (qusit) and time (qutit) of which all space (time) intervals are an integer multiple.*

Proof: According to the Theorem 18 of the Consistent Universe, the set S of all possible intervals of space in which the laws of physics apply can only be finite and discrete, and then with a minimum value m (Definition 13 and Theorems 10 and 11). And the extension of any of these intervals has to be an integer multiple of m. Indeed, suppose that one of those intervals L is not an integer multiple of m. We would have:

$$L = n \times m + \delta; \; \delta < m \qquad (8)$$

where n is a positive integer. Since the physical laws apply on the intervals of extension m, $n \times m$ and L, they must also apply on the interval of extension δ, which is impossible because $\delta < m$, and m is the minimum extension of an interval in which the physical laws apply. Exactly the same argument proves that every time interval in which physical laws apply must be an integer multiple of the minimum time interval in which these physical laws apply. □

The existence of these indivisible intervals of space and time, derived in purely formal terms, supports the idea that Planck length and Planck time, although deduced from dimensional equations, could also indicate the existence of an indivisible minimum of space and an indivisible minimum of time. These minima impose certain restrictions on experimental and theoretical physics. For example, one could not measure the speed of light over a distance $d < l_p$, nor the duration t of events such that $t < t_p$. These limitations have been confirmed theoretically and tested experimentally (see, for instance, [10, 159, 273, 254]). And here, it is formally proved in the form of the following:

Corollary 23 (of Discrete Threshold) *The laws of physics do not apply in spaces smaller than the indivisible unit of space nor in times* *smaller than the indivisible unit of time, both being of non-zero extension (duration).*

Proof: It is an immediate consequence of the Theorem 35. □

As noted above, although the Corollary 23 of the Discrete Threshold is not explicitly stated in contemporary physics, its statement has broad theoretical and empirical support. It is a fundamental result for the construction of discrete models of the universe. From now on, the indivisible minima of space and time will be called qusits (discrete space units) and qutits (discrete time units), respectively. Finally, let us prove the following:

Theorem 36 (of Adjacency) *No space exists between any two successive qusits and no time elapses between two successive qutits.*

Proof: Let AB and CD be two successive qusits (simplified to a one-dimensional version) and assume they are not adjacent, i.e assume that $0 < BC$. BC must be less than the minimum space unit (a qusit), otherwise AB and CD would not be two successive qusits. Therefore, the physical laws would not apply in BC (Corollary 23 of the of Discrete Threshold), nor in AD, which goes against the Theorem 18 of the Consistent Universe because BC and AD are intervals of space. In consequence, it must be $BC = 0$, and no space exists between two successive qusits. So, the successive qusits can only be adjacent. A similar argument applied to successive qutits proves they also must be adjacent. □

Corollary 24 (of Finite Space and Time) *Every space interval (or time interval) is finite and can only be divided into an integer number of adjacent qusits (qutits).*

Proof: It is an immediate consequence of Theorems 29, 30 and 36. □

Note that in a discrete reality, where both space and time are discrete, no accessible object (e.g. elementary particles) could be smaller than a qusit, nor last less than a qutit. The existence of these limits for the intervals of space (Planck length) and time (Planck time) is also assumed in quantum mechanics, and their existence can be proved semi-formally from Heisenberg's Uncertainty Principle [208, pp. 269-272]. There would also be a maximum velocity of one qusit per qutit, so that the second principle of special relativity would not be necessary, but a direct consequence of the discrete nature of space and time.

The primary and secondary literature of physics is replete with expressions that reveal the scant interest of physicists in the formalism of their infinitist mathematical language: adjacent points, contiguous points, point to point, and so on. But in the spacetime continuum there are no contiguous (adjacent) points, nor is it possible to go "point to point over all points". Between any two points of this spacetime continuum there is always the same uncountable infinite number of different points, exactly $2^{\aleph_o}$, as many as in the whole three-dimensional universe.

An immediate consequence of the above theorems and corollaries is that physical objects, if they have size (can a physical object not have size?), cannot be point-like, nor have the same structure as a point, because points have neither size nor, consequently, structure. Nor can something physical be concentrated in a point without making it disappear, because points do not occupy space: something concentrated in a point would not occupy any space. However, authors who need no introduction have written texts such as:

> ... a beam of light emanating from a point... consists of a finite number of energy quanta, localized at points of space... (A. Einstein, quoted in [46, pp. 45-46].

> ... the interpretation of the photon as a pointlike structure... (A. Einstein, quoted in [46, p. 46].

> ... in the position to concentrate energy upon a single point in space... (M. Planck, quoted in [46, p. 46].

I believe that physics should be more rigorous in the use of ordinary language, which would help to consider the discrete alternative of space and time.

18.3 The problem of change

The protagonist of this section is the old problem of change, which was raised more than twenty-five centuries ago [253, 315, 30, 31, 239, 151]. A problem that, although forgotten in modern science, has not yet been solved. An oversight that is particularly important in physics, the science of change (the science of the regular succession of events in Maxwell words [235, p.98]); the science that should be the most interested in its solution. Indeed, it is not a satisfactory fact for a science as physics to have to admit its inability to explain the problem of change; its inability to explain, for example, how a simple change of position of an object in uniform motion occurs.

As is well known, the problem of change is related to Zeno Dichotomy [195], a dichotomy to which several solutions have been proposed from different areas of mathematics and physics [139, 140, 374, 141, 143, 142, 238, 237, 269, 6, 288, 312, 165, 323], but the problem of change itself remains unsolved. Forgetting a problem is no way to solve it. And the inability to explain how a simple change of position of an object in uniform motion occurs should point to a fundamental flaw in the model used to explain the physical world. That model has classically been based on an essentially continuous time and space (see chapter 16), which basically coincides with the modern spacetime continuum, the supposedly infinite structure of space and time in which all solutions to the problem of change have been sought.

The simplicity of the problem and the sterile search for solutions for more than twenty-five centuries have certainly contributed to the abandonment of the search and the pretense that the problem does not exist. But the problem does exist, and it is very fundamental. The attitude of science towards the problem of change is truly shameful. The only justification is that our sensory perception of the physical world deceives us: it shows us a continuity that could be discontinuous (think of a movie and its frames). But in any case, this discontinuity should have been explored. Especially since the invention of cinema. As we will see in this section, the problem of change could be solved in a model where space and time are discrete instead of continuous, which would of course confirm the conclusions drawn above.

As just indicated, physics models the physical world as a continuous world, probably because our very senses perceive it as a continuous world. The problem is that this perceived continuity is actually illusory because of the way our brain constructs the images we see: it takes a time greater than zero ($\approx$13 ms [282]) to process each visual image (α, β, γ and δ motions and ϕ phenomenon [102]), so a continuum of visual images is physiologically impossible, they will always be separated by a time interval greater than zero. Just as a movie is a discontinuous sequence of images that is perceived as a continuum, natural motion could be a discontinuous sequence of position changes that, for the same reason as a movie, is perceived as continuous by our brain. This illusory continuum (the impossibility of sensory perception of nature's discrete nature) is surely behind our attempts to explain the physical world in terms of the spacetime continuum. The discrete nature of space and time would certainly open the door to a discrete interpretation of special relativity (the science of the spacetime continuum) in terms of apparent, not real, space contractions, time dilations, and local simultaneities [208].

The problem of change has been forgotten by physics, certainly because of its unconditional submission to the spacetime continuum. In fact, the development of contemporary theoretical physics has taken place exclusively within the framework of infinitist mathematics, and with a total lack of interest in the formal consistency of the Hypothesis of the Actual Infinity that underlies infinitist mathematics. Fortunately, experimental physics can only be discrete and finitist, and requires that theoretical models be adapted to its results. Addressing the problem of change would have had a double benefit: checking the inconsistency of its infinitist mathematical language, and solving the problem of change itself. Confronting the problem of change would surely have had the consequence of discovering that the most appropriate language for physics is not infinitist mathematics, but computational language, or a new kind of undeveloped discrete mathematics. But it's never too late.

After formally posing the problem of change, it will be proved here that it cannot be solved within the spacetime continuum, the only framework in which its solution has been sought so far. And it cannot be solved precisely because of the lack of immediate successiveness (lack of adjacency between points and between instants) in the spacetime continuum. It is then proved that the problem of change can be solved within the framework of a discrete space and time, which here will be a model similar to cellular automata (CALM, cellular automata like models). The discrete solution of the old pre-Socratic problem would confirm the need for a discrete model to explain the physical world, because the physical world is essentially a consistent and constantly changing world.

(The text of the following two sections is an up-to-date summary of [212, pp. 329-338] and [208, pp. 571-583].)

18.4 Canonical changes

If Ob is a physical object, we will say Ob changes causally from the state S_a to the state S_b if there exist a set of (physical) laws L such that, under the same conditions C, and as a consequence of those laws and conditions, the state of Ob is S_a at the instant t_a, and S_b at an ulterior instant t_b, symbolically:

$$S_a \mapsto S_b : L(S_a, C, t_a) = (S_b, t_b) \qquad (9)$$

Here we will only deal with causal changes defined according to (9). They will be referred to simply as changes.

The change $S_a \mapsto S_b$ can be direct, without intermediate states. In such a case, it will be referred to as *canonical* change. It can also be the result of an ordered sequence of canonical changes:

$$\langle S_a \mapsto S_b \rangle : S_a \mapsto S_1 \mapsto S_2 \mapsto \cdots \mapsto S_v \mapsto S_b \qquad (10)$$

Note that, except S_1, each element S_n of $\{S_i\}$ must have an immediate predecessor S_{n-1} (symbolically $S_{n-1} < S_n$) so that S_n can be causally derived from S_{n-1}:

$$\forall S_{1 < n \leq b} : L(S_{n-1}, C_{n-1}, t_{n-1}) = (S_n, t_n) \qquad (11)$$

The objective of the discussion that follows is the analysis of the canonical changes, whether or not they are part of a sequence of canonical changes. We will begin by proving the following two theorems:

Theorem 37 (of the Canonical Changes) *Every change is either a canonical change of a discrete and finite sequence of canonical changes.*

Proof: Let $S_a \mapsto S_b$ be any change. If it is not a canonical change it will be a sequence of changes. A sequence that cannot be densely ordered (Theorem 7, page 22). Therefore, it will be a sequence with a first change; a last change; and each change (except the first) will be immediately preceded by another change and will be immediately followed by another change (except the last). It will therefore be a discrete sequence of canonical changes, which can only be finite (Theorem 9, page 23). □

Theorem 38 (of Change) *Canonical changes are instantaneous and then impossible in the spacetime continuum.*

Proof: Let $S_a \mapsto S_b$ any change of any object Ob and suppose it lasts for any time $t > 0$. Let t' be any instant in the interval $(0, t)$. If at t' the state of Ob is S_a, the change has not yet begun, and its duration will be less than t. If the state of Ob at t' is S_b, the change will have already ended and its duration will also be less than t. Therefore the duration of the change will be less than any real number t greater than zero. The duration of the change cannot be negative either because in that case S_b would be prior to S_a. Therefore the duration of the change has to be non-negative and less than any real number greater than zero. That is, it must be zero. The change must be instantaneous, which in spacetime continuum is only possible if both states coexist in the same instant, because in the spacetime continuum between any two different instants always elapses a time greater than zero. Now, if both initial and final states coexist in the same instant it is not possible to establish which state is the cause of the other. Therefore, canonical change is impossible in the spacetime continuum. □

Since change is so pervasive in the physical world, the above theorem of change may indicate that the spacetime continuum is inappropriate for representing physical space and time. Space and time may in fact be of a discrete nature. And, as we will see in the next section, instantaneous changes are possible in such a discrete space and time, although it is very difficult to grasp the idea of instantaneous changes.

Indeed, how can a change be instantaneous? If the change results from a process (the process of change) and that process has zero duration, the process has no existence and the change remains impossible. We arrive at the starting point of Zeno paradoxes, immediate consequences of the impossibility of change. But changes exist, they do not stop happening. Therefore, everything indicates that we need a new paradigm about the intimate constitution and functioning of the physical world at its most essential scale, even beyond the atomic scale.

The directional evolution of the universe (Principle 1 of Directional Evolution, page 41) shows that this evolution is subject to a consistent set of physical laws (Theorem 18, page 42). The changes in the universe must be consistent, otherwise no directionality would be possible, because these changes would occur in all directions, and it would not be possible to progress in any of them. We conclude that all changes in the universe are consistent, and if they are consistent, they must be instantaneous. The problem is that we have no idea how this is possible. In the last section of this chapter, we will analyze whether canonical changes are possible in a discrete space and time model. And before that, the next section raises some interesting questions about continuous versus discrete reality.

18.5 Discrete versus continuous

As indicated above (and especially in chapter 16 of this book), our theories about the physical world have certainly been influenced in some way by the fact that we ourselves perceive that physical world as a continuous reality. We now know that this continuous perception of the world is a deception of our brain (similar to that of cinematography). Moreover, a discrete reality would be much simpler than the current models of the physical world based on the spacetime continuum, which have been virtually the only models considered by science throughout its history.

Leaving aside the fact that the universe is consistent (Theorem 18 of the Consistent Universe, page 42) and the spacetime continuum inconsistent (Theorem 8 of the Inconsistent Continuum, page 23), we will have to admit that it is not the same to explain the evolution of $\approx 2.66 \times 10^{185}$ qusits[1] as that of $2^{\aleph_0}$ points, with the additional difficulty that any region of the universe, however small, has the same number of points as the entire universe. In addition, some interesting questions may be raised that physics has ignored throughout its history:

1. If space can deform, expand, and vibrate, is it not a real physical object? Under these conditions, how does physical space relate to the geometric points of the spacetime continuum?

2. If physical space is real, what is the physical reality of points? And how does ordinary matter relate to these points?

---

[1] The total number of qusits in the observable universe if they where of a Planck volume.

3. How is it possible, for example, that in a linear space of one millionth of a millimeter and in one millionth of a second as many virtual particles are created as in the entire three-dimensional universe in its entire history of more than 13.7 billion years?

4. How is the irreversible and directional geological record possible in a reversible and non-directional spacetime continuum?

And above all:

5. How can a consistent universe be constantly changing if change is inconsistent in the spacetime continuum?

6. How is it possible for physics to pretend to explain a physical world in constant change without first solving the problem of change?

7. Will these questions finally deserve the attention of the hegemonic infinitist streams of thought in modern physics?

As will be seen in the next section, most of these questions, and many others not properly addressed by modern physics, could find simple answers in discrete and finite models of space and time.

18.6 A discrete model: cellular automata

Cellular automata like models (CALMs) provide a new interesting perspective to analyze the way the universe could be evolving. In particular it provides a discrete space-time model in which a new analysis of the incomprehensible oddities of contemporary physics, including change, would be possible. As we will see in the next short discussion, twenty five centuries after it was posed, the old problem of change could find a first consistent solution in the discrete spacetime of CALMs.

In CALMs, space consists exclusively of minimal indivisible units: cells (qusits). Time also consists of a sequence of successive indivisible minimal units: qutits. There is no extension between a qusit and its immediate successor in any spatial direction. Similarly, no time elapses between a qutit and its immediate successor. Each qusit can have different states, each defined by a certain set of variables. The states of all qusits change simultaneously in each successive qutit according to the laws that drive the evolution of the automaton. Once changed, the state of each qusit remains unchanged for a qutit. In the following, we will assume that this is the case, although instead of one qutit, the state of each qusit could remain unchanged for a certain integer number of qutits. Note that the problem of change is not yet solved: it will be necessary to explain how the successive changes of state of each qusit occur. A consistent way to explain how such changes could occur is suggested on page 218.

18.6 A discrete model: cellular automata

Let $u, v, c, \ldots z$ be the set of variables that define the state of each qusit of a certain CALM A. Let us represent the nth state of each qusit x_i by $x_i(u_{i,n}, v_{i,n}, \ldots z_{i,n})$, where $u_{i,n}, v_{i,n} \ldots z_{i,n}$ denote the particular values of the state variables of x_i at the nth qutit. Let finally L be the set of laws driving the evolution of the automaton, including the laws that relate the different state variables to each other. L determines the way each qusit x_i changes from a qutit to the next one, taking into account the state of x_i as well as the state of any other qusit with which it interacts, which may include all qusits. All these current states define the conditions C_i under which the laws L determine the state of each qusit in the next qutit, that is, the laws that determine the change that each qusit undergoes in each successive qutit.

The automaton engine changes the state of each qusit at each qutit and maintains it just for one qutit. Thus we can write for each particular qusit x_i:

$$L(x_i(u_{i,n} \ldots, z_{i,n}), C_n, t_n) = (x_i(u_{i,n+1} \ldots, z_{i,n+1}), t_{n+1})$$
$$L(x_i(u_{i,n+1} \ldots, z_{i,n+1}), C_{n+1}, t_{n+1}) = (x_i(u_{i,n+2} \ldots, z_{i,n+2}), t_{n+2})$$
$$L(x_i(u_{i,n+2} \ldots, z_{i,n+2}), C_{n+2}, t_{n+2}) = (x_i(i, u_{n+3} \ldots, z_{i,n+3}), t_{n+3})$$
$$L(x_i(u_{i,n+3} \ldots, z_{i,n+3}), C_{n+3}, t_{n+3}) = (x_i(u_{i,n+4} \ldots, z_{i,n+4}), t_{n+4})$$
$$\ldots$$

Certain sets of qusits might remain grouped with the same configuration through successive qutits. They could be called CALM's objects. It is significant that the operation of a CALM is similar to that of a computer: its internal clock defines the indivisible units of time in which all operations and updates take place. And remember that computers are man-made machines capable of simulating physical phenomena [337].

Since both space and time are discrete, each qutit t_n has an immediate predecessor t_{n-1} and an immediate successor t_{n+1}, so that no other qutit passes between t_{n-1} and t_n, nor between t_n and t_{n+1}. In other words, no time elapses between any two consecutive qutits. This simple property of CALMs (together with the perceptible and interacting modes explained later in this chapter) is sufficient to solve the logical problem of change, because discrete space-time allows instantaneous changes: the state A_n at qutit t_n changes to A_{n+1} at the next qutit t_{n+1}, where the time elapsed between t_n and t_{n+1} is zero. One could say that all qusits of a CALM are updated simultaneously at every qutit. In the case of the points and instants of the space-time continuum, things are different, because between any two of its points (instants), whatever they are, there are other $2^{\aleph_o}$ different points (instants), so that none of them has an immediate successor, which makes it impossible for a change to occur (Theorem 38 of Change).

On the other hand, we must remember that our sensory perception of the physical world is continuous. And so we are used to thinking in terms of a spacetime continuum. So far, our only way of thinking. All of our models of the physical world have assumed that the physical world is a continuous world. It is then almost inevitable to extrapolate this way of thinking to any new discrete paradigm, which would obviously be catastrophic. Thinking in (physically) discrete terms will certainly require a long process of re-education.

For example, an electron could be in the state S_1 at a certain instant t_1 and in a different state S_2 at a later instant t_2, without ever having been in an intermediate state between S_1 and S_2 (quantum jump). So it is a canonical change. In the spacetime continuum, the interval (t_1, t_2) must always be greater than zero, and during this time the electron can neither be in the state S_1, nor in the state S_2, nor in any other conceivable state. Therefore, it must cease to exist for a time greater than zero. It must disappear at t_1 and reappear at t_2. In the discrete space and time of a CALM, all we have to do is consider two successive qutits, t_1 and t_2. At t_1 our electron would be in the state S_1, and at t_2 in the state S_2, so that no other qutit passes between t_1 and t_2. But we must recognize that this is an incomplete explanation.

The directional evolution of the universe shows that this evolution is subject to a consistent set of rules, physical laws, (Theorem 18 of the Consistent Universe). So, canonical changes must be consistent processes, and then instantaneous. The problem is that we have no idea how this is possible. As a very adventurous hypothesis, one could propose that qusits have two modes of existence:

1. Permanence mode: The state of each qusit remains unchanged at least for one qutit. This would be the only perceptible state of qusits.

2. Interacting mode: All qusits update synchronically their respective states through appropriate processes driven by the laws of the automaton, which has to last at least one qutit.

Although, in accordance with what has been said above, the problem of change will now appear in terms of these changes of modes. So we would have to admit that the interactive mode is simultaneous with the permanence mode, although it remains in an imperceptible background (such as computer applications running in the background) that changes to the permanence mode at each successive qutit (or something like that). In other words, the perceptible state of the qusits would coexist with the interactive mode of updating (changing) that defines the next perceptible state, which will become perceptible in the next qutit, as a kind of binary flip-flop. The newly substituted perceptible states are the new source of interactions of the new interactive

18.6 A discrete model: cellular automata

mode.

It is interesting to remember that in computers, very rapid changes occur in the contents of their discrete memory units (their qusits) at the rate set by the successive discrete time units of their internal clocks (their qutits). The contents of these discrete units of memory are updated by successive qutits, and with them the various devices controlled by each computer.

One could argue that the same could happen in the spacetime continuum: the content of each point (or group of points) is updated at each instant (or group of instants). The problem here is that no point (or group of points) has adjacent points (or adjacent groups of points): there are neither adjacent points nor groups of adjacent points (no group has a last point adjacent to the first point of the next group). And the same is true for instants (or groups of instants). Thus, the above possible discrete solution of the problem of change cannot be applied to the spacetime continuum.

What has just been presented is not the solution to the problem of change, but a way of solving it based on the discreteness of space and time. The only thing that is clear is that in the spacetime continuum this solution is not possible. Nor is any other, because in the spacetime continuum canonical changes are impossible. (Theorem 38).

From the point of view of CALM model, it would be interesting to analyze its compatibility, or even its formal relationships, with the implicit order proposed by David Bohm [42]. In any case, and by way of example, assume that: In any case, and by way of example, assume that:

- The universe has 7.6564×10^{196} qusits.
- The universe contains 10^{80} elementary particles.
- Each particle is defined by p variables
- Each particle is, somehow, present in each qusit.

Let U be a tridimensional CALM of 7.6564×10^{196} qusits in which the state of each qusit is defined by $p \times 10^{80}$ state variables. If it were possible to simulate U, perhaps we would observe the self-organizing and evolution of an object similar to our universe.

U would be incomparably less complex than, say, any matrix of infinite elements (which are common in mathematics and theoretical physics). We could model the universe, provided that we know the basic laws that make it evolve. Under these circumstances, simulating does not mean reproducing the exact history of the universe: recursive interactions between qusits and the resulting nonlinear dynamics open the door to unexpectedness and creativity, as in the terrestrial biosphere. In any case, we could theorize about U, we could use it as a theoretical reference to grasp the nature, size, and possibilities of real universes. Colossal as it may seem, U would be a finite object, and then composed of a number of elements incomparably smaller than the number of points ($2^{\aleph_0}$), a simple interval of, say, a trillionth of a millimeter of continuous space. Moreover, while the points of the space continuum are abstract artifacts devoid of intrinsic physical properties, each element of U would have plenty of intrinsic physical meaning: each of them represents a real part of a real object contained in another real and unique object: the discrete space of U.

19. Universal preinertia

19.1 Introduction

If the enigma of parallels in geometry was called "the shame of elementary geometry" in the 19th century (see chapter 13), so in the 21st century one could also call the concept of preinertia "the shame of elementary physics". Probably the most fundamental and universal concept in physics, which physics has not yet discovered, although it is implicitly and unintentionally used in a wide variety of arguments. Preinertia is the first goal of this chapter; the second is to prove the impossibility of detecting absolute motion, if any, precisely because of the preinertia of all physical objects (including photons). Except perhaps in one situation, which is also introduced at the end of this chapter and discussed in the last two chapters of this book.

19.2 Definition of Preinertia

Although only implicitly, preinertia appears in Galileo's first relativistic discussions. Specifically, in the discussion of the fall of the lead ball thrown from the top of the vertical mast of a ship moving at uniform velocity v. In this discussion Galileo refutes the Aristotelian conception of motion [124, p. 106-275]. In fact, and contrary to the hegemonic Aristotelian view, Galileo defended the ball hitting the base of the mast. According to Galileo, the ball follows a vertical trajectory for observers on the ship, while it follows a non-vertical trajectory for observers on the dock.

But there is one fact that will be observed in the same way by all observers, those on the ship and those on the dock:

The ball always moves parallel to the mast of the ship.

And here is where preinertia appears: for the observers in the ship, the ball moves parallel to the mast, because no force other than gravity acts on the ball as it falls; this is also true for the observers in the dock, but for these observers the ball can move parallel to the mast of the ship only if it continues to move at the same relative velocity v of the mast of the ship (figure 19.1). In other words, IF THE BALL INHERITS AND MAINTAINS THE RELATIVE VELOCITY v AS IT FALLS DOWN.

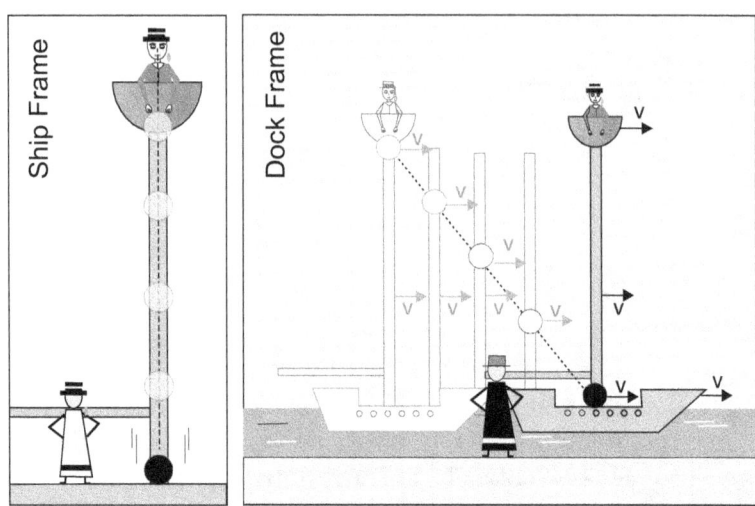

Figure 19.1 – Galileo's mast: the ball always falls parallel to the ship's mast, both in the ship's reference frame (left) and in the dock's reference frame (right).

We could replace Galileo's ball with a modern visible laser beam, emitted on Galileo's ship in the vertical direction, parallel to the ship's mast (in the direction from bottom to top, or from top to bottom). Whatever the velocity v of Galileo's ship, both inside and outside the ship, we will always see a visible laser beam (e.g. green) parallel to the ship's mast. If the ship is observed moving at velocity v, the laser beam will be seen as a thick line of green light moving in solidarity with all elements of the ship. It will be observed as a moving vertical laser beam, always parallel to the ship's mast, with the same velocity v as the mast; i.e., it will be observed as a vertical green light mast moving with the same velocity v as any other object on the ship. (Figure 19.2).

As in the case of Galileo's ball, the observation of Galileo's laser is only possible if each of its photons inherits, when emitted, the ship's velocity vector $\vec{v}$ as one of the components of its own velocity vector $\vec{c}$. In the primary and secondary literature of modern physics, the idea that the speed of light is independent of the speed of its emitting source is widespread. But that is true only in scalar terms; in vector terms -and velocity is a vector magnitude- photons always inherit the velocity vector of their emitting source as one of the components of their own velocity vector. We will prove this formally in the next section. For now, we will finish this introduction to the concept of preinertia, defining it as follows:

19.2 Definition of Preinertia

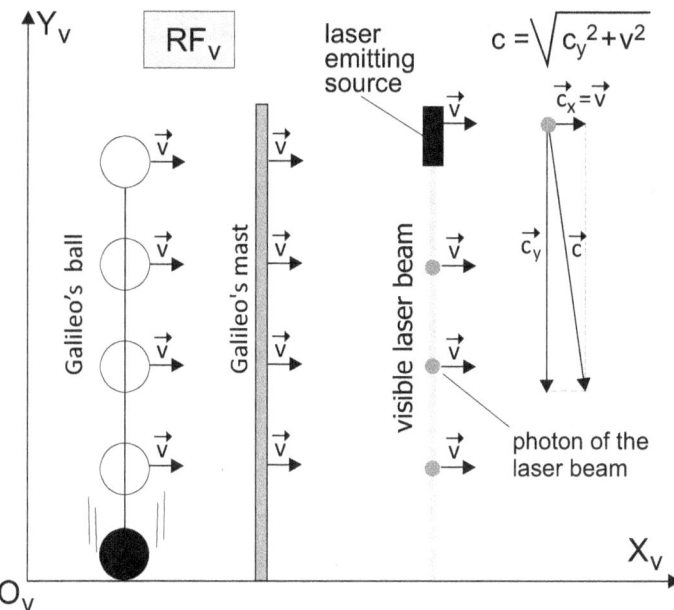

Figure 19.2 – The photons of Galileo laser beam also inherit the relative velocity vector of their emitting source as a component of their own velocity vector.

Definition 18 (of Preinertial Objects) *Preinertia is the property of all physical objects to inherit the relative velocity vector of the reference frame in which they are set in motion.*

It is evident that without preinertia the physical world would be completely different: every time an object, for example on Earth, momentarily leaves its physical contact with the Earth it would shoot out at a speed of 367 km/s (more than one million three hundred thousand kilometers per hour!) in the same direction and in the opposite sense to that in which the Earth moves with respect to the Cosmic Microwave Background. The next section will prove that all physical objects, including (supposedly) massless objects like photons, are preinertial. Preinertia is a universal property of all physical objects with the highest empirical evidence. A universal property that opens the door to some interesting discussions.

Since preinertia is a universal property of all physical objects, how is it possible that 21st century physics has not yet discovered it? Perhaps it has been influenced by the widely publicized relativistic conclusion that the speed of light is independent of the speed of its emitting source, without adding that it is so only in scalar terms. In vector terms, light (like any other physical object) always inherits the relative velocity vector of its emitting source. And when something is widely published in the official media, it tends to be perpetuated. Also, it has always been, and still is, difficult to maintain dissident positions with academic officialdom. In the case of preinertia, the omission is so obvious, and the consequences so serious, that it should make us reflect on our way of constructing science. It seems to me that we have too much ego and lack humility and critical spirit.

19.3 Photons are preinertial

Let us consider a photon reflecting vertically on two horizontal mirrors (Einstein's clock of light) in the proper reference frame RF_o of the mirrors. In RF_v, from whose perspective RF_o moves from left to right at a velocity $v = kc$, $(0 < k < 1)$, parallel to the direction of the increasing X_v, the photon follows a trajectory inclined with respect to the vertical by an angle β_v (Figure 19.3, left) given by:

$$\sin \beta_v = \frac{vt_v}{ct_v} = k \tag{1}$$

$$\beta_v = \arcsin k \tag{2}$$

In RF_v the motion of the reflecting photon can also be referred to the vertical walls of the clock, as Figure 19.3 (right) shows. This way of referring the motion of the reflecting photon illustrates that both, the clock and the photon, move with the same relative velocity $v = kc$ with respect to RF_v. Now we will prove that photons inherit the relative

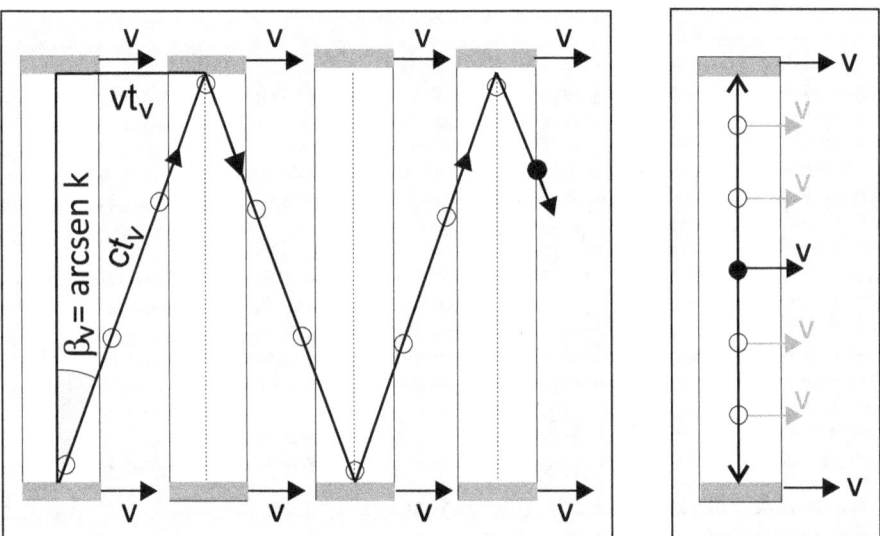

Figure 19.3 – Two ways of observing Einstein clock of light in relative motion from the perspective of RF_v.

velocity vector of their emitting source, as a component of its own velocity vector, whatsoever be the proper inclination at which they are emitted. For this, consider a photon a^* that is emitted at any angle α_o with respect to the X_o axis of the reference frame RF_o of its emitting

19.3 Photons are preinertial

source. Obviously, after a proper time t_o this photon will have traversed a horizontal distance d_{ox} and a vertical distance d_{oy} such that:

$$d_{ox} = t_o c \cos \alpha_o \tag{3}$$

$$d_{oy} = t_o c \sin \alpha_o \tag{4}$$

The corresponding components c_{ox}, c_{oy} of its velocity vector will be:

$$c_{ox} = c \cos \alpha_o \tag{5}$$

$$c_{oy} = c \sin \alpha_o \tag{6}$$

Assume the frame RF_v coincides with RF_o at the precise instant $t_{oo} = t_{vo} = 0$ when the photon a^* is emitted in RF_o, being t_{oo} and t_{vo} respectively measured in RF_o and RF_v. From the perspective of RF_v, the frame RF_o moves from left to right parallel to X_v at a uniform velocity $v = kc$, $(0 < k < 1)$. Thus, for the observers in RF_v the photon a^* travels a vertical distance d_{vy}:

$$d_{vy} = d_{oy} = c t_o \sin \alpha_o \tag{7}$$

in a time t_v:

$$t_v = \gamma t_o + \frac{\gamma (t_o c \cos \alpha_o) kc}{c^2} \tag{8}$$

$$= \gamma t_o (1 + k \cos \alpha_o) \tag{9}$$

Therefore, and taking into account that $\gamma^{-1} = \sqrt{1 - k^2}$, the vertical component c_{vy} of the velocity vector of the photon a^* will be:

$$c_{vy} = \frac{c t_o \sin \alpha_o}{\gamma t_o (1 + k \cos \alpha_o)} \tag{10}$$

$$= \frac{c \sin \alpha_o}{\gamma (1 + k \cos \alpha_o)} \tag{11}$$

$$= \frac{c \sqrt{1 - k^2} \sin \alpha_o}{1 + k \cos \alpha_o} \tag{12}$$

To calculate the horizontal component c_{vx} of the velocity vector of the photon a^* in RF_v we will assume the universality of the speed (modulus of the velocity vector) of light. Accordingly, we can write:

$$c_{vx}^2 = c^2 - c_{vy}^2$$

$$= c^2 - \frac{c^2(1-k^2)\sin^2\alpha_o}{(1+k\cos\alpha_o)^2}$$

$$= \frac{c^2(1+k^2\cos^2\alpha_o + 2k\cos\alpha_o) - c^2\sin^2\alpha_o + k^2c^2\sin^2\alpha_o}{(1+k\cos\alpha_o)^2}$$

$$= \frac{c^2 + c^2k^2\cos^2\alpha_o + 2c^2k\cos\alpha_o - c^2\sin^2\alpha_o + k^2c^2\sin^2\alpha_o}{(1+k\cos\alpha_o)^2}$$

$$= \frac{c^2(1-\sin^2\alpha_o) + k^2c^2 + 2c^2k\cos\alpha_o}{(1+k\cos\alpha_o)^2}$$

$$= \frac{c^2(\cos^2\alpha_o + k^2 + 2k\cos\alpha_o)}{(1+k\cos\alpha_o)^2}$$

$$= \frac{c^2(k+\cos\alpha_o)^2}{(1+k\cos\alpha_o)^2}$$

And then:
$$c_{vx} = \frac{c(k+\cos\alpha_o)}{1+k\cos\alpha_o} \tag{13}$$

A little algebra suffices now to prove that the horizontal component c_{vx} of the photon velocity with respect to RF_v given by (13) implies that our photon a^* inherited (in vector terms) the relative velocity kc of its emitting source (and then of RF_o) with respect to RF_v:

$$c_{vx} = \frac{c(k+\cos\alpha_o)}{1+k\cos\alpha_o} \tag{14}$$

$$= \frac{kc + c\cos\alpha_o + k^2c\cos\alpha_o - k^2c\cos\alpha_o}{1+k\cos\alpha_o} \tag{15}$$

$$= \frac{(1-k^2)c\cos\alpha_o + kc + k^2c\cos\alpha_o}{1+k\cos\alpha_o} \tag{16}$$

$$= \frac{(1-k^2)c\cos\alpha_o}{1+k\cos\alpha_o} + kc \tag{17}$$

$$= \frac{(1-k^2)^{1/2}c\cos\alpha_o}{(1-k^2)^{-1/2}(1+k\cos\alpha_o)} + kc \tag{18}$$

$$= \frac{\gamma^{-1}c\cos\alpha_o}{\gamma(1+k\cos\alpha_o)} + kc \tag{19}$$

$$= \frac{\gamma^{-1}t_o c\cos\alpha_o}{\gamma(t_o + kt_o\cos\alpha_o)} + kc \tag{20}$$

19.3 Photons are preinertial

$$= \frac{\gamma^{-1}t_o c \cos\alpha_o}{\gamma t_o + \dfrac{\gamma k t_o c^2 \cos\alpha_o}{c^2}} + kc \tag{21}$$

$$= \frac{\gamma^{-1}t_o c \cos\alpha_o}{\gamma t_o + \dfrac{\gamma(t_o c \cos\alpha_o)kc}{c^2}} + kc \tag{22}$$

$$= \frac{\gamma^{-1}t_o c \cos\alpha_o}{t_v} + kc \tag{23}$$

$$= \frac{\gamma^{-1}t_o c \cos\alpha_o + kct_v}{t_v} \tag{24}$$

Therefore, during the time t_v, and with respect to RF_v, the photon a^* runs through a horizontal distance d_{vx}:

$$d_{vx} = \gamma^{-1}t_o c \cos\alpha_o + kct_v \tag{25}$$

The right side of (25) has two terms:

1) According to (3), the first term $\gamma^{-1}t_o c \cos\alpha_o$ is the horizontal distance a^* moves with respect to RF_o for the time t_o, although contracted by the relativistic factor γ^{-1}. This would be the horizontal distance our photon a^* would have traversed with respect to RF_v if it had not inherited the relative velocity vector of its emitting source.

2) The second factor kct_v is the distance the emitting source moves with respect to RF_v during the time t_v with a velocity kc.

All of which suggests that the velocity vector of the photon a^* inherited the relative velocity vector as part of its component parallel to the direction of relative motion. And indeed, equation (25) allows us to prove that this is the case:

$$c_{vx} = \frac{d_{vx}}{t_v} \tag{26}$$

$$= \frac{\gamma^{-1}t_o c \cos\alpha_o + kct_v}{t_v} \tag{27}$$

$$= \frac{t_o}{\gamma t_v} c \cos\alpha_o + kc \tag{28}$$

$$= \frac{t_o}{\gamma^2 \left(t_o + \dfrac{kt_o c \cos\alpha_o}{c}\right)} c_{ox} + kc \tag{29}$$

$$= \frac{1}{\gamma^2(1+k\cos\alpha_o)}c_{ox} + kc \qquad (30)$$

$$= \frac{1-k^2}{1+k\cos\alpha_o}c_{ox} + kc \qquad (31)$$

This conclusion is confirmed by the following argument: Assume that with respect to the X_v axis of RF_v the photon a^* only travels the distance:

$$d_{vx} = \gamma^{-1}d_{ox} \qquad (32)$$

Since $d_{ox} = t_o c \cos\alpha_o$ (3), equation (23) would give rise to:

$$c_{vx} = \frac{\gamma^{-1}t_o c \cos\alpha_o}{t_v} + kc \qquad (33)$$

$$= \frac{\gamma^{-1}d_{ox}}{t_v} + kc \qquad (34)$$

$$= \frac{d_{vx}}{t_v} + kc \qquad (35)$$

$$= c_{vx} + kc \qquad (36)$$

$$\therefore k = 0 \qquad (37)$$

which is not the case because $k > 0$.

In consequence, once emitted and from the perspective of RF_v, our photon a^* moves in the direction of the relative motion (apart from the projection of its inclined trajectory on that direction) the same distance and for the same time as its emitting source (25), and it inherits the relative velocity vector as a part of its vector component parallel to the direction of the relative motion (31). The other part of this component is a fraction of c_{ox} defined by the complementarity factor f_{vx}, which according to (31) is:

$$f_{vx} = \frac{1-k^2}{1+k\cos\alpha_o}; \quad 0 \leq f \leq 1 \qquad (38)$$

which, obviously, decreases with the relative velocity kc and increases with α_o (Figure 19.4). The other vector component c_{vy} will be defined according to:

$$c^2 = c_{vx}^2 + c_{vy}^2 \qquad (39)$$

$$c_{vy}^2 = c^2 - c_{vx}^2 \qquad (40)$$

$$= c^2(1 - f_{vx}^2 \cos^2\alpha_o) \qquad (41)$$

19.3 Photons are preinertial

It is then clear that, from the perspective of the reference frame RF_v, the velocity vector of the photon a^* inherits, as part of one of its components, the relative velocity vector of its emitting source. At this point, and according to all theoretical and experimental evidence, we have no choice but to accept that photons are preinertial; that they do indeed inherit the relative velocity vectors of their emitting sources as a component of their own velocity vectors (except, perhaps, in the case where both vectors are parallel, which would be a kind of test of the prevalence of preinertia on the universality of the speed of light, or vice versa).

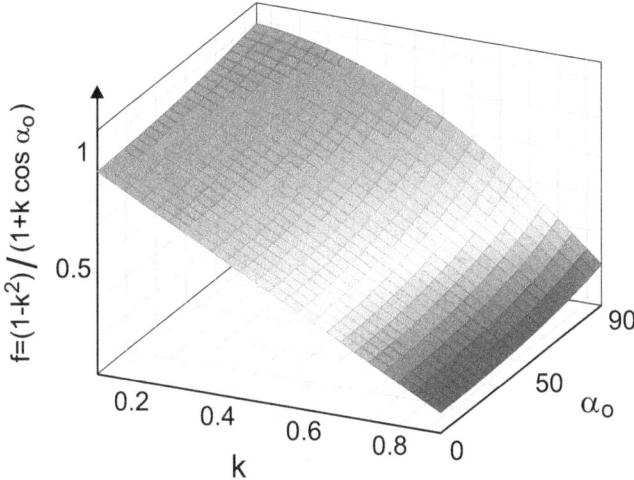

Figure 19.4 – Complementarity factor $f = (1 - k^2)/(1 + k \cos \alpha_o)$ in terms of k and α_o.

We must conclude that whenever the direction of the relative velocity v of a photon source is different from the emitting direction d_e, each emitted photon inherits the relative velocity v of its emitting source as a component (or part of a component) vector of its velocity vector, the other component in the plane defined by v and d_e being such that the module of the resulting vector is the universal speed of light c.

The above conclusion applies to all directions except the direction of the relative motion of its emitting source. This exception has an immediate (and axiomatic) explanation: the Second Principle of Relativity. If the photon is emitted in the same direction and sense as the relative motion, then the exception could also have a physical explanation in a discrete space-time: the existence of a maximum insurmountable speed of one unit of space per one unit of time. If the photon is fired in the same direction as the relative motion, but in the opposite sense, then only the axiomatic explanation remains. Thus, for all directions and senses, the inherited relative velocity has a physical explanation, except in the case just indicated, which is the case that motivates Santiago del Collado's experiment [208, pp. 463-488]. Obviously, the above argument about photons can also be applied to any other physical object, whether it is an elementary particle or not, and whether it is a massive object or not. So it proves the following:

Theorem 39 (of Preinertia) *Every physical object inherits in one of its vector components the relative velocity vector of the reference frame where it is set in motion, provided that the resulting speed does not exceed the possible maximum limit.*

We have made use of the Lorentz Transformation, and therefore of the special relativity, to demonstrate the preinertial nature of photons. So it can be said that special relativity implies preinertia. But preinertia being such a basic and universal property of matter, and there being so much empirical evidence confirming it (for example, each time an object falls to the ground), we may wonder whether the principles of relativity are necessary to demonstrate preinertia, or is preinertia an aspect of the Principle of Inertia that has gone unnoticed, perhaps because of its excessive evidence. This enormous empirical evidence of preinertia recommends making it independent of special relativity (SR) and incorporating it into the statement of the Principle of Inertia by simply adding three words, namely:

> IS, PREINERTIAL, AND.

Thus, in order to make preinertia explicit, the Principle of Inertia could be stated as follows:

Principle 3 (of Inertia) *Every physical object is preinertial and remains at rest or moves at a constant uniform velocity, unless an external force acts upon it.*

We accept the truth of the principle of inertia because of its extraordinary empirical evidence. It is an inductive principle that we use, along with others, to begin to construct an explanation of the physical world, the construction of physics. Now, should that be the starting point, or can we try to solve some even more basic questions? For example:

Let A and B be any two physical object of the same type at rest in their common inertial reference frame RF_o. Let now A be set in linear uniform motion with respect to RF_o.

- What determines and controls the linear trajectory of A, its successive positions along the successive instants?

- How does A remember that it was set in motion? Where lies the imprint of that action?

- What changed, if any, in the internal structure of A as a consequence of being set in motion?

- What distinguishes an object that has been set in motion from another that was not?
- Is space and time somehow affected by an object set in motion?
- Knowing that A was set in motion and B was not, is it the same to say that A moves with respect to B as to say that B moves with respect to A?
- If A is a massless photon, what quality of its nature determines its preinertia? Or is it not a massless object?
- To set A in motion is the same as to set the rest of the universe in motion?
- Is there any absolute describable reality?
- If there is no reality describable in absolute terms, are there as many realities as there are relative forms of observing it? To observe what?
- Could the universe be described, as such an object, from outside the universe?
- Are we living beings endowed with the capacity to reason but not to observe reality?
- Is the theory of special relativity the ultimate theory?
- etc.

As Feynman said, we know how objects move but not why they do (why they move in a straight line) [116, p. 18]. But should science give up answering the above questions?

19.4 Preinertia and absolute motion

The impossibility of measuring absolute velocities has been confirmed experimentally, but its derivation from the principles of relativity is axiomatic and then devoid of physical meaning. The impossibility of measuring absolute velocities is better explained physically by the preinertial nature of photons and, according to the Principle 3 of Inertia, of all physical objects. There would be a kind of mechanical entanglement between all physical objects of an inertial reference frame (even if they are created in that frame when set in motion, as photons). And this entanglement is maintained forever, unless some force modifies it. As will be seen below, this mechanical entanglement of all objects in an inertial reference frame makes it impossible to use just these elements, once set in motion, to detect the absolute motion of the frame in which they are set in motion. Although, as indicated above, there may be an

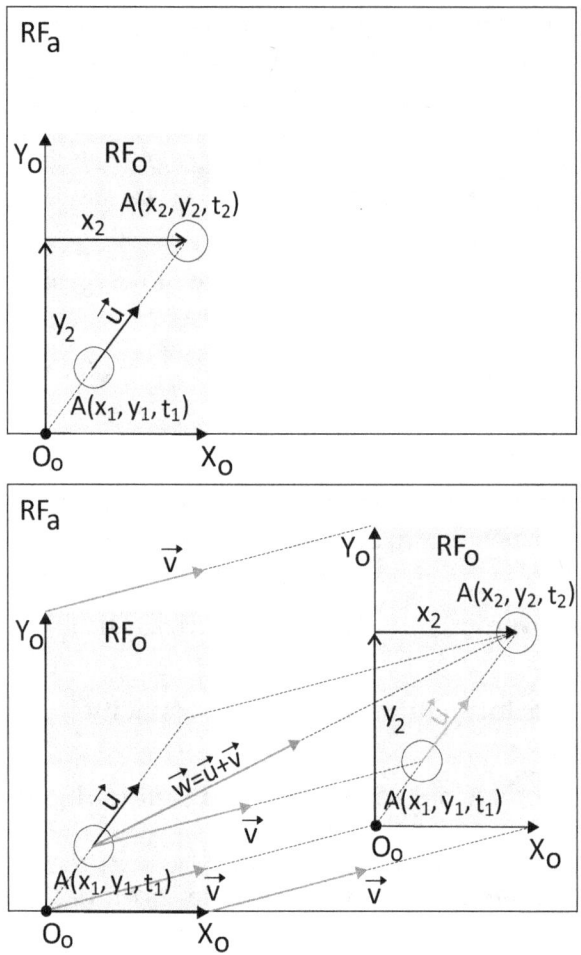

Figure 19.5 – Absolute motion is undetectable. Left: the physical object b is set in motion in its proper reference frame RF_o at rest in RF_a. Right: the physical object b is set in motion in its proper reference frame RF_o, which moves through RF_a with an absolute velocity $\vec{v}$.

exception.

Assume, just for a moment! that there exist an absolute reference frame RF_a (perhaps made of indivisible space units, qusits) through which physical objects can move in absolute terms. Let RF_o be a reference frame at rest in RF_a, and let b be any physical object at rest in its proper reference frame RF_o, where at instant t_1 it is placed at rest in the position (x_1, y_1, t_1) of RF_o (for simplicity, we dispense with the z-coordinate). Let b be set in motion at t_1 with a uniform velocity $\vec{u_b}$ so that at the instant t_2 it is placed in the position (x_2, y_2, t_2) (Figure 19.5, left).

Consider now that RF_o moves in RF_a with an absolute and uniform velocity $\vec{v}$, and let b be set in motion under the same above conditions when RF_o was at rest in RF_a. Thanks to preinertia, b *inherits* the abso-

19.4 Preinertia and absolute motion

lute velocity $\vec{v}$ of RF_o with respect to RF_a, and thanks to the Principle of Inertia, b *maintains* $\vec{v}$ along its own motion with respect to RF_o (Figure 19.5, right). Let O be the origin of coordinates of RF_v. This point O moves respect to RF_a at a velocity $\vec{v}$, while b moves with respect to RF_a at a velocity:

$$\vec{w_b} = \vec{u_b} + \vec{v} \qquad (42)$$

The object b (that moves with respect to RF_a at the velocity $\vec{w_b}$ given by (42)) will move with respect to O (that moves with respect to RF_a at the velocity $\vec{v}$) at a velocity $\vec{u'_b}$ given by:

$$\vec{u'_b} = \vec{w_b} - \vec{v} \qquad (43)$$

$$= \vec{u_b} + \vec{v} - \vec{v} \qquad (44)$$

$$= \vec{u_b} \qquad (45)$$

which is the same velocity as if RF_o were at rest with respect to RF_a. In consequence, the coordinates of b in RF_o at t_2 will be the same as in the first case when RF_o was at rest in RF_a. So, the coordinates of b at t_2 will also be (x_2, y_2, t_2), and they cannot be used to detect the absolute motion of RF_o.

Assume now that with the intention to measure the absolute velocity $\vec{v}$ of RF_o with respect to RF_a, two physical objects b and c are set in motion in RF_o with the respective uniform velocities $\vec{u_b}$ and $\vec{u_c}$. The velocity $\vec{w_b}$ of b with respect to RF_a, and the velocity $\vec{w_c}$ of c with respect to RF_a will be:

$$\vec{w_b} = (\vec{u_b} + \vec{v}) \qquad (46)$$

$$\vec{w_c} = (\vec{u_c} + \vec{v}) \qquad (47)$$

and then:

$$\vec{w_b} - \vec{w_c} = (\vec{u_b} + \vec{v}) - (\vec{u_c} + \vec{v}) \qquad (48)$$

$$= \vec{u_b} - \vec{u_c} \qquad (49)$$

So, the difference $\vec{w_b} - \vec{w_c}$ of velocities between b and c with respect to RF_a will be the same whatsoever be $\vec{v}$, and then it cannot be used to determine $\vec{v}$.

Since RF_o is any reference frame, b and c any two physical objects initially at rest in RF_o, and $\vec{u_b}$ and $\vec{u_c}$ any two velocities, we must conclude that the absolute motion of a reference frame is undetectable by setting into motion any physical object (or objects) of that reference frame. Let me now recall again with admiration Newton's words [260,

Corollary V, p 144]:

> The motions of bodies included in a given space are the same among themselves, whether that space is at rest, or moves uniformly forwards in a right line without any circular motion.

The above argument also applies to the case of elementary particles set in motion in the same conditions as b and c (or even created and set in motion in those conditions). It could be argued, however, that the argument only applies to elementary particles as such particles, but not to their corresponding associated waves. Evidently, if this were the case, particle-wave decoupling would occur, which is unknown in modern physics (as far as I know). It seems reasonable to conclude that positive results should not be expected in experiments a la Michelson-Morley (except, maybe, in the case of Santiago del Collado experiment [208, p. 371-378]).

19.5 Two key questions

There is enormous empirical evidence that preinertia is a universal property of all physical objects, even (supposedly) massless physical objects such as photons. We can then ask what is the reason for preinertia. If photons have spin 1 but no electric charge, no color charge, and no mass, what essential property of photons (and of all other particles, since they are all preinertial) is responsible for their preinertia? This is our first key question.

To pose our second key question, consider a photon a^* emitted parallel to the Y_o axis of the proper reference frame RF_o of its source S (Figure 19.6). According to special relativity (SR) and preinertia, the trajectory of the photon a^* will be different in different references frames, depending on their relative velocities respect to RF_o. According to SR, each of these trajectories is as real as any other. Recall Einstein's own words:

> ... and shows that the clock goes slower than if it were at rest relatively to K'. These two consequences, which hold, mutatis mutandis, for every system of reference, ... [91, p. 38].
> ... From all of these considerations, space and time data have a real, and not a mere fictitious, significance; [91, p. 30].
> ... It is clear that the same results hold good of bodies at rest in the "stationary" system, viewed from a system in uniform motion.[94, p. 49]

So, preinertia and SR make it inevitable to face the following two alternatives:

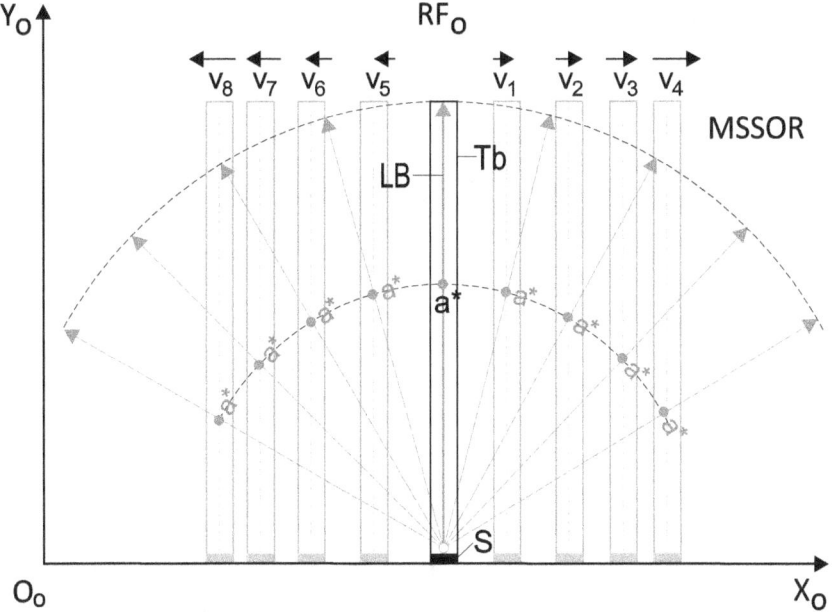

Figure 19.6 – The vertical trajectory of the photon in the proper reference frame RF_o of its source S is seen at different inclinations depending on the relative velocity at which it is observed. MSSOR: multiple superposed and simultaneous objective realities.

a) There are multiple superposed and simultaneous objective realities, as many of them as possible relative velocities at which any physical object or event can be observed.

b) There is a unique objective reality in which all objects moves THROUGH the same absolute space frame.

The second alternative seems much more simple, and it is not incompatible with relative motion: relative motion is an inevitable consequence of moving with different velocities through the absolute space frame. The problem is that, until now, absolute motion is undetectable. And for the reasons given above, maybe it will always remain undetectable. According to this alternative, SR is a mathematical theory on apparent realities, as apparent as the deformed rod partially submerged in water. Because, in fact, the rod is not deformed no matter how many experimental measurements we make checking its observed deformation.

19.6 Preinertia and the nature of light

Almost every general physics textbook includes a chapter on the nature of light [346, 347, 117, 362, 321] etc., not to say publications specifically devoted to the matter, for instance [232, 61, 304, 295, 331, 115, 161, 104, 242] etc. In page 291 of the Oxford Dictionary of Physics [74] we can read (the italic is mine):

light The form of electromagnetic radiation to which the human eye is sensitive and on which our visual awareness of the universe and its content relies.

... In 1905 Einstein showed that the photoelectric effect could only be explained on the *assumption* that light consists of a stream of discrete photons of electromagnetic energy.

... While it is not easy to construct a model that has both wave and particle characteristic, it is *accepted*, according to the Principle of Complementarity proposed by Niels Bohr, that in some experiments light will *appear* wavelike, while in other it will *appear* to be corpuscular.

Assumption, accepted, appear, ... it seems that, in the end, we don't know what light really is. Indeed, let us see how quickly the answers to successive pertinent questions are exhausted:

1) What is light? Answer: A set of electromagnetic waves.
2) What is an electromagnetic wave? Answer: An oscillation of the electric field associated with an orthogonal oscillation of the magnetic field.
3) What is a field? Answer: A region of space in which certain forces manifest.
4) What is space? Answer:
5) What is a force? Answer:

And given the preinertial nature of light, does preinertia intervene in the gravitational interactions of photons with massive objects? Could these hypothetical interactions (forces) explain the observed gravitational curvature of space without having to curve space? And, as noted above, if it is not mass, what fundamental property of the nature of light could be responsible for its preinertia?

We can describe light and its propagation through various media up to a certain level of essentiality. And for the same reasons as with the principle of inertia, we should not stop at this level of essentiality. But in order to continue asking and answering more essential questions about the nature of light, we would surely first have to clarify the nature of space and time. This, in turn, will not be possible if we continue to ignore the inconsistent nature of actual infinity and thus the inconsistency of the spacetime continuum. Many things will change when we accept and prove the discrete nature of space and time. Perhaps the CALM models (see Chapter 18) would be of great help in initiating this new physical exploration of the physical world.

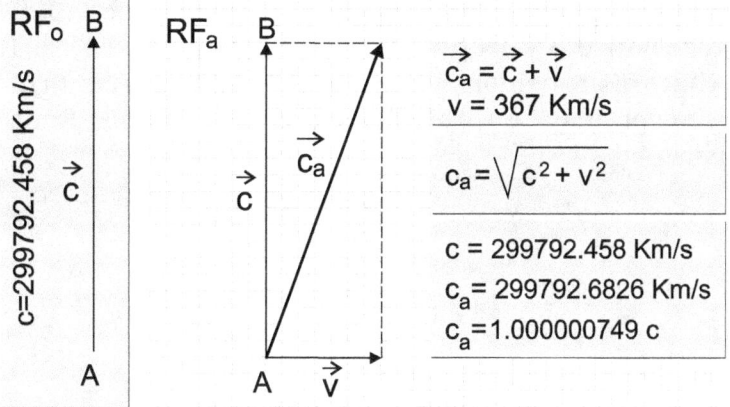

Figure 19.7 – Speed of light c and absolute speed of light c_a. The velocity $v = 367 Km/s$ is the velocity of the Earth through RF_a, the reference frame of the isotropic cosmic microwave background, assumed as an absolute reference frame.

19.7 The speed of light and absolute motion

The speed of light has been measured on different occasions, in different places and with different methods. Today it is unanimously accepted that this speed is 299792458 m/s. Now, since photons are of a preinertial nature, it makes sense to ask about the possible consequences on their velocity of inheriting the relative velocity vector of their corresponding emitting sources. In this sense, assume we measure the speed of light in our terrestrial reference frame (Figure 19.7, left). We must consider the following two alternatives:

a) Light inherits simultaneously a potentially infinite number of relative velocities, as many as relative velocities at which it can be observed from any possible inertial reference frame.

b) Light inherits a unique absolute velocity, which is the absolute velocity of the reference frame where it is created and set in motion.

The first alternative is very uncomfortable from the physical point of view. According to the second alternative, the measured velocity c of light in our planet would include the absolute velocity v of our planet through the same absolute reference frame RF_a (Figure 19.7, right). So, we could write for the absolute velocity of light c_a:

$$\vec{c_a} = \vec{c} + \vec{v} \tag{50}$$

$$c_a = \sqrt{c^2 + v^2} \tag{51}$$

$$= \sqrt{299792.458^2 + 367^2} \tag{52}$$

$$= 299792.6826 Km/s \tag{53}$$

In consequence, the absolute speed of light c_a would be 1.000000749 times greater than the velocity c measured here, in the Earth. The problem is that this conclusion could be impossible to confirm here in the Earth. Except, may be through the above mentioned Santiago del Collado experiment.

20. Zeno Dichotomies

(A chapter of the book Infinity put to the test [212])

20.1 Introduction

More than 2500 years ago, Zeno of Elea posed his famous paradoxes about the nature of space and its (supposed) infinite divisibility. These paradoxes are still paradoxical today, although they have lost their relevance and interest in contemporary science. Zeno challenges are recalled here, albeit in the modern terms of infinitist mathematics. As will be seen in this chapter, infinitist mathematics provides the formal tools to attempt a solution to Zeno paradoxes: it is possible to prove that these enigmas are true inconsistencies (although so far the echo of these demonstrations has been very slight). Recognizing Zeno inconsistencies is an inevitable first step in any discussion on the nature of physical space.

20.2 Introductory definitions

This chapter introduces a formalized version of Zeno Dichotomy in its two variants (here referred to as Dichotomy I and Dichotomy II) based on the successiveness and discontinuity of ω-order (Dichotomy I) and of ω^*-order (Dichotomy II). Each of these formalized versions leads to a contradiction pointing to the inconsistency of the Hypothesis of the Actual Infinity (the existence of the "*totality of finite cardinal numbers*", in Cantor's words [56, p. 103]) from which the first transfinite ordinal number ω is deduced [56, p. 160, Theorem §15 A].

In the second half of the XX century, several solutions to some of Zeno paradoxes were proposed with the aid of Cantor's transfinite arithmetic, topology, measure theory and, more recently, internal set theory (a branch of non-standard analysis) [139, 140, 374, 141, 143, 142, 238, 237]. It is also worth noting the solutions proposed by P. Lynds [226, 227] within classical and quantum mechanics frameworks. Some of these solutions, however, have been contested. And in most cases, the proposed solutions do not explain where Zeno arguments fail. Moreover, some of the proposed solutions gave rise to a new collection of problems so exciting as Zeno paradoxes [269, 6, 288, 312, 165, 323].

In the discussion that follows I propose a new way to discuss Zeno Dichotomies based on the notion of ω-order, the type of order of the well-ordered sets whose ordinal number is ω, the least transfinite ordinal [56, p. 160, Theorem §15 A]. The set $\mathbb{N}$ of the natural numbers in their natural order of precedence is an example of ω-ordered set. In this type of ordering, each element n has an immediate successor $n+1$ with no elements between n and $n+1$, that is why $n+1$ is called immediate successor of n, and n the immediate predecessor of $n+1$. Immediate succession is also called adjacency or contiguity.

A sequence $\langle a_i \rangle$ indexed by the ω-ordered set $\mathbb{N}$ of the natural numbers is also ω-ordered by the relation of precedence of their indexes (Theorem 4 of Indexation, page 19 of this book and Theorem of the Indexed Sets [212, p. 55]), which can be the same, or not, as their natural precedence, if any. As is well known, in an ω-ordered sequence there is a first element but not a last one, and each element has an immediate successor and an immediate predecessor, except the first one, which has no predecessor. So, assuming the set of the natural numbers exist as a complete infinite totality (Hypothesis of the Actual Infinity subsumed into the Axiom of Infinity) means that any ω-ordered sequence can also exist as a complete infinite totality, despite the fact that no last element completes the sequence. Recall that a complete totality is a set defined by comprehension in which every element that should be in the set, is in the set (Definition 3, page 11).

An ω^*-ordered sequence is one in which there exists a last element but not a first one, and each element has an immediate predecessor and an immediate successor, except the last one that has no successor. Since there is not a first element these sequences are non-well-ordered. From the same infinitist perspective, ω^*-ordered sequences are also complete infinite totalities, in spite of the fact that there is not a first element to begin with. The *increasing* sequence of negative integers, $\mathbb{Z}^* = \ldots, -3, -2, -1$, is an example of ω^*-ordered sequence.

That said, let us consider a point particle P moving through the X axis (of a Cartesian coordinate system) from the point -1 to the point 2 at a constant finite velocity v (Figure 20.1). Assume P is in the point 0 just at the precise instant t_0. At instant $t_1 = t_0 + 1/v$ it will be exactly in the point 1. Consider now the following ω^*-ordered sequence of Z*-points $\langle z_i^* \rangle$ within the real interval $(0, 1)$, defined by [358]:

$$z_{n*}^* = \frac{1}{2^n}, \ \forall n \in \mathbb{N} \tag{1}$$

20.2 Introductory definitions

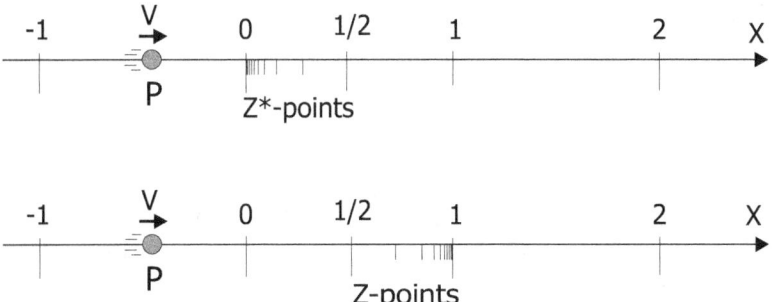

Figure 20.1 – Z^*-points and Z-points.

where z_{n*}^* stands for the last but $n-1$ element of the ω^*-ordered sequence $\langle z_i^* \rangle$ of Z^*-points. Consider also the sequence of Z-points $\langle z_i \rangle$ within the real interval $(0,1)$ defined by:

$$z_n = \frac{2^n - 1}{2^n}, \; \forall n \in \mathbb{N} \tag{2}$$

Although the points of the X axis are densely ordered (between any two of its points infinitely many other points do exist), Z^*-points and Z-points are not. Between any two successive Z^*-points $z_{(n+1)*}^*$, z_{n*}^* there is no other Z^*-point (ω^*-discontinuity), and a distance greater than zero $z_{n*}^* - z_{(n+1)*}^* > 0$ always exists. Because of ω^*-discontinuity, Z^*-points can only be traversed (by a point object as P) in a successive way, one at a time, one after the other, and in such a way that between any two successive Z^*-points, a distance greater than zero $z_{n*}^* - z_{(n+1)*}^* > 0$ must always be traversed. The traversal will take a time greater than zero if it is traversed at a finite velocity. The same applies to Z-points, which exhibit ω-discontinuity.

As P passes over the points of the closed real interval $[0,1]$ of the X axis, it must traverse the successive Z^*-points and the successive Z-points. It makes no sense to wonder about the instant at which P begins to traverse the successive Z^*-points because there is not a first Z^*-point to be traversed. The same can be said on the instant at which P ends to traverse the Z-points, in this case because there is not a last Z-point to be traversed. For this reason, we will focus our attention on the number of Z^*-points P has already traversed and on the number of Z-points it must still traverse at any instant t within the closed real interval $[t_o, t_1]$.

In this sense, and being t any instant within $[t_o, t_1]$, let $Z^*(t)$ be the number of Z^*-points P has traversed just at instant t. And let $Z(t)$ be the number of Z-points to be traversed by P at the instant t. The discussion that follows examines the evolution of $Z^*(t)$ and $Z(t)$ as P moves from the point 0 to the point 1. Both discussions are formalized versions of Zeno Dichotomy II and I respectively. See, for instance, [52, 53, 359, 312, 165, 363, 71, 236].

The strategy of pairing off the Z*-points (or the Z-points) with the successive instants of a strictly increasing infinite sequence of instants was firstly used (in a broad sense) by Aristotle [16, Books-III-VI] when trying to solve Zeno dichotomies. Although Aristotle ended up by rejecting his original strategy, it is still the preferred one to discuss on both paradoxes. As we will see, however, the discontinuity and separation of Z*-points and Z-points leads to a conflicting conclusion.

20.3 Zeno Dichotomy II

P3 Let us begin by analyzing the way P passes over the Z*-points. Since the sequence of Z*-points is ω^*-ordered, its first point does not exist, and consequently its first n points, for any finite number n, do not exist either. Thus, and taking into account that P is in the point 0 at t_0 and in the point 1 at t_1, it holds:

$$\forall t \in [t_0, t_1] \begin{cases} t = t_0: & Z^*(t) = 0 \\ t > t_0: & Z^*(t) = \aleph_o \end{cases} \quad (3)$$

According to (3), no instant t exists within $[t_0, t_1]$ at which $Z^*(t) = n$, whatever be the finite number n, otherwise there would exist the impossible first n elements of an ω^*-ordered sequence. Notice $Z^*(t)$ is well defined in the whole interval $[t_0, t_1]$. Thus, equation (3) represents a dichotomy, ω^*-dichotomy: $Z^*(t)$ can only take two values along the whole closed interval $[t_0, t_1]$: 0 and $\aleph_o$. □

In agreement with P3 and regarding the number of traversed Z*-points, P can only have two successive states: the state $P^*(0)$ at which it has traversed zero Z*-points, and the state $P^*(\aleph_o)$ at which it has traversed aleph-null Z*-points. The number of traversed Z*-points change directly from zero to $\aleph_o$ (ω^*-dichotomy), without finite intermediate states at which P has traversed only a finite number of Z*-points.

P4 Taking into account the ω^*-discontinuity of Z*-points and the fact that between any two successive Z*-points a distance greater than zero always exists, to traverse two successive Z*-points $z^*_{(n+1)*}, z^*_{n*}$, whatsoever they be, means to traverse a distance greater than zero:

$$z^*_{n*} - z^*_{(n+1)*} > 0, \forall n \in \mathbb{N} \quad (4)$$

In consequence, to traverse $\aleph_o$ of such successive Z*-points in the same direction means to traverse a distance greater than zero. And to traverse a distance greater than zero at the finite velocity v of P means the traversal has to last a time greater than zero. □

20.3 Zeno Dichotomy II

Although it is impossible to calculate neither the exact duration of the transition $P^*(0) \to P^*(\aleph_o)$ nor the distance P must traverse while performing such a transition (there is neither a first instant nor a first point at which the transition begins), we have proved in P4 that, indeterminable as they might be, that duration and that distance must be greater than zero. It will now be proved they cannot be greater than zero.

P5 Let d be any real number greater than zero and consider the real interval $(0, d)$. According to the ω^*-dichotomy (4), at any point x within $(0, d)$ our point-particle P have already traversed $\aleph_o$ Z^*-points. In consequence the distance P must traverse while performing the transition $P^*(0) \to P^*(\aleph_o)$ is less than d. And since d is any real number greater than zero, we must conclude the distance P must traverse while performing the transition $P^*(0) \to P^*(\aleph_o)$ is less than any real number greater than zero. □

So then, according to P4, the distance P must traverse while performing the transition $P^*(0) \to P^*(\aleph_o)$ is greater than zero. And according to P5 that distance must be less than any number greater than zero. But there is no real number greater than zero and less than any real number greater than zero. So, it is impossible for the distance P must traverse while performing the transition $P^*(0) \to P^*(\aleph_o)$ to be greater than zero. The same conclusion, and for the same reasons, applies to the time elapsed while performing the transition $P^*(0) \to P^*(\aleph_o)$.

In line with P4 and P5, the point particle P needs to traverse a distance greater than zero for a time greater than zero to perform the transition $P^*(0) \to P^*(\aleph_o)$, but neither that distance nor that time can be greater than zero. Note this is not a question of indeterminacy but of impossibility. If it were a question of indeterminacy there would exist a nonempty set of possible solutions, although we could not determine which of them is the correct one. In our case the set of possible solutions is the empty set, because the set of the real numbers greater than zero and less than any real number greater than zero is the empty set. In short:

A) According to the actual infinity hypothesis, the transition $P^*(0) \to P^*(\aleph_o)$ takes place.

B) The transition $P^*(0) \to P^*(\aleph_o)$ can only take place along a distance and a time greater than zero, because of the ω^*-discontinuity and to the distance greater than zero that P must traverse at its finite velocity v.

C) The transition $P^*(0) \to P^*(\aleph_o)$ cannot take place along a distance and a time greater than zero, because of the ω^*-dichotomy, and because no real number greater than zero is less than all real numbers greater than zero.

D) Zeno Dichotomy II is, therefore, a contradiction derived from ω^*-order.

20.4 Zeno Dichotomy I

P6 We will now examine the way P traverses the Z-points between the point 0 and the point 1. Being $Z(t)$ the number of Z-points to be traversed by P at the precise instant t in $[t_0, t_1]$, that number can only take two values: $\aleph_o$ and 0. In fact, assume that at any instant t within $[t_0, t_1]$ the number of Z-points to be traversed by P is a finite number $n > 0$. This would imply the impossible existence of the last n points of an ω-ordered sequence of points. Thus, we have a new dichotomy that can be expressed as follows:

$$\forall t \in [t_0, t_1] \begin{cases} t < t_1 : & Z(t) = \aleph_o \\ t = t_1 : & Z(t) = 0 \end{cases} \tag{5}$$

Therefore, no instant t exists in $[t_0, t_1]$ at which $Z(t) = n$, whatever be the finite number n. Notice $Z(t)$ is well defined in the whole interval $[t_0, t_1]$. Thus, equation (5) expresses a new dichotomy, ω-dichotomy: $Z(t)$ can only take two values: $\aleph_o$ and 0. □

In accord with P6 and regarding the number of Z-points to be traversed, P can only have two successive states: the state $P(\aleph_o)$ at which that number is $\aleph_o$, and the state $P(0)$ at which that number is 0. The number of Z-points to be traversed by P decreases directly from $\aleph_o$ to 0, without finite intermediate states at which it has to traverse only a finite number of Z-points.

P7 Taking into account the ω-discontinuity of Z-points and the fact that between any two successive Z-points a distance greater than zero always exists, to traverse two successive Z-points, whatsoever they be, means to traverse a distance greater than zero:

$$z_{n+1} - z_n > 0, \forall n \in \mathbb{N} \tag{6}$$

In consequence, to traverse $\aleph_o$ of such successive Z-points in the same direction means to traverse a distance greater than zero. And to traverse a distance greater than zero at the finite velocity v of P means the traversal has to last a time greater than zero. □

Although it is impossible to calculate neither the exact duration of the transition $P(\aleph_o) \to P(0)$ nor the distance P must traverse while

20.5 Conclusion

performing such a transition (there is neither a last instant nor a last point at which the transition ends), we have proved in P7 that, indeterminable as they might be, that duration and that distance must be greater than zero. It will now be proved they cannot be greater than zero.

P8 Let τ be any real number greater than zero, and consider the real interval $(0, \tau)$. According to the ω-dichotomy (5), for any instant t within $(0, \tau)$ the number of Z-points that P must still traverse at the instant t is $\aleph_o$. In consequence, the time P needs to perform the transition $P(\aleph_o) \to P(0)$ is less than τ. And since τ is any real number greater than zero, we must conclude the time P needs to perform the transition $P(\aleph_o) \to P(0)$ is less than any real number greater than zero. □

So then, according to P7, the time P needs to perform the transition $P(\aleph_o) \to P(0)$ is greater than zero. And according to P8 that time must be less than any real number greater than zero. But there is no real number greater than zero and less than any real number greater than zero. So, it is impossible for the transition $P(\aleph_o) \to P(0)$ to last a time greater than zero. The same conclusion, and for the same reasons, applies to the distance P must traverse while performing the transition $P(\aleph_o) \to P(0)$.

In line with P7 and P8, P needs to traverse a distance greater than zero for a time greater than zero to perform the transition $P(\aleph_o) \to P(0)$, but neither that distance nor that time can be greater than zero. Note this is not a question of indeterminacy but of impossibility. If it were a question of indeterminacy there would exist a nonempty set of possible solutions, although we could not determine which of them is the correct one. In our case the set of possible solutions is the empty set because the set the of real numbers greater than zero and less than any real number greater than zero is, in fact, the empty set.

In short:

a) According to the actual infinity hypothesis, the transition $P(\aleph_o) \to P(0)$ takes place.

b) The transition $P(\aleph_o) \to P(0)$ can only take place along a distance and a time greater than zero, because of the ω-discontinuity and of the distance greater than zero P must traverse at its finite velocity v.

c) The transition $P(\aleph_o) \to P(0)$ cannot take place along a distance and a time greater than zero because of the ω-dichotomy, and because no real number greater than zero is less than all real numbers greater than zero.

d) Zeno Dichotomy I is, therefore, a contradiction derived from ω-order.

20.5 Conclusion

According to the Hypothesis of the Actual Infinity, the set of Z-points and the set of Z*-points do exist as complete totalities. Therefore the transitions $P^*(0) \to P^*(\aleph_o)$ and $P(\aleph_o) \to P(0)$ take place while P moves from the point 0 to the point 1. Now then, the transitions $P^*(0) \to P^*(\aleph_o)$ and $P(\aleph_o) \to P(0)$ can only take place along a distance and a time greater than zero. The problem is that they cannot take place along a distance and a time greater than zero because that time and that distance is less than any real number greater than zero, and no real number greater than zero and less than any real number greater than zero do exist.

The above contradictions are direct consequences of assuming that ω-ordered and ω^*-ordered sets, as the sets of Z-points and of Z*-points, exist as complete infinite totalities (Definition 3, page 11), which in turn is a consequence of assuming the existence of all finite natural numbers as a complete totality [56, p. 103-104], which is the Hypothesis of the Actual Infinity subsumed into the Axiom of Infinity in modern set theories. An hypothesis that, consequently, should be put to the test.

21. Achilles, the tortoise and the speed of light

Abstract.-A photon replaces Achilles and the Tortoise in a variant of the famous Zeno paradox discussed in this paper. The discussion takes place in the infinitist scenario of the spacetime continuum. In this theoretical scenario, and thanks to the dense order of real numbers, the paradox becomes first a dichotomy and then a contradiction, of which there are only two coherent solutions: either the speed of light is NOT finite, or the Hypothesis of the Actual Infinity is NOT consistent. An inconsistency that would change almost everything in modern mathematics and then in the formal foundation of modern physics. The article ends by pointing to a new finite and discrete scenario for space and time in which this and many other contradictions and paradoxes dissolve immediately. But it also warns of the extraordinary difficulties that will be involved in exchanging the hegemonic infinitist paradigm of our days for an alternative based on the finitist discreteness of space and time.

Keywords: actual infinity, discrete solution of Zeno Contradiction, dense order, discrete space and time, finiteness of the speed of light, foundation of physics, power of the continuum, spacetime continuum, Zeno Contradiction, Zeno Dichotomy, Zeno Paradoxes.

21.1 Introduction

On Zeno paradoxes there is, for obvious reasons, an abundant literature, including the alleged solutions proposed in different areas of contemporary mathematics such as transfinite arithmetic, topology, measure theory, and internal set theory [139, 140, 374, 141, 143, 142, 238, 237], even solutions proposed within classical mechanics and quantum mechanics [226, 227]. Some of those solutions have been contested, and none of them convincingly explains where Zeno original arguments fail. In any case, the new discussions have given rise to new problems as challenging as the paradoxes themselves.

[269, 6, 288, 312, 165, 323].

The most famous of these paradoxes is undoubtedly that of Achilles and the Tortoise. In this article I discuss a formalized variant of that paradox in which instead of a tortoise and an athlete only a photon intervenes. The fastest object in the universe moving along a straight line on which certain points have been defined, which I will call AT points in memory of the paradox of Achilles and the Tortoise. The scenario of the new theoretical discussion, in which a photon moves through the AT-points, will be the contemporary infinitist mathematics built on the Hypothesis of the Actual Infinity subsumed in the Axiom of Infinity (Theorem 1, page 17). In that scenario, absolutely hegemonic in contemporary mathematics, any infinite set exists as a COMPLETE TOTALITY: a set defined by comprehension in which every element that should be in the set, is in the set (Definition 3, page 11). This will be the case of the densely ordered set of AT-points that our photon has to travel.

Once the AT-points are defined and the photon is fired, it will be proved that in this variant of Zeno paradox, and due to the dense order of AT-points, the paradox becomes a dichotomy: Zeno Dichotomy. And being zero the only non-negative real number that is less than all real numbers greater than zero, the dichotomy becomes a contradiction: Zeno Contradiction. Since the only possible causes of Zeno Contradiction are the finiteness of the speed of light and the Hypothesis of the Actual Infinity, it seems reasonable to assume that, taking into account the overwhelming empirical confirmation of the finiteness of the speed of light, the Hypothesis of the Actual Infinity subsumed in the Axiom of Infinity must be the formal cause of Zeno Contradiction. If so, we would have an inconsistent axiom in the infinitist foundation of modern mathematics (through set theory), and then of modern physics.

In the ordinary language of the primary and secondary physics literature, most physicists ignore this infinitist formalism and describe the physical world in a reasonably correct way, although using expressions that are not compatible with the foundational infinitism of their theories (some examples are given in Section 21.6). Practically all the strangeness derived from the actual infinity (as, for example, that a 10^{-32}mm line segment has the same number of points as the whole three-dimensional universe) is absent from the ordinary language with which physicists describe physical phenomena. For that reason they do not find it necessary to revise the infinitist formalism that underlies their theories, in practice they act as if those foundations were something else. But they are not something else, they are what they are, and it is convenient to revise them in order to ensure that physics is built on consistent fundamentals, which may not be the case if, for example, the formal proof offered in this article is well constructed and the Axiom of Infinity is inconsistent.

21.2 AT-points

Consider any straight line in the spacetime continuum. For example the X_o-axis of an inertial reference frame RF_o. Let us define as AT-points all points of the open interval $(0,1)$ of the X_o-axis of RF_o, in which any metric is defined, for example, the Euclidean metric in the SI, so that the considered interval corresponds, for example, to 1 meter: $(0m, 1m)$. Although, as usual, we will not indicate neither the unit of measurement nor the coordinates that are not involved in the discussion. The set of AT-points $(0,1)$ is an infinite non-numerable set of points containing exactly $2^{\aleph_o}$ points, the same number of points as the whole X_o-axis, or the entire observable three-dimensional universe (Dimension Problem proved by G. Cantor [19, 78, 324, 363, 133, 76, 59, 66]). Although $(0,1)$ is a complete totality, there does not exist a first AT-point following point 0, nor a last AT-point preceding point 1. Moreover, if p stands for any natural number, or even for the first infinite cardinal $\aleph_o$, there do not exist in $(0,1)$ the firsts p points following 0. Nor do the last p points preceding 1 exist. Between point 0 and ANY point within $(0,1)$ there are always the same infinite number of points, just $2^{\aleph_o}$ points. And the same applies to point 1. A very appropriate set, then, to discuss on Zeno paradoxes.

21.3 Zeno Dichotomy

Suppose that at the point $x = -1$ of the X_o-axis of RF_o, a photon γ (considered as a point particle[1]) is emitted along the X_o-axis, in the direction of its increasing values. Being c the speed of light, at the instant $1/c$ the photon γ will be exactly on the point 0, and has not yet begun to travel the AT-points of $(0,1)$. Let us represent this state of the photon by $\gamma(0)$, where 0 indicates that γ has traveled exactly 0 AT-points. At any instant after $1/c$ the photon γ will already be inside $(0,1)$, and taking into account that any subinterval of $(0,1)$ has the same number of points as the whole interval $(0,1)$, exactly $2^{\aleph_o}$ points, we can conclude that at any instant after $1/c$ the photon γ has already traveled $2^{\aleph_o}$ AT-points. Thus, from the point of view of the number of traveled AT-points, the photon γ can only have two states: the state $\gamma(0)$ at which it has traveled 0 AT-points, and the state $\gamma(2^{\aleph_o})$ at which it has already traveled $2^{\aleph_o}$ points. No intermediate state is possible for

---
[1]The following theoretical argument can be adapted immediately for the center of mass of any object moving with any finite uniform velocity.

the reason given at the end of the previous section: for any cardinal $p < 2^{\aleph_0}$, the first p points following the 0 point do not exist. Therefore, the infinitist dense order of real numbers turns Zeno Paradox into Zeno Dichotomy: the number of AT-points traveled by the photon γ can only be 0 or $2^{\aleph_0}$. Note that this is not an indeterminacy, but an impossibility: the set of positions of γ within the open real interval $(0,1)$ for which the photon γ has traversed a number of AT-points other than $2^{\aleph_0}$ is the empty set. This is infinitist mathematics!

21.4 Zeno Contradiction

Let τ now be any time interval greater than zero, and suppose the photon γ takes a time τ to change from the state $\gamma(0)$ to the state $\gamma(2^{\aleph_0})$. At any instant t in the interval τ such that $0 < t < \tau$, the photon γ is already within $(0,1)$; therefore it has already traveled $2^{\aleph_0}$ AT-points, and has already reached the state $\gamma(2^{\aleph_0})$. Therefore, in changing from $\gamma(0)$ to $\gamma(2^{\aleph_0})$, the photon γ takes less time than any time interval τ greater than zero. But there is only one non-negative (γ travels into the future) real number less than all real numbers greater than zero: just zero. We have to conclude, therefore, that γ takes zero time to change from the state $\gamma(0)$ in which it has not traveled any AT-point, to the state $\gamma(2^{\aleph_0})$ in which it has already traveled $2^{\aleph_0}$ AT-points. Now, since at point 0 the photon is in the state $\gamma(0)$, to change to the state $\gamma(2^{\aleph_0})$ the photon must necessarily change its position. And it has to do it in zero time, which is not possible with its finite velocity c. Therefore, the photon γ changes its position during a zero time, an does not change its position during a zero time. The fact that zero is the only non-negative real number less than any real number greater than zero causes Zeno Dichotomy to become Zeno Contradiction.

21.5 The Axiom of Infinity and Zeno Contradiction

Obviously, no contradiction can be admitted in a scientific theory, otherwise, anything could be proved within it. And to eliminate the contradiction in a theory it is necessary to analyze the cause of that contradiction, which in a correct argument cannot be other than the inconsistency of at least one of its foundational hypotheses. In the case of Zeno Contradiction, if the above argument is correct, there are only two hypotheses that could be the reason of the contradiction: either the speed of light is NOT finite; or the Hypothesis of the Actual Infinity is NOT consistent. The finiteness of the speed of light has an overwhelming empirical confirmation, and moreover it can be confirmed experimentally in an instant, never better said. On the contrary, the inconsistency of Hypothesis of the Actual Infinity would be confirmed by the more than forty proofs that the interested reader can find in [212, Link]. It seems then reasonable to propose the inconsistency of the Hypothesis of the Actual Infinity (subsumed in the Axiom of Infinity) as the cause of Zeno Contradiction.

21.6 A discrete solution to Zeno Contradiction

Zeno Contradiction, and many other contradictions and paradoxes, dissolve immediately in a discrete scenario with indivisible minimal units for space (qusits) and time (qutits). Discreteness already empirically confirmed and universally accepted in the cases of matter, energy and all kinds of charges, electrical and non-electrical. [213]. Indeed, in our case, the photon will be in qusit 0 in a certain qutit, and in the next contiguous qutit it will be in the next contiguous qusit in the direction of its motion, moving at the maximum speed of one qusit per qutit. Changing the infinitist dense order by the finitist contiguity (adjacency), Zeno Contradiction disappears.

But the solution is not so immediate because first we have to solve the old problem of change (particularized for changes of position), a problem that has been posed for 26 centuries and has not yet been resolved. In fact, it has been completely forgotten by physics, the science of change (!), the science of the regular succession of events [235, p. 98]. It can be proved that the problem of change has no solution in the spacetime continuum [197, Link], but it can be solved in a discrete universe functioning in a similar way to CALMs (Cellular Automata Like Models [212, Link]).

The proposed discrete scenario may seem novel and extravagant, but it is actually very old: the early pre-Socratics already considered points as indivisible units with a non-zero extent [231]. It is a pity that soon after they discovered the irrational numbers and with them the impossibility of non-zero extent points. And in the ninth and tenth centuries, the Arab philosophical and theological current known as Kalam developed a discrete cosmology that denied irrational numbers, and in which, matter, space and time were constituted by minimal indivisible units greater than zero; and motion had to occur in leaps and bounds separated by discrete units of time; the fewer units of time, the faster the motion [176, p. 62-68].

But, discrete models for space and time have at least two major drawbacks. The first is our sensory perception of the physical world as a continuous world. Although it is a deceptive perception because the human brain takes a certain amount of time (≈ 13 milliseconds [282]) to process each image (the base of the well known α, β, γ and δ movements, and of ϕ-phenomenon [102]). Therefore, it can process only a finite number of images per unit of time, although that time is so short that we perceive the discontinuous succession of those images as if it were a continuous succession, just as we perceive the succession of frames in a movie. So, this first drawback is not really a drawback but a suggestive warning that if the physical world were discrete, discontinuous, with sufficiently short qutits (Planck time?) we would perceive it as continuous.

The second drawback is much more difficult to overcome. It is the existence of (absolutely) hegemonic streams of thought that leave few options for dissent. The mathematical infinitist stream of our days is one of them. It has become, moreover, the mathematical basis on which physics is formally founded. A symptom that this infinitist foundation of physics is not the most appropriate is the fact that this infinitist formalism is never consistently reflected in the ordinary language of physicists. Indeed, in many issues the ordinary language of physicists is incompatible with the infinitist formalism that underlies physical theories. The origin of this serious incoherence between what physicists say and what they should say according to the mathematical foundations of their theories, is precisely one of those foundational items: the Hypothesis of the Actual Infinity with its sequel of densely ordered continuums. Let me at this point quote the words of the renowned philosopher of physics T. Maudlin [234, p. xiv]:

> Unfortunately, physics has become infected with very low standards of clarity and precision on foundational questions, and physicists have become accustomed (and even encouraged) to just "shut up and calculate," to consciously refrain from asking for a clear understanding of the ontological import of their theories.

This is an untenable situation that physicists should consider because, among other things, it perpetuates the foundational infinitism of physical theories in exchange for a schizophrenic use of ordinary language that, on the one hand, describes physical phenomena reasonably well and, on the other, is incompatible with that foundational infinitism. Indeed, in the primary and secondary literature of physics we can find thousands of expressions such as:

... points are small rectangles of infinitesimal extent ...

... particles small enough to be considered as points ...

... it propagates through the adjacent points ...

... through each of the contiguous points ...

... in the immediately next instant ...

... at each successive point ...

21.6 A discrete solution to Zeno Contradiction

... it is not only possible but absolutely certain that points will gradually coalesce.

... the more spacetime points there are in a region ...

... to an infinitesimal spacetime point ...

etc. etc.

which are incompatible with the infinitist foundation of physical theories, because the spacetime points have a null extent [213], are densely ordered, they cannot be contiguous, and the same number of them exist in any region, linear or not, of the spacetime continuum.

But, on the other hand, the infinitist formalism that underlies, for example, the relativistic spacetime continuum is so widely accepted that reputable philosophers of physics can write [234, p. xiii]:

... all there is to the physical world, at a fundamental level, is accounted for by the theory of space-time and the theory of matter.

But if the Hypothesis of the Actual Infinity is inconsistent, the spacetime continuum (our present and unique theory of space-time) will also be inconsistent, so that both space and time will have to be "discretized." And the same would have to be done with those theories that, as special relativity, have been built on the basis of the space-time continuum. In this case all the inertial relativistic deformations of space and time would disappear. Deformations which, on the other hand, could be only apparent, as apparent as the refractive deformations which, however empirically confirmed Snell's Law may be, are not real. indexSnell's Law

In any case, I will end this work with an urgent question: What would have to happen, and how could one collaborate, for contemporary science to begin to consider the possibility that the Hypothesis of the Actual Infinity is inconsistent? An inconsistency that would be anything but irrelevant to science, philosophy, and even theology.

22. Infinity, physics and language

In the primary and secondary literature of physics, an ordinary language is used that is not compatible with some of the mathematical assumptions of physical theories, nor with most of the equations with which these theories are constructed. I denounce here, with numerous examples, this logical abuse between the formal and the ordinary language of physics. And the question arises what would happen if physicists forced themselves to maintain the necessary consistency between what they say and what they should say according to the mathematical (infinitist) foundations of their theories.

22.1 Introduction

The purpose of this chapter is not to expose the limitations of ordinary language with respect to the mathematical language of physical theories. The purpose of this chapter is TO DENOUNCE THE FORMAL INCONSISTENCY of the ordinary language of contemporary physics with respect to its infinitist mathematical foundations, as a consequence of the assumption of the Axiom of Infinity (the existence of an actually infinite set as a complete ordered totality, even though there is no last element to complete the ordered totality). The chapter also recalls the problems arising from the potentially infinite regress of arguments, definitions, and causes, which can only be resolved by admitting the necessity of indemonstrable statements, primitive concepts, and arbitrary first causes that cannot be explained in terms of other causes, all of which set severe limits to human knowledge. Limits to which science (not only physics) does not pay the necessary attention, as if they did not exist. For example, no science has established what its primitive concepts are.

In the primary and secondary physics literature, the reader can find thousands of sentences with the same errors as the sentences cited in this chapter. Therefore, it does not seem fair to name the authors of the erroneous sentences contained in this chapter: it would be like singling out some authors and not others for the same serious errors. I invite the reader to briefly review what he has just read and ask himself how we have arrived at this situation. For those who are interested in the history of science, I suggest to study this kind of errors in the most important authors of physics of the 20th and 21st centuries. And I repeat: the errors denounced here are not those that could arise from the limitations of ordinary language to express mathematical language [60], but real contradictions between the two languages that can be avoided either by changing the corresponding ordinary expressions or by replacing the current infinite and continuous paradigm by a finite and discrete one, which is, in my opinion, what should happen.

22.2 Infinity and ordinary language

Some of the requirements of the actual infinity subsumed in the Axiom of Infinity (Theorem 6 of the Axiom of Infinity, page 21) are incompatible with ordinary language descriptions of physical objects and phenomena. Contemporary physics resolves these incompatibilities by ignoring them. The immediate consequence is the perpetuation of these incompatibilities. Among the most common forms of this incompatibility between ordinary language and formal language in physics we find [208]:

1. Expressions of ordinary language that contradict the mathematical foundations of physical theories.

2. Expressions of ordinary language that are compatible with its contradictory use.

3. Omission of certain infinitist consequences that are difficult to fit into the physical world.

And, of course, one can ask oneself: What would happen if physicists forced themselves to maintain the necessary formal consistency between what they say and what they should say according to the mathematical foundations of their theories? What would happen if they said certain things that are never said but should be said? Would they end up discovering that their infinitist foundations are inappropriate? Would they end up discovering the discrete and real nature of space and time? Would they end up discovering the inconsistency of actual infinity? An affirmative answer to the last three questions would mean that it is not necessary to change the ordinary language of physics, but only its infinitist foundations, which is exactly what this book proposes.

22.3 As firm as a rock

According to the Theorem 6 of the Axiom of Infinity that was proved in Chapter 3, the infinity subsumed in the Axiom of Infinity is the actual infinity, not the potential infinity. As is well known, the axiom of infinity is one of the foundations of modern set theory and then of a good part of contemporary mathematics. This infinitist mathematics is practically the only mathematics of our time. It is worth recalling some quotations from G. Cantor (1845-1918), the great founder (together with R. Dedekind (1831-1916) and G. Frege (1848-1925)) of this omnipresent infinitism:

> ...in my opinion the absolute reality and legality of the natural numbers is much higher than that of the sensory world. This is so because of a unique and very simple reason, namely, that natural numbers exist in the highest degree of reality, both separately and collectively in their actual infinitude, in the form of eternal ideas in Intellectus Divinus. ([244]; reference and (Spanish) text in [114])

> ...I am only an instrument of a higher power, which will continue to work after me in the same way as it manifested itself thousands of years ago in Euclid and Archimedes ... ([58, pp 104-105])

> ...I cannot regards them [the atoms] as existent either in concept or in reality no matter how many useful things have up to a certain limit been accomplished by means of this fiction. ([57, p 78], English translation of [54])

> My theory stands as firm as a rock; every arrow directed against it will return quickly to its archer. How do I know this? Because I have studied it from all sides for many years; because I have examined all objections which have ever been made against the infinite numbers; *and above all because I have followed its roots, so to speak, to the first infallible cause of all created things.* [82, p. 283] (the italic is mine).

This rock-solid belief has spread to the vast majority of contemporary mathematicians and physicists, so that infinitism has become virtually the only stream of physical-mathematical thought. The finitist alternative is not even considered. And as is often the case with unique streams of thought, this one does not tolerate criticism. Dissent is scorned, insulted, and ostracized. Contemporary infinitism is not questioned. Nothing more anti-scientific can be imagined. And as a result, the following happens.

22.4 Points and instants of the spacetime continuum

As will be seen below, the points and instants of the spacetime continuum can only be of zero extension and densely ordered. However, contemporary physics expresses itself as if this were not the case, which is an unacceptable violation of the First Law of Logic: *Things are what they are, and they are not what they are not.* The following subsections highlight specific instances of this violation.

THE SIZE AND DENSE ORDER OF POINTS AND INSTANTS

As far as I know, point and instant are primitive concepts whose intuitive notion almost never agrees with their most important infinitist properties: their null extension and their dense order. In this subsection both statements are discussed, and examples of their inconsistent usage in the ordinary language of contemporary physics are given, but first let us prove the following:

Theorem 40 (of non-extensive points) *The points (instants) of the spacetime continuum have not extent (duration).*

Proof: Suppose that all points of the spacetime continuum have the same extent of δ meters in a given metric, being δ any real number greater than zero. Let AB and CD be the lengths of any two lineal intervals of that continuum such that, in the same metric, the first is less than the second:

$$AB < CD \tag{1}$$

On the other hand, since the number of points of AB and CD is the same, just $2^{\aleph_0}$, we can write:

$$AB = 2^{\aleph_0} \times \delta \text{ m} = 2^{\aleph_0} \text{ m} \tag{2}$$

$$CD = 2^{\aleph_0} \times \delta \text{ m} = 2^{\aleph_0} \text{ m} \tag{3}$$

$$\therefore AB = CD \tag{4}$$

which contradicts (1). Therefore, points cannot have an extent greater than zero. For the same reasons, instants cannot have a duration greater than zero. □

On the other hand, recall that an element of a strictly ordered set is between two given elements of that set if, in the ordering relation of the set, that element is a successor of one of the given elements and a predecessor of the other. And that this set is densely ordered if between any two of its elements there exists at least one other different element. Recall also the densely ordered nature of the set of real numbers $\mathbb{R}$, of the rational numbers $\mathbb{Q}$, of the real line, and of all vector (cross or cartesian) products of $\mathbb{R}$ and $\mathbb{Q}$, such as the spacetime continuum $\mathbb{R}^4$. All continuous lines (straight or not) of the spacetime continuum

22.4 Points and instants of the spacetime continuum

are also densely ordered, because otherwise there would be at least two different points (real numbers) between which no other point (real number) of the line exists, which of course violates the conditions of mathematical continuity. Furthermore, and by definition, in densely ordered sets their elements cannot be adjacent (contiguous). This is the case of the points and instants of the spacetime continuum (as Aristotle would say [16, Book V, 228a], points cannot touch each other).

In this chapter we assume the Axiom of Infinity, and therefore we will not consider any of the demonstrations of its inconsistency already given in previous chapters and in [212]. However, some rather unpleasant questions arise. For example, it is impossible to describe a simple continuous motion along a continuous line: Suppose a point particle P moves with uniform velocity along the X-axis of a reference frame in $\mathbb{R}^3$, in the direction of the ascent of this axis, and so that at time t_o it is exactly at the origin (point 0) of this axis, symbolically $P(0)$. If we wanted to describe the motion of P along the interval $(0, 1)$, we could not do so, because there is no first point where P begins its path along this interval, there is no position following $P(0)$. At every moment after t_o the particle P has already passed through $2^{\aleph_0}$ points of $(0, 1)$, symbolically $P(2^{\aleph_0})$. The particle must go directly from $P(0)$ to $P(2^{\aleph_0})$, with no intermediate positions where it has passed only a finite or countable infinite number of points in the interval $(0, 1)$. This is what happens if we assume the existence of an ordered list of elements without a first element that starts the list. Obviously, this conclusion is closely related to Zeno paradoxes (see Chapter 20).

The above elementary results about the spacetime continuum are invariably violated in the ordinary language that physics uses today to describe the physical world. In any of his books and articles, or on the Internet, the reader can find hundreds of expressions such as the following:

... mass points of a true *infinitesimal* size ...

... it propagates to the *adjacent* point ...

... varies *from point to point* ...

... through each of the *contiguous* points ...

... it is distributed *point to point* ...

... a *collision* of particles occurring *at the same point* in space ...

... all the energy in the universe contained *in one point* ...

... in the *next* instant ... in the *previous* instant ...

... to the *next* point ... from the *previous* point ...

... their spacing increases *uniformly* ...

... a point moving *uniformly* along a straight line ...

... *continuous motion* of a point ...

... which *moves uniformly* along its trajectory ...

... the universe began at an *infinitesimal point* ...

... just in the *next instant after* the Big-Bang ...

... etc.

Similar expressions were also not uncommon in classical authors before G. Cantor such as B. Cavalieri, I. Newton or G.W. Leibniz:

... area as formed by a number of very small rectangles or "points".

... points as small rectangles of infinitesimal extension ...

... particles small enough to be considered as points.

... etc.

But in these cases they are justified because there was not yet a mathematical theory of infinity. Infinity was still "a form of speech" as the Prince of Mathematics C. F. Gauss said [127, Vol. II, p. 268]). But once the Axiom of Infinity was accepted, infinity ceased to be a form of speech and became a formal requirement of the corresponding mathematical and physical theories. Nevertheless, modern physicists continue to express themselves in unacceptable ordinary terms similar to those above.

The Dimension Problem

Once the existence of infinite sets is accepted (Axiom of Infinity), and with the help of one-to-one correspondences and ellipsis (...), it is possible to demonstrate very shocking results (for some authors it seems that the extravagance and the shocking add value to scientific theories). For example, it is easy to show that any linear segment, say of Planck's length (1.6×10^{-32} millimeters), has the same number of points as the entire observable three-dimensional universe. Or that any interval of time, e.g. the duration of Planck time (5.39×10^{-44} s), has the same number of instants as the entire history of the observable universe (more than 17.8 billion years).

As is well known, this is the so called Dimension Problem proved by Cantor [19, 78, 324, 363, 133, 76, 59, 66]. It is only necessary to define the appropriate one-to-one correspondence between the two sets and to make appropriate use of ellipsis (the magic wands of infinitism) to prove it. The physical consequences of this result are as unacceptable to physics as they are little discussed among physicists. For example, that a linear segment of Planck length contains as many point entities (point charges, point masses, point virtual particles, etc.) as the entire

22.4 Points and instants of the spacetime continuum

universe. This is one of those topics that is never discussed in contemporary physics, lest someone happen to make a critical analysis of the actual infinity.

THE ZERO POINT ENERGY

The so-called Zero-Point Energy is a consequence of the well-known Heisenberg Uncertainty Principle. It can be described in several ways, one of which defines it as *the energy associated with empty space.* The physical problem arises for all those contemporary physicists (the vast majority) for whom neither space nor time are real, they are mere fictions useful for representing and discussing relative spatial and temporal positions of material objects (according to them, absolute positions do not exist). If space were real, all material objects would move **THROUGH** the same real space, and both space and motion would be absolute, i.e. anathema to contemporary physics. In consequence, on the one hand, we can read things like:

1. Spacetime is not a fabric, it is not material ...
2. Space is just an illusion, time is just an illusion ...
3. ... space is a good way of simplifying the concept of general relativity to the public.
4. ... space and time, like society, are in the end also empty conceptions ...
5. For me it is absurd to attribute physical properties to "space."
6. ... space and time lose the last remnant of physical reality ...
7. etc.

And at the same time and in the same physical literature we find:

1. ... virtual point particles are continuously being created and annihilated ...
2. ... any finite region of empty space is filled with energy ...
3. Every thing came from nothing ...
4. The universe began with a fluctuation out of nothingness ...
5. The universe could have emerged out of the vacuum ...
6. ... space grows between material objects ...
7. ... a stage at which space expanded exponentially ...
8. ... space and time emerged out of a quantum buble ...
9. etc.

Being consistent, these expressions consider the energy of something that does not exist; births of things that do not exist; growths and expansions of things that do not exist; fluctuations of things that do not exist; and so on. But that which does not exist contains nothing, cannot be born, cannot expand, cannot grow, cannot fluctuate, precisely because it does not exist.

There is a finite and discrete alternative in which ordinary language is expressed in a manner consistent with formal language. Moreover, using the experimental detection of gravitational waves, it can be shown that space is a real physical object; and using the inconsistency of actual infinity, it can also be proved that this real object is formed by a finite number of indivisible and contiguous units of non-null extension, which contains and possibly generates all the material objects of the universe, to which it offers no resistance in their motions and makes their mutual interactions possible [214].

22.4.1 ω-Asymmetry

The reader interested in the Hypothesis of the Actual Infinity (an issue of paramount importance to physics) can find more than 40 different proofs of its formal inconsistency in [212]. Here we will disregard all of them and assume the formal consistency of that hypothesis, as contemporary infinitists do. Under these conditions, let us recall that ω is the first infinite ordinal number. It is, for example, the ordinal of the set $\mathbb{N}$ of natural numbers in their natural order of precedence and considered as a complete totality (Definition 3, 11) in which there is no last element that completes the totality. It is also the ordinal of any sequence of elements indexed by $\mathbb{N}$, for example $\langle x_i \rangle = x_1, x_2, x_3, \ldots$.

Let then AB be a straight line of 90 billion light years (perhaps the diameter of the visible universe). Suppose that AB is partitioned into an ω-ordered sequence of contiguous parts by means of an ω-ordered sequence of successive points $\langle P_i \rangle = P_1, P_2, P_3, \ldots$, in which the successive points $P_1, P_2, P_3 \ldots$ define the successive parts of the partition $AP_1, P_1P_2, P_2P_3 \ldots$, thus being $P_1 < P_2 < P_3 \ldots$, where $P_n < P_{n+1}$ means that point P_n (for all n in $\mathbb{N}$) precedes point P_{n+1} in the direction from A to B. Naturally the sequence $\langle P_i \rangle$ must be convergent, being its limit the point B of the straight line AB. In these conditions, let C now be a point in the interior of AB arbitrarily close to the endpoint B, for example a point C such that CB has the Planck length (1.616199×10^{-32} millimeters). Since C is not the limit of $\langle P_i \rangle$, there will exist in CB at least one point P_v of $\langle P_i \rangle$. Consequently, the segment AC, of ≈ 9.9 billion light-years contains only a finite number (less than v) of the partition points, and hence a finite number of parts of the partition. Consequently, the remaining segment CB of Planck length must contain an infinite number of such parts.

In short, practically all parts of the partition fall within a ridiculously small final segment CB, while only a finite number, and therefore ridiculously small compared to the infinite number of parts within CB, fall within AC no matter how many billions of light-years long AC

22.4 Points and instants of the spacetime continuum

is, and no matter how tiny the segment CB is. There is no way to undo this indescribable asymmetry. It will always occur, and for the same reason, in all ω-ordered partitions. An asymmetry that is never talked about, probably because it is so ugly; or because we have failed to recognize its inevitable and crazy dissymmetry; or simply because no one has discovered it yet.

ON THE UNIVERSAL CONSTANTS

Natural numbers are all finite, but they can be enormous. Leaving aside hyperfactorial and n-hyperfactorial numbers [212], we can imagine numbers written in standard text at 5 mm per digit that are longer than the diameter of the observable universe (90 billion light-years). They would be finite natural numbers because they would have a finite number of digits (the integer obtained by dividing 9.9 billion light-years expressed in millimeters by 5), which is obviously a finite number. Traveling at the speed of light, it would take us 90,000 million years to go through all of its digits. Now imagine that we put a zero and a decimal point in front of the first digit of one of these numbers. We would have a rational number q with a finite number of digits after the decimal point. We will use q as a reference in the next brief discussion.

Some physical constants are defined by mathematical expressions containing irrational numbers, such as π, $\sqrt{2}$, e, etc. These numbers have an infinite number of (ω-ordered) decimal places. Therefore, their full written expressions would contain a gigantic number of decimal places compared to our q, whose written expression is larger than the diameter of the visible universe. If universal constants with q decimal places were necessary to describe the workings of the universe, we would say that the universe is quite grotesque: a few decimal places should be enough. But much worse would be the case if universal constants with an infinite number of decimal places were necessary, as happens with universal constants defined with the intervention of irrational numbers interpreted in the infinitist continuum scenario.

It is therefore shocking to see the jubilation with which measurements with an accuracy of a few tens of decimal places are celebrated:

1. A distance of less than 10^{-17} m has been measured.
2. The critical density has important consequences if its 27th decimal place is altered.
3. The shortest measured time is approximately 10^{-26} seconds.
4. The stretching/shortening of space caused by gravitational waves is of order 10^{-21} m.
5. A very approximate value of the Planck time is $5.39106(32) - 10^{-44}$ seconds.

6. The radius of a proton is $8.41235641483227 - 10^{-16} m$.

7. etc.

All of these, and many others, are very important and significant from an operational point of view, but insignificant from the fundamental point of view of the infinitist paradigm of the continuum.

22.5 The problem of change

I have dealt with the problem of change on several occasions, to which I refer the reader (see Chapter 18). Here I will only mention some important aspects related to the content of the chapter. In this case related to the lack of attention paid by physicists to a fundamental problem, without the solution of which it will be impossible to explain the physical world. Let us then recall that:

1. The problem of change was posed 26 centuries ago.
2. It has not yet been solved.
3. It is completely forgotten by physics.
4. Physics has not been able to explain how a simple change of position occurs.
5. Change is inconsistent in the spacetime continuum (Theorem 8, page 23).
6. Change could find a solution in a discrete and finite model of the universe.
7. The physical world cannot be explained in fundamental terms until the problem of change is solved.

22.6 Anything but discrete

In contemporary physics literature one finds such strange demonstrations as:

> The reason why space cannot be discrete is because it would go against the Principle of Relativity.

The demonstration given by its author (a university professor of physics and astrophysics, and a science popularizer) is based on the FitzGerald-Lorentz contraction, which he considers to be real, not apparent, a matter on which there is clear disagreement among physicists. (see, for instance, the example of the elastic cord in Chapter 15 and in [208]). In any case, that contraction is a formal consequence of the Lorentz Transformation, which in turn is a formal consequence of the Principle of Relativity and the Principle of the Constancy of the Speed of Light. Whereas its author only implies the Principle of Relativity:

> The laws of physics have the same form in all inertial reference frames.

and considering that in relativistic physics all reference systems are defined in the spacetime continuum, the author of the sentence quoted above actually says:

> The reason why space cannot be discrete is because it would go against the spacetime continuum.

An example of what happens when you confuse speed with bacon (a classic Spanish saying). The correct expression, and much more humble, would be:

> For those of us who assume that space is continuous, we assume that it is not discrete.

22.7 The Aristotelian infinite regress

The propositions that make up scientific theories say things about the world. But it is necessary to demonstrate the truth of what they say, because propositions do not themselves demonstrate the truth of what they say; they are not self-demonstrable; except for tautologies, they are not self-true. In some cases, their truth can be confirmed by observing the physical world and checking the correspondence between the proposition and the observed facts. But in this (inductive) way we can never be sure of the universal truth of the proposition. In other cases, even this local empirical confirmation is not possible.

Thus, to confirm the universal truth of a proposition, we would have to use a second proposition of which the first is a logical consequence. But this second proposition is not self-demonstrable either, and we are left with the same situation as with the first. This opens up a potentially infinite regress of propositions (denounced by Aristotle [12, Book 1, p. 1-16; 71a-76a]), which has no other solution than to admit a first proposition whose truth we assume without proof. These propositions whose universal truth we accept without proof are the axioms in the case of formal sciences and the fundamental principles and laws in the case of experimental sciences.

The same situation occurs with concepts, which are also not self-defining. To define a concept, we must use at least one other different concept. And the same is true of that other different concept. As with propositions, we cannot avoid the potentially infinite regress of definitions. Again, the only solution is to allow the use of undefined concepts. These are the primitive concepts that are inevitable in all areas of human knowledge. Moreover, these primitive concepts are the most basic ones in any discipline: set, point, time, force, etc.

Very occasionally, we have the possibility to define a concept considered primitive until then, as is the case of the concept of space, whose definition is now possible thanks to the detection of gravitational waves [214]. The need for axioms and primitive concepts considerably limits human knowledge, although this limitation, much more serious and less controversial than certain theorems that are much talked about [201], is usually not mentioned. As B. Russell would say, science is that of which we will never know what we are talking about and whether what we are talking about is true [306, p. 959] [309]. Although some scientists speak so convincingly and emphatically that they seem to know what they are talking about.

There is a third area of knowledge in which a potentially infinite regress is also involved: causes. It has just been pointed out that propositions do not prove themselves, nor do concepts define themselves. It happens that, for the same reasons as propositions and concepts, physical objects and physical phenomena are not the cause of themselves. But if the first two territories are rarely visited by modern science, this third territory is anathema to modern science, which, with its excessive ego and its spirit of revenge, excommunicates and ostracizes dissidents.

And the fact is that this third territory is much more compromising than the other two: if objects cannot originate from themselves, they must originate from a cause. A cause that can only be external to the object itself. And the universe being an object... What follows I leave to the reader. But do not forget what was said on the page 122 of this book:

> What if the universe were eternal? Well, in that case its duration would be infinite, it would have, for example, an infinite number of seconds or any other arbitrary unit of time. That is, it would have an inconsistent duration (Theorem 4). What if the universe had arisen from a fluctuation of nothing? Well, then nothing would not be nothing, but something with the ability to fluctuate, and we would have to apply the Corollary 17 of the First Cause to that something with the ability to fluctuate. What if the present universe were a stage in a cyclic succession of universes being continuously created and destroyed? Well, in this case the number of cycles could only be finite (Theorem 4) and therefore there would be a first universe (Theorem 23 of the First Element) in the cyclic succession of universes to which the Corollary 17 of the First Cause could be applied.

22.8 Conclusion

There is a clear inconsistency between what physicists say about the physical world when they say it in ordinary language and when they say it in the infinitist formalism of their mathematical language. If they were to force themselves to put an end to this inconsistency, they might end up discovering that the assumed mathematical infinitism underlying their theories is not the most appropriate assumption. This is what these linguistic inconsistencies point to.

23. Gravitational Waves as Empirical Proofs of Space Reality

Abstract.-This article presents an unexpected formal consequence of the experimental detection of gravitational waves. It proves that space is not a theoretical fiction useful for discussing the relative positions of physical bodies, but a real physical object. It also proves that the physical space is not formed by a non-denumerable infinitude of inextensive points, but by a finite number of discrete, indivisible and contiguous units of non-zero extension. The demonstration is based both on the detectable reality of gravitational waves, and on the inconsistency of the Hypothesis of the Actual Infinity subsumed in the Axiom of Infinity, inconsistency that is very briefly demonstrated in Chapter 3 of this book. I invite you to examine that very short demonstration, and if you do not find it correct, you can stop reading this chapter right there.

Keywords: gravitational waves, spacetime continuum, dense order, discreteness, physical space, physical time, real objects versus useful fictions, fundamental physics, operational physics.

23.1 Introduction: Gravitational waves

Some precedents for the idea of gravitational waves may be found somewhat forcibly in the suggestion made by M. Faraday in 1847 that gravity might involve some kind of radiating phenomenon [107], and somewhat more explicitly in those made in 1870 by W.K. Clifford on the vibrations of space caused by mass [67, p. 158]:

> (1) That small portions of space are in fact of a nature analogous to little hills on a surface which is on the average flat; namely, that the ordinary laws of geometry are not valid in them.

> (2) That this property of being curved or distorted is continually being passed on from one portion of space to another after the manner of a wave.

(3) That this variation of the curvature of space is what really happens in that phenomenon which we call the motion of matter, whether ponderable or etherial.

(4) That in the physical world nothing else takes place but this variation, subject (possibly) to the law of continuity.

In June 1905, H. Poincaré also predicted the existence of gravitational waves [281, p. 1507]:

... I was first led to assume that gravitational propagation is not instantaneous, but takes place at the speed of light.

... in this case, we're talking about the position or speed at the instant when the gravitational wave left the body.

Gravitational waves also appear in 1916 as a deducible possibility of general relativity [90], and become an experimentally confirmed reality in 2015. The history of gravitational waves during those 100 years was full of lights and shadows, with more shadows than lights (Einstein himself was not sure of their existence, although he was sure of their interest [100]). But confidence in its existence was never completely lost, from the failed detection by J. Weber with his detector (electromechanical antenna) in 1969 [365, 366], to its definitive confirmation in 2015. In recent months, primordial gravitational waves possibly originated in the first events in the history of the universe have been detected. The detection has made headlines in print and broadcast media around the world.

An acceptable way to explain the nature of gravitational waves is to compare them with the well-known electromagnetic waves generated by interactions between electric charges: As the saying goes, gravitational waves are to gravitational interactions what electromagnetic waves, e.g. light, are to electromagnetic interactions. If interactions between electric charges produce electromagnetic waves, interactions between gravitational masses produce gravitational waves. The enormous difference between the intensities of the two interactions explains the enormous difficulties that have had to be overcome in order to detect gravitational waves experimentally: the gravitational interaction is 10^{30} times weaker than the electromagnetic interaction. For this reason, we can only hope, at least for the moment, to detect the gravitational waves generated by the gravitational interactions between the most massive objects in the universe, such as neutron stars or black holes.

Unlike the electromagnetic interaction, which can be both attractive and repulsive, the gravitational interaction is always attractive, so although much more weaker, its additivity extends it over enormous areas of the universe, which does not occur with the electromagnetic interaction due to the approximate cancellation between electromagnetic attractions and repulsions. What they do resemble is that both types of waves are transverse (they vibrate in directions perpendicular to the direction of their propagation), although gravitational waves are exclusively quadrupole waves.[1]. Both propagate at the same speed, the speed of light: 299792.458 Km/s.

Obviously, the gravitational interactions capable of generating detectable gravitational waves must be those of the highest intensity, which in turn originate from interactions between the most massive objects in the universe. Sources of gravitational waves could be, for example:

1. Supernova explosions.
2. Binary stars (revolving around each other).
3. Binary pulsars.
4. Irregular neutron stars rotation.
5. Neutron star collisions.
6. Rotating black holes.
7. Binary black holes in spiral approach.
8. Collision of black holes.

As usual in these cases, there is an abundant primary and secondary literature on gravitational waves, including that related to their experimental detection and to the latest discoveries on primordial gravitational waves, e.g. [1, 23, 40, 49, 50, 62, 100, 119, 146, 154, 250, 261, 299, 333, 372]. The interested reader can also visit on the Internet the sites of some important gravitational wave research projects:

> https://www.ligo.org
> http://www.geo600.org
> https://www.virgo-gw.eu

23.2 Space deformations

In addition to the corresponding mathematical equations, the gravitational deformations of space are usually explained in ordinary language with the help of some metaphors (almost always the same ones) and more or less simplifying drawings (also almost always the same). What is never done is to describe these deformations in ordinary language in terms consistent with the dense order of the points that supposedly constitute space. This dense order is that of the set $\mathbb{R}^3$ of all

---

[1] Like the sides of a square that contract and stretch alternately in the two orthogonal directions of its orthogonal sides

3-tuples of real numbers that model space. One of the most important consequences of this dense order is that it is impossible for two points in space to be contiguous, i.e. adjacent: between any two points in space there must always be the same infinite number ($2^{\aleph_0}$) of different points. A really uncomfortable infinitist requirement for the explanation of the physics of space and its deformations (local curvatures, extensions, vibrations).

Moreover, points have neither extension nor shape, which can be proved almost immediately (Theorem 15, page 34) and [213, p. 61]. And the extension of a non-denumerable number of densely ordered points of null extension is a mathematical indeterminable ($0 \times 2^{\aleph_0}$). So, it is an indeterminable extension. Whatever region of space we consider has the same number of points as the whole universe, and its extension will always be indeterminable if it is to be expressed as a consequence of the extension of its individual constituents (points). For that reason arbitrary metrics have to be defined without any relation to the inextensive points that make up space.

When all the peculiarities of points and dense order are taken into account, practically insurmountable difficulties appear in explaining in physical and consistent terms how a space formed by points can be deformed. For example, the followings:

1. Points cannot be deformed because they have no shape to deform. Therefore a space of points cannot be deformed by deforming its points.

2. Points have no extension:

    a) Therefore a space of points cannot contract by the contraction of its points: points have no extension to contract.

    b) Neither can the space be stretched by stretching its points: those points would cease to be inextensive, and therefore would cease to be points.

3. A space of points cannot be deformed by sliding its points: they would leave gaps of indeterminable extension and space would no longer be a continuum of points.

4. Furthermore, contacts between points would occur, which is impossible because the collided points would have to be contiguous, which is not possible because contacting points are impossible in a densely ordered set of points.

5. The deformation of a space of points cannot occur either by destruction of points or by creation of new points:

    a) In the first case, gaps of indeterminable extension would be created and the continuum would cease to be a continuum.

b) The second case is also impossible, since there would have to be previous holes for the new points, which is impossible in a continuum of points.

6. A deformed space would be indistinguishable from an non-deformed space: in any direction of space, whether or not deformed, there would be the same number of inextensive, formless, non-deformable and non-adjacent points: always $2^{\aleph_0}$ of such points.

Consequently, it seems impossible to deform a space constituted by a continuum of points without size, without form, without contiguity, and of which the same number would exist in any region of the universe, whatever its size, from a Planck volume ($l_p^3\ m^3$) to the whole three-dimensional universe.

It is formally objectionable, therefore, the ease and lack of infinitist consistency with which contemporary physicists speak in ordinary language of the deformations of the spacetime continuum of points and instants. It could be said that they make a schizophrenic use of language: consistent in their mathematical language, inconsistent in their ordinary language. I have the impression that by forcing physicists to express themselves in ordinary language consistent with the infinitism of their mathematical language, they would end up rejecting that mathematical infinitism.

23.3 Physical space is a real physical object

I agree with T. Maudlin that it is impossible to exaggerate the importance of explaining what space is for physics [233, p. 25]. Although I disagree with the majority opinion of contemporary physicists, according to which space is only an unreal instrument useful for describing the relative positions of physical objects (see Appendix 23 of this chapter). An opinion that will be seriously compromised by recent experimental confirmations of some of its physical properties. Indeed, the experimental detection of gravitational waves has an immediate formal consequence on the nature of physical space:

> Physical space can vibrate and be the transmitting medium of its own vibrations, which are of transverse, quadripolar type, of a great variety of frequencies and with a velocity of propagation through space itself of $299792458\ ms^{-1}$. This implies that the space must have the necessary physical properties to enable these vibratory and transmitting capabilities.

But a necessary condition for an object to have empirically detectable physical properties is that the object actually exists, for what does not exist cannot have empirically detectable physical properties. Vibrations of something that does not exist also do not exist, and therefore cannot be empirically detected. Consequently, if they are empirically detectable, they do exist. This simple argument (Modus Tollens) leads us to the conclusion that physical space is real.

Since gravitational waves are transmitted through the enormous empty spaces between material objects, we must conclude that space is a physical object, not a property of material objects. In addition, gravitational waves are vibrations of space that interact with the arms of the interferometers that detect them, modifying, even minimally, their length. So they can only be real vibrations of real stuff. We have, therefore, proved the following:

e-Theorem 1 (of the Physical Space) *Physical space is a real physical object with certain physical properties that can be tested and measured in experimental terms.*

Where an e-theorem is a statement whose veracity is supported by both empirical data (in this case the empirical detection of gravitational waves) and logical inferences.

To some readers, the above e-Theorem may seem superfluous. They may even think that its deduction and emphasis is unnecessary. But it turns out that for most contemporary physicists, neither space nor time are real. As noted above, for them they are mere fictions useful for explaining the RELATIVE spatial and chronological positions of the physical objects included in the observable universe. Therefore, after the above brief analysis of the immediate consequences that the detection of gravitational waves will have on the (real or fictitious) nature of physical space, it is worth taking a brief look at Appendix 23 of this chapter, which summarizes the position of some relevant physicists on the ontological nature of space.

23.4 Physical space is discrete

It is formally proved in this section that if physical space is real and formally consistent, then it must be discrete, i.e. it must consist of contiguous units of a non-zero extension, rather than inextensive and densely ordered points. The proof makes use of the inconsistency of the actual infinity, a formal result demonstrated by the author (more than 25 years ago) with more than 40 different proofs that the interested reader can find in [212].

23.4 Physical space is discrete

With a few exceptions, whom I thank for their support[2], the hegemonic infinitist stream in modern mathematics ignores all these proofs. One of them is reproduced is reproduced in Chapter 3 of this book. You can examine it a few minutes, and if it does not seem to be a correct argument, you can stop reading this article right there.

Let us prove that the real physical space cannot be continuous, but discrete, i.e. consisting of basic units of non-zero extension, hereafter called discrete units or qusits. Indeed, if space were a continuum of inextensive points, then it would be formally equal to the continuum R^3, which is a non-denumerable infinite set. And being non-denumerable it contains a non-denumerable infinitude of denumerable infinite proper subsets. We may therefore consider the following argument:

(a) If R^3 is consistent, then there exist consistent denumerable sets.

(b) If there exist consistent denumerable sets, then the Axiom of Infinity is consistent.

(c) The Axiom of Infinity is not consistent (Theorem 6 of the Inconsistent Axiom of Infinity and [212, p. 24-25])

(d) Therefore, the continuum R^3 is not consistent (Modus Tollens a-b-c)

(e) So, the continuum of points of physical space, which formally equals R^3, is inconsistent.

(f) Consequently the real physical space, if consistent, cannot be a continuum of points.

On the other hand, physical space cannot be constituted by an infinite number (denumerable or non-denumerable) of discrete units either, since it would be formally equivalent to an infinite set, and since infinite sets are inconsistent (Corollary1 of the Inconsistent Set, page 18 of this book, and [212]), so would be physical space. The finiteness of the number of discrete units of physical space has at least three consequences:

(g) As noted above, the extension of any infinite and densely ordered set of points with a null extension (as would be the case of the entire three-dimensional space of points) is a mathematical indetermination: $0 \times 2^{\aleph_0}$. On the other hand, a totality formed by a finite number of discrete units of null extension, has exactly a null extension, which is not the case of physical space. Therefore, the discrete units of physical space cannot have a null extension. So, they can only have an extension greater than zero.

(h) The discrete units of the physical space cannot be densely ordered,

---

[2] For mental health reasons (social anxiety disorder) I do not correspond with anyone, but I thank the messages.

because in that case the physical space would be constituted by an infinite number of such units. Therefore, and being inconsistent the actual infinity (Theorem 6, page 21), in any considered spacial direction the discrete units forming the physical space must be contiguous, adjacent.

(i) The discrete units of the real physical space, as such discrete units, cannot be divided, since the new discrete units arising from the division would be equally divisible and we would have an infinite number of such discrete units, and therefore the real physical space would be inconsistent.

The above argument (a)-(i) proves the following:

e-Theorem 2 (of the Discrete Space) *Physical space can only consist of discrete units which, as such units, are of a non-null extension, indivisible and contiguous in all directions.*

This is the same formal conclusion that was reached in the theorems and corollaries in Chapters 3-5 of this book. As extravagant as the e-Theorem of the Discrete Space may seem to some readers, let us remember that this scenario of the physical world was already considered by the early pre-Socratics [231]. And that the Arabic philosophical-theological school of thought known as Kalām (IX-X centuries) developed a discrete cosmology, with discrete units for mass, space, and time, so that motion had to occur in leaps and bounds separated by a certain number of discrete units of time, less time units the faster the motion is [176, p. 62-68].

Although overly speculative, one might consider the possibility that the universe functions in a manner similar to the functioning of cellular automata. In such a case, its discrete units would have different states that could give rise to the fundamental interactions and elementary particles from which all material objects would arise. In certain extreme states of gravitational interactions the discrete units of space could deform (which, for the reasons given above, is not possible with points) and propagate the deformations as gravitational waves; or they could originate some change in the content of the discrete units which would also propagate as do the detected gravitational waves, with the same consequences on the transmitting space.

On the other hand, it does not seem to be a coincidence that gravitational waves propagate with exactly the same velocity as electromagnetic waves: $299792458\,m/s$. The coincidence could be due to the fact that both velocities represent the maximum possible velocity in a discrete space and time: one discrete unit of space per one discrete unit of time (the discrete nature of time is demonstrated in the next section).

For obvious reasons, I do not intend to explain here gravitational waves from the perspective of a discrete space (simply because I would not know how to do it), but to point out that the discrete scenario, so far only considered by a few authors (curiously making use of infinitist mathematics) is the only consistent scenario if the actual infinity is inconsistent (I remind you the proof of that inconsistency included in this book, and the more than 40 others included in [212]). At most, I dare to indicate that in a discrete and finitist scenario things could be very different in the fundamentals, while being compatible with all known experimental data.

23.5 Time is a discrete magnitude

Physical objects can be defined in terms of their composition, structure and properties. This is the case, as seen in the two previous sections, of space, once it has been proved that it is a real physical object (e-Theorem 1), reason for which it would be possible to give a true definition of space (See the proposed Definition 19 at the end of this chapter).

Time, on the contrary, does not seem to be a physical object but a magnitude that measures a basic and universal property of all physical objects: their ability to persist in a given state. But it is very likely that time, being a magnitude that measures such a basic and universal property, is a primitive concept that can only be defined in operational terms, which in this case would be related to the permanence and evolution of the successive states of physical objects, including their changes in position.

Since the states and changes of state of the physical objects are real, they can be empirically detected, and their permanence can be measured in comparative terms, for example by comparison with any arbitrarily chosen permanence, it makes sense to assign a magnitude that measures the permanence of each state of each material object. That magnitude would be time, and it would be real in the same sense that any other empirically detectable property of material objects is real. On the other hand, the relativistic deformations of time (even if they are only apparent, such as refractive deformations [208]) prove that time is not a universal physical object independent of the rest of the physical objects, but a magnitude linked to the state of the particular physical objects.

Time, as such a magnitude, can be measured in objects in different states of motion, and confirm that it is different in each object, depending on the state of motion with which that object is observed, which confirms that it is a universal property of material objects and not a particular and independent physical object. Although, I repeat, these variations depending on the way in which an object is observed (relativistic inertial time dilations) could be only apparent.

The difficulty, perhaps insurmountable, of defining time in non-operational terms suggests that time is not, in fact, a physical object like space, for if it were, we would expect a non-circular and descriptive definition of that object, which has never happened, nor does it seem likely to happen. What can be given, as will be shown below, is an operational definition of time which clearly indicates that it is a measurable magnitude of a fundamental and universal property of all material objects. From its operational definition one could define operational units of time and other related concepts, such as time interval, simultaneity, flow of time, etc.

On the other hand, modern physics considers that time has the mathematical structure of a one-dimensional continuum of instants (the "points" of time) equivalent in formal terms to the continuum $\mathbb{R}$ of the real numbers. Exactly the same argument (a)-(i) developed in the previous sections for the case of space can be made for the case of time, which leads us to state the following:

e-Theorem 3 (of Discrete Time) *Time is a discrete magnitude whose discrete units, as such units, have to be indivisible, contiguous and of a non-null extension.*

The discrete units of space could all have the same extension, or not. If they did not all have the same extension it could happen that the general physical laws would vary with spatial directions, which does not seem to be the case. It is therefore reasonable to assume that in terms of the extension of its discrete units (which we could call *qusits*, quantum space units) physical space is isotropic. The same is true for the discrete units of time taken as a reference to measure time, although here there are only two directions (past and future), all the discrete units of time (which we could call *qutits*, quantum time units) should have the same extension, otherwise we would have to consider the possibility that the physical laws vary with the evolution of the universe.

It is interesting to note at this point that there is an exclusively formal way, independent of the empirical detection of gravitational waves, which allows us to demonstrate the existence of the above discrete and real units of space and discrete real units of time. An argumentation whose starting point is the Principle 1 of the Directional Evolution (page 41):

> *The observable universe always evolves independently of its rational observers and in the same direction of increasing its global entropy.*

A principle that, as we know, has enormous inductive support. The following results, among others, can be immediately deduced from it (Chapters 3-5 of this book and [205, 207]):

1. The universe evolves under the control of a unique set of invariant and consistent physical laws.

2. There is an indivisible minimum of space (time) of which all space (time) intervals are an integer multiple.

3. The indivisible units of space (qusits) and time (qutits) are physical, and then real and absolute.

4. Every space interval (or time interval) is finite and can only be divided into an integer number of adjacent qusits (qutits).

5. The continuum densely ordered spacetime cannot be used to model uniform motion.

6. The laws of physics do not apply in spaces smaller than the indivisible unit of space nor in times smaller than the indivisible unit of time, both being of non-zero extension (duration).

23.6 Absolute motion

The fact that we cannot observe qusits does not mean that they do not exist (some authors, as G. Cantor, said the same thing about atoms in the first half of the 20th century). According to the above arguments they must exist if the universe is consistent. And it can be proved that the universe is formally consistent on the basis of the above Principle of Directional Evolution. If we could observe qusits, it would be possible to choose any of them and refer to it the motion of all material objects in the universe.

Naturally, all these motions would be absolute: motions THROUGH a unique, real and absolute physical space. The differences of these absolute motions would be the cause of the relative motions of all material objects to each other. These relative motions are the only motions we can observe due to PREINERTIA,[3] and the fact that, as mentioned above, it is not possible, at least for the time being, to observe qusits and establish an absolute reference frame with them.

But the absolute or relative nature of natural motions should not depend on the biological fact that we humans can or cannot perceive qusits: motion existed long before living things appeared in the universe. So, although it is anathema in contemporary physics, we would

---
[3] All physical objects, including photons, inherit the relative velocity vector (with respect to any other inertial reference frame, including absolute space) of the inertial reference frame in which they are set in motion [208].

have to consider the possibility that, indeed, all motions of all material objects are absolute in nature, motions THROUGH a finite, discrete, real and absolute space. Contemporary physics would then be operational physics, not fundamental physics. To develop a fundamental physics one would have to change the current paradigm of the infinitist continuum to a new paradigm based on the finite discreteness of space and time.

It is really significant that material objects are not altered by the physical space they continuously pass through, and that space is altered by the presence of material bodies, which in relativity is called spacetime curvature, and which in reality could only be an alteration in the content of the qusits closest to the material objects, an alteration produced by the close presence of these material objects.

The preinertia of electromagnetic waves and this alteration of qusits would be sufficient to explain gravity without having to modify the shape of space, perhaps too much of an apparatus. Although the gravitational deformation of the discrete qusits would also be possible, what is not possible, for the reasons given above, is the deformation of the spacetime continuum of densely ordered points. A model to start thinking about the new discrete paradigm could be the cellular automata like models (CALMs) [197].

An additional attraction of finitist and discrete models, such as CALMs, is that they could solve the old problem of change, which has been posed for 26 centuries without having found a solution. A problem that physics has been ignoring for the last centuries. But the physical world will not be properly explained (fundamental physics) until the problem of change has been resolved. With all its mathematical apparatus (always infinitist), the physics of our days has not yet been able to explain how a simple change in position of any material object occurs.

Moreover, it can be formally proved that in the spacetime continuum, change is an inconsistent process [197], which had already been anticipated by some philosophers such as H. W. F. Hegel [148, 152, 240, 270, 293, 361], while others, as J. M. E. McTaggart, came to the same conclusion as Parmenides [271] on the impossibility of change [239]. All of which points to the fact that it would be convenient to start considering the possibility of changing the infinite and continuous scenario of contemporary physics for a finite and discrete one.

At this point, it is interesting to recall the following (very simplified) Aristotelian argument about the non-existence of the vacuum [16, Book IV, 209a, p. 216], in force for more than 2000 years:

If the vacuum were something and we placed anything else in it, we would have two somethings in the same place, which is impossible.

The reality of gravitational waves and the interpretation of the universe as a CALM invalidates this Aristotelian conclusion: Space is something that contains and generates all material objects. Therefore, an object at a place in space is a material object generated by space at that place. There is then no inconsistent superposition of *somethings*.

I end this paper with the following definition of physical space as a recapitulation of all that has been said and all that has been proven up to this point in this paper:

Definition 19 (of Physical Space) *Space is a real physical object formed by a finite number of indivisible and contiguous units of a non-null extension that contains, and possibly generates, all the material objects of the universe, to which it offers no resistance to their motions and makes possible their mutual interactions.*

APPENDIX

Physicists and physical space

(Text taken from [213, p. 119-121]) The dominant idea in contemporary physics is that neither space nor time are real physical objects. Answers like the next one can be found on some physics well-known FAQ websites (obviously answered by 'expert' physicists):

> Spacetime is not a fabric, it is not material. Space is just an illusion, time is just an illusion therefore spacetime is just an illusion and a good way of simplifying the concept of general relativity to the public.

This has also been the opinion of many relevant authors in the history of science and thought (particularly empiricists): G. Leibniz, D. Hume, C. Huygens, E. Mach, H. Poincaré, E. Borel, L. Wittgenstein etc. And of the vast majority of contemporary physicists. For example [325, p. 266]:

> ... space and time, like society, are in the end also empty conceptions. They have meaning only to the extent that they stand for the complexity of the relationships between the things that happen in the world.

Although, on the other hand, we can also read the contrary opinion. For instance, according to:

1. A. Einstein

- I agree with you that the general theory of relativity is closer to the ether hypothesis than the special theory [186, p. 68].
- According to the general theory of relativity, space is endowed with physical qualities... [186, p. 98].

2. F. Wilczek:
   - Spacetime is also a form of matter [371, p. 180].
   - Spacetime has a life of its own [371, p. 180].
   - According to general relativity, spacetime is extremely rigid [371, p. 181].
   - Dark energy could be a universal density of space itself [371, p. 194].
   - What appears to our eyes as empty space is revealed to our minds as a complex medium full of spontaneous activity [370, p. 1].

3. N. A. Tambakis:
   - It seems to me that in this way we can confirm the well-known epistemological assumption that space and time are not fictions but rather modes of the dynamic existence of matter [342, p. 146].

4. M. Kaku:
   - In a sense, gravity does not exist; it is the distortion of space and time that moves the planets and stars (cited in [40, p. 63]).

The case of A. Einstein is a bit more complex. Let's remember some of his words through the years about space and time:

> 1905: The introduction of a "luminiferous ether" will prove to be superfluous inasmuch as the view here to be developed does no longer need an absolute space, at absolute rest, with physical properties [87, p. 891].

> 1913: For me it is absurd to attribute physical properties to "space" (Letter to E. Mach cited in [185, p. 135]).

> 1914: As much I am not disposed to believe in ghosts so I do not believe un the enormous thing about which you are talking and which you call space [89, p. 345].

> 1915: Thereby (through the general covariance of the field equations) space and time lose the last remnant of physical reality (Letter to M. Schlick cited in [185, p. 134]).

However, in 1916 Einstein changed his mind about the physical nature of space and the existence of the ether:

23.6 Absolute motion

> 1916: I agree with you that the general theory of relativity is closer to the ether hypothesis than the H. A. Lorentz cited in [185, p. 135]).
>
> 1919: Thus, once again "empty" space appears as endowed with physical properties, i.e. no longer as physical empty, as seemed to be the case according to special relativity. One can thus say that the ether is resurrected in the general theory of relativity, though in a more sublimated form (Morgan Manuscript, cited in [185, p. 137]).
>
> 1938: Our only way out seems to be to take for granted that space has the physical property of transmitting electromagnetic waves... We may still use the world ether, but only to express some physical property of space... At the moment it no longer stands for a medium built up of particles [97, pp. 159-160] [98, p. 115]

In later writings he defended that the physical notion of space is linked to the existence of rigid bodies, but he rejects the idea that space is an *a priori form of intuition* [37], as Kant defended [179]. Einstein *"always supported an objective description of physical reality, without interference of the observer"* [177, p. 128].

24. The substance of physical space

Abstract.-Once the detection of gravitational waves is accepted, the fact that physical space must be a real object with the necessary properties to vibrate and transmit its own vibrations must be admitted. This short article starts the discussion on what must be the substance that constitutes this real physical space, taking into account its vibrational properties and, above all, its universal relationship with ordinary matter, which is undoubtedly of extraordinary importance for physics.
Keywords: space deformations, real space, space matter, discrete space.

24.1 Introduction

Once the detection of gravitational waves is accepted (see chapter 23), the fact that physical space must be a real object with the necessary properties to vibrate and transmit its own vibrations must be admitted. This short chapter begins the discussion on what must be the substance that constitutes this real physical space, taking into account its vibrational properties and, above all, its universal relationship with ordinary matter, which is undoubtedly of extraordinary importance for physics.

After a brief introductory discussion of the attitude of physicists toward the ontological problem of space, i.e., the debate about its existence as a real object or as a fiction, it is recalled that the experimental detection of gravitational waves can only lead to the conclusion that space must be a real physical object. We will then discuss the properties that this space substance might have, taking into account the mutual interaction between this space substance and ordinary matter.

Finally, a starting point for the construction of a theory of space substance is proposed. Given the material, finite, and discrete nature of space and its role as the container (and possibly generator) of all material objects, the proposal simply points to a model well known in computer science: cellular automata. In particular, because of the way in which in these theoretical artifacts a large variety of complex objects are generated and evolve under the control of a small set of very simple laws.

24.2 Physical space is real and discrete

As almost everyone knows, gravitational waves have already been detected several times since 2015 [23, 49, 261, 372]. It is also well known that they are transverse quadrupole waves traveling at the speed of light through space itself. For the reasons given in [214], the empirical detection of these ondulatory deformations of space demonstrates unequivocally that space can only be a real physical object with the necessary properties to vibrate, to be the transmitting medium of its own vibrations and to physically interact with the interferometers that detect them by changing the distances between their mirrors. A conclusion that seems unquestionable if it is accepted that what does not exist cannot vibrate or have empirically detectable properties.

But it happens that a considerable part of modern (20th and 21st century) physicists have defended the idea that neither space nor time are real, that they are only theoretical instruments useful for expressing the relative positions of physical objects in that fiction known as the spacetime continuum. For example, in 1902, E. Mach wrote (quoted in [176, p. 142]):

> Concerning the conceptual monstrosities of absolute space and absolute time, I could take nothing back.

And in our days we find on some physics well-known FAQ websites (obviously answered by *expert* physicists):

> Spacetime is not a fabric, it is not material. Space is just an illusion, time is just an illusion therefore spacetime is just an illusion and a good way of simplifying the concept of general relativity to the public.

This has also been the opinion of many relevant authors in the history of science and thought (particularly empiricists): G. Leibniz, D. Hume, C. Huygens, E. Mach, H. Poincaré, E. Borel, L. Wittgenstein etc. And of the vast majority of contemporary physicists. For example [325, p. 266]:

> ... space and time, like society, are in the end also empty conceptions. They have meaning only to the extent that they stand for the complexity of the relationships between the things that happen in the world.

Presumably, this position on the ontological nature of space will even-

24.2 Physical space is real and discrete

tually disappear, and the reality of physical space will be universally accepted, since observation and experimentation are the only valid ways to confirm scientific theories.

Relativistic physics considers other deformations of space: the inertial deformation (FitzGerald-Lorentz contraction) and the deformation caused by the presence of massive bodies (gravitational deformation). The former could only be apparent, because if it were real, material objects would have to have different sizes at the same time; or, for example, an elastic band could have stretched and contracted zones at the same time without any force acting on it (Chapter 15, page 166 of this book, and and [208]). The second type of deformation could also be explained in an alternative way, without the need to curve space, in this case by using preinertia, a universal property of all physical objects, including photons, by virtue of which they all inherit the relative velocity vector of the reference frame in which they are set in motion [208, Part IV].

According to this preinertia alternative, it would be the trajectories of photons that would be curved due to their gravitational interaction with massive objects, instead of having to curve the three-dimensional space itself, which is much more complex from all points of view (remember that physical nature bears the signature of simplicity, as has been recognized at least since the time of Galileo Galilei [125, p. 183-184]). This alternative would imply that photons would not be the massless particles they are supposed to be. They could, for example, have a quantum rest mass m_q defined by:

$$m_q = \sqrt{\frac{G\hbar^3 R_\infty^4}{c^5}} = \hbar t_p R_\infty^2 = 6.845023 \times 10^{-64} Kg \qquad (1)$$

where t_p is the Planck time and R_∞ is the universal Rydberg constant, which is specific to each chemical element and varies slightly with its mass.

But for the deformations caused by gravitational waves, there seems to be no other explanation than the actual existence of the object vibrating and transmitting its own vibrations. The reality of physical space is an important first step for physics, which is followed and will be followed by others. To begin with, once the real nature of physical space has been accepted, it can be proved (now using the inconsistency of the actual infinity (Chapter 3 of this book and [212]) that such space cannot consist of inextensive and densely ordered points, but of discrete, indivisible, contiguous units of non-zero extension (*qusits*) [214].

A similar reasoning applied to time shows that it must also be formed by discrete, indivisible, contiguous units of non-zero extension (*qutits*), although time is not a physical object like space, but a universal magnitude related to the stability of the different states of the different objects, maybe related to the vibrational frequency of qusits [214]. The next step is to discuss what the substance of physical space might be. To initiate this discussion is the purpose of this chapter.

24.3 On the substantiality of physical space

The formal consistency of the universe can perhaps be accepted as an inductive principle, but it can also be formally demonstrated as a theorem from another even more fundamental principle, which establishes the directional evolution of the universe (in the direction that increases its entropy). A principle that has overwhelming inductive evidence (see Chapter 5). In this consistent universe scenario, a first fundamental discussion begins here about what the substance of space should be, given what we already know about ordinary matter, about space itself, and about its interaction with ordinary matter. It also takes into account the inconsistency of the actual infinity, a key issue for physics that is virtually never considered.

On the basis of these varied supports, it is possible to draw the first conclusions about the materiality of real physical space and some of its most important physical consequences related to ordinary matter, time, motion, and the own evolution of the universe. Thus, without being exhaustive, the following points can be emphasized:

1. Taking into account the Principle of Inertia, we must conclude that the substance of space (from now on space matter) is completely transparent to ordinary matter. Therefore, the space matter cannot be of the same type as ordinary matter, because ordinary matter is not transparent to ordinary matter.

   Comment: Since gravitational waves change the length of the arms of their detectors (interferometers), one would have to admit some interaction of the vibrations of physical space on ordinary matter. Or some other explanation of the phenomenology of detection would have to be given, perhaps based on the discrete nature of space and time.

2. On the contrary, space matter is sensitive to the presence of ordinary matter: its properties are altered in the vicinity of material objects, the more so the more massive those objects are, and the closer the regions of space under consideration are to those massive objects.

   Comment: These space deformations (gravitational deformations) affect other material bodies that are different from the material body that produced them. Therefore, we can say that space matter allows material objects to interact with each other at a distance.

3. The space substance can vibrate and transmit its vibrations at a speed of 299792458 m/s (gravitational waves).

   Comment: Vibrations of the space matter could be due to reversible periodic deformations of its discrete units (see item 10 below) or to periodic changes of state of these same units.

4. Electromagnetic waves propagate through the space matter, so this matter must have the necessary physical properties to be able to transmit gravitational and electromagnetic waves at the same speed of 299792458 m/s.

   Comment: The coincidence of these two velocities could indicate the existence of a maximum velocity of one qusit (discrete unit of space) per qutit (discrete unit of time).

5. The space matter is virtually transparent to the entire electromagnetic spectrum, from gamma rays to radio waves.

6. The space matter has certain electrical and magnetic properties, such as magnetic permeability μ_o and electrical permittivity ϵ_o, which define the speed of light c in (supposedly) empty physical space: $c = (\mu_o \epsilon_o)^{-1/2}$.

7. The space matter must be able to exhibit and propagate different values of electric and magnetic charges.

8. The substance that makes up physical space must have the properties necessary to manifest in its bosom variable intensities of the four fundamental forces (force fields) that make possible all the physical interactions of all material objects (remember that the interior of atoms is occupied in its practical totality by physical space).

   Comment: The physical properties of space matter make possible the formation, maintenance, and evolution of all material objects.

9. Does space matter have anything to do with dark energy?

10. For the reasons given in [214] (inconsistency of actual infinity), and also taking into account the consistent nature of the universe, physical space must be formed by minimal indivisible units, contiguous in all directions and of non-zero extension, units which we will provisionally call qusits. Thus, as in the case of ordinary matter, energy and all kind of charges, the space matter is also quantum in nature.

    Comment: The same reasons for the formal consistency of the observable universe allow us to prove that time must also consist of minimal indivisible and contiguous units (Chapter 23), which we will call qutits for the time being. And for the same reasons, motion must also be discrete.

11. The consistency and universality of physical laws requires the universal existence of some relation between the discreteness of space matter and the discreteness of ordinary matter, electromagnetic energy, electric and non-electric charges.

    Comment: qusits, the discrete units of real physical space, could not only contain the discrete units of ordinary matter, but be their generative cause, in a similar way to the objects of cellular automata.

12. The fundamental properties of qusits must be the same in all directions of space, and their spatial distribution must be homogeneous, both of which are necessary conditions for the universality of physical laws in an isotropic and formally consistent universe. The space matter must therefore be homogeneous, isotropic, and stable as such a substance.

13. The real physical space, consisting of the same homogeneous and isotropic space matter, will be the same for all material objects contained within it. Thus, physical space is universal and absolute.

14. If a qusit were the size of a Planck volume, the observable universe would consist of $\approx 1.85 \times 10^{186}$ qusits, and a proton would occupy $\approx 1.44 \times 10^{59}$ qusits. All of them finite numbers.

    Comment: In a consistent universe, there can be no infinite sets (actual infinity, not potential infinity). Consequently, and whatever it is, the number of qusits in space will always be finite, which greatly simplifies their evolution.

15. Since all the ordinary matter of the universe is contained in physical space, it makes sense to speak of the material content of qusits. Even of the generation of ordinary matter within the space qusits.

16. Since ordinary matter evolves (changes), so must the material content of qusits.

    Comment: A magnitude can be defined to measure the persistence of the changing material content of the different qusits of physical space: TIME. The same magnitude that, from the point of view of the material objects, would measure the persistence (stability) of their states, is intimately related to the persistence (stability) of the material content of qusits (their vibrational frequency, one could say). In this sense, time would be perhaps the most essential magnitude in the evolution of the universe.

17. For the same reasons of formal consistency as in the case of space, time must also be discrete, and its units as such units must be

24.3 On the substantiality of physical space

indivisible, contiguous and of non-zero extension. It seems appropriate to call them qutits (quantum space units).

18. The universality of the physical laws applied to the evolution (history) of the universe requires the homogeneity of the qutits towards the past and towards the future. The discrete units of time should therefore be homogeneous and isotropic in both directions.

    Comment: The homogeneity of space and time is what makes possible the geological Principle of Actualism-Uniformism [225], which is insistently and without exception recorded in all the rocks of the Earth. This record, in turn, is an impressive proof of the Principle 1 of Directional Evolution of the universe (page 41) and of all its formal consequences. Unfortunately, it is also a proof of the insufficient connectivity between the different sciences.

19. The isotropic sequence of homogeneous qutits is the same for all material objects. Time is therefore an absolute and universal magnitude that measures the stability of the state of the different qusits of the physical space and the ordinary matter they contain.

20. For the same reasons of formal consistency as in the case of space and time, motion must also be discrete, which means that, at least for one qutit, nothing moves in the universe.

    Comment: See page 293 at the end of the next section.

21. The motion of material objects occurs THROUGH the same space matter, or what is the same, THROUGH a universal and absolute physical space. It is an absolute motion because time is also an absolute magnitude.

22. Preinertia and the lack of sensory and instrumental perception of qusits (at least for now) make the detection of absolute motion impossible.

    Comment: Preinertia is a universal property of all physical objects (including photons), by virtue of which all objects inherit the velocity VECTOR! of their proper reference frame when they are set in motion, which is why it is impossible to detect the absolute motion of a reference frame with the sole aid of the objects set in motion in that reference frame. The different absolute velocities of the different material objects give rise to the different relative velocities of the different material objects. Relative velocities are the only velocities that can be observed, detected and measured (except perhaps in the special situation explored in the Santiago del Collado experiment [208, p. 371-378]).

23. The inertial deformations of space and time described by special relativity would be only apparent, as is the case with refractive deformations.

24. The existence of a maximum speed of one qusit per qutit determines the existence of a maximum insurmountable speed for all material objects, including photons. This could be the speed of gravitational waves and electromagnetic waves through the real physical space.

25. The principles of special relativity are therefore operational, but not fundamental. They are tools for local explanations of the local observations in a local reference frame, and with mathematics based on the infinitist continuum, which is also inconsistent (Chapter 3).

26. From the point of view of fundamental physics, special relativity is neither necessary nor consistent.

27. The real physical space constituted by qusits could be reversibly deformed if these qusits were reversibly deformable, and their deformation could be propagated through the successive qusits.

    Comment: It can be proved that the points and instants of the spacetime continuum have no extension (duration) and are densely ordered (between any two of them there is always the same number $2^{\aleph_0}$ of other different points). There is no contiguity neither between points nor between instants (these are the formal causes of Zeno's dichotomies). Under these conditions it can be proved (apart from the inconsistency of the continuum itself) that the spacetime continuum cannot be deformed (Chapter 25 of this book and [214]).

28. The gravitational deformations of space may not be such deformations, but consequences of the gravitational interactions of photons with massive objects, which would produce the deformation of photon trajectories, a deformation much simpler than the deformation of three-dimensional space.

    Comment: A physical theory such as general relativity that is based on an inconsistent mathematical concept (such as the actual infinity that defines the spacetime continuum) cannot be consistent. It could be an operational theory, but not a fundamental theory. At best, it would be an operational theory that explains local observations made and interpreted in a particular mathematical framework, which, being inconsistent, cannot be applied to a complete universe that is formally consistent (Theorem 18 of the Consistent Universe, page 42).

24.4 Cellular Automata Like Models

As is well known, in the late nineteenth and early twentieth centuries the Hypothesis of the Actual Infinity was accepted as subsumed in the Axiom of Infinity, one of the axioms of the various set theories that were developed from those years on. This acceptance was, in my opinion, one of the great errors in the history of modern science. Espe-

24.4 Cellular Automata Like Models

cially because of the universal, uncritical and dogmatic way in which this axiom was accepted and imposed; and because of the hostility towards the few dissenters. The fact is that the only mathematics since then has been infinitist mathematics, that is, the mathematics that assumes the existence of complete ordered lists with an infinite number of elements without a last element completing the list.

One of the emblematic objects of infinitist mathematics is the relativistic spacetime continuum, with which physics became irrevocably linked to mathematical infinitism: the theories of relativity (special and general) are theories essentially linked to the spacetime continuum. Consequently, and since this continuum is an inconsistent set, the theories of relativity can only be provisional operational theories that justify certain empirical observations, but not fundamental theories that explain the real physical phenomena that give rise to the observed real facts (assuming, as it can be formally proved, that the universe is a formally consistent object).

Since the beginning of the twentieth century, there have been new computational-logical instruments made up of discrete, indivisible units, contiguous in all directions, called cells, distributed in space in a homogeneous and isotropic manner and subject to certain evolutionary rules, generally very simple. These theoretical artifacts are called CELLULAR AUTOMATA. What is unexpected is the behavior of the objects that these automata give rise to, which can become really stable and at the same time very complex, given the simplicity of the rules that give rise to them and make them evolve, evolution that takes place under the control of new emerging laws that act on these objects.

Given the discrete nature of physical space, ordinary matter, and time, and the existence of rules (laws) governing their evolution, and given the need to change the current infinitist and continuous paradigm of the physical world to one of finite and discrete nature, one could begin to consider formal options such as those offered by Cellular Automata-like Models (CALMs). In these discrete models, the problem of change (posed 26 centuries ago) may find a solution, although the solution will not be as immediate as one might think [208, p. 579, Link]:

> As a very adventurous hypothesis, it could be proposed that qusits have two simultaneous mode of existence:
>
> 1. Permanence mode: The state of each qusit remains unchanged at least for one qutit. This would be the only perceptible state of qusits.
>
> 2. Interacting mode: All qusits update synchronically their respective states through appropriate processes driven

by the laws of the automaton, which has to last at least one qutit.

Although, in accordance with what has been said above, the problem of change will now appear in terms of these changes of modes. So we would have to admit that the interacting mode is simultaneous with the permanence mode, although it remains in an imperceptible background (such as computer applications running in the background), which changes to the permanence mode at each successive qutit (or something similar).

In any case, remember that in order to explain the physical world it is absolutely necessary to solve the problem of change beforehand.

24.5 Additional reasons for the paradigm shift

The paradigm shift I referred to in the previous section is first and foremost a change in the conception of physical space. And the main reason for proposing it is twofold: first, the empirical detection of gravitational waves, and second, the inconsistency of the actual infinity that underlies the mathematical language of contemporary physics. But, as will be seen below, there are additional reasons that make the idea of proposing the change convenient. On the other hand, it is to be expected that there will be many difficulties and resistance to changing a stream of thought, the infinitist and continuous (analogical) paradigm, which has been an absolutely hegemonic stream of thought for more than a century. Moreover, it coincides with our sensory perception of the physical world as a continuous reality, in the same way and for the same reasons that the projection of a film (in reality formed by a discontinuous series of images) appears to us as continuous.

As we know, computers can create an great variety of virtual worlds, but all of them must be, and are, discrete. Computers cannot create or simulate virtual worlds of an analog nature (simulations will always be truncated by the finite number of ciphers they can actually handle). This is an important limitation. On the one hand, it may explain why we did not seriously confront the analog-digital dilemma before the development of computer science. On the other hand, it raises the question of whether only discrete worlds can really exist. In any case, it might be interesting to end this chapter by listing some of the most important (at least apparent) advantages of CALMs:

1. The actual infinity is not necessary (and recall that it is inconsistent).

2. The problem of change could be finally resolved by admitting that each cell (qusit) has two superimposed and simultaneous modes

24.5 Additional reasons for the paradigm shift

of existence: the permanence mode and the interacting mode.

3. Contrary to points and instants, qusits and qutits are plenty of physical significance: qusits are distinguishable physical parts of a real physical object: absolute space (which could be the generator of all other physical objects). Qutits are the minimal units of a magnitude that measures a universal property of all physical objects: the minimal permanence of their corresponding states.

4. While knowing the position of a real object in the spacetime continuum does not provide relevant physical information, the position of a real object in discrete space represents the real position of that object within a unique real object, discrete space, which contains (and surely generates) the rest of the real objects. It thus represents the position of a part within a single whole.

5. CALMs are much more simple than the continuum model: while between any two points of the spacetime continuum uncountably many other ($2^{\aleph_0}$) points exist, the number of qusits in the whole visible universe would be finite ($\approx 8.29 \times 10^{184}$ if they were cubes of a Planck's volume l_p^3).

6. In the analog models of nature, extent and shape loss their physical meaning at the elementary particle level. This meaning could be found within the discrete space and time of CALMs.

7. Since each variable defining the state of a qusit is updated at each successive qutit, the content of qusits can oscillate in multiple forms.

8. Nothing can last a time less than one qutit nor move a distance less than one qusit. The maximum speed in a CALM is therefore one qusit per qutit. That could be the speed of light in a vacuum, though not necessarily.

9. The vibrations of all physical fields propagate in the vacuum with the same speed of 299792.458 km/s, which could be the maximum speed of one qusit per qutit.

10. The speed of light, or of any other physical object, does not depend upon the relative motion of the observer that perform the measurement: whatsoever be the number of qusits an object traverse in a given number of qutits, both the number and the particular qusits the object traverses will be the same for all observers.

11. Motion, and then physical laws, have not to be referred to abstract reference frames but to the actual fabric of space qusits (although for practical reasons we could also make use of symbolic reference frames).

12. The theory of special relativity could be reinterpreted as a theory of apparent deformations, making it unnecessary most of its inconveniences.

13. Certain qusit states could have organizing effects and then could give rise to emergent objects and properties.

14. Quantum entanglement and quantum non-locality could be a natural consequence of CALMs synchronized way of functioning.

15. The flow of time and its irreversible arrow, enigmatic from a spacetime continuum perspective, is naturally explained in CALM terms.

16. The slippery concept of *now* could also be easily explained in CALM terms.

17. The incessant quantum activity of "free space" could be better explained in CALM terms than it is in the inconsistent model of the spacetime continuum.

18. The place and role of information in the physical world could also be explained in terms of CALMs: information is discrete, digital.

19. The universe would be discrete not only for matter, electromagnetic energy, information and all types of charges, but also for the space and time in which all physical objects evolve.

20. Although quantum mechanics appears to be a complete science, its mathematical language, like that of the rest of physics, is infinitist. Surely with a finite and discrete language some of the mysteries associated with this branch of physics would be better explained and understood.

21. Observers, instruments and observed objects would form part, all of them, of the same CALM. Their current mutual interactions, including recursive interactions, would determine their irreversible future.

22. Synchronicity [337] and quantum entanglement, could be explained in terms of CALMs.

It is reasonable to suspect that the most interesting consequences of discrete paradigms, such as CALMs, are not to be found in the above list, but rather in the minds of some of its readers, if any.

25. On space deformations

This chapter is an article of the series Towards a Discrete Cosmology [213].

Abstract.-This paper in the series examines certain problems related to the contraction-expansion of space that have not yet been considered by contemporary physics. It examines how these contractions and expansions would have to be carried out given the formal properties of the set $\mathbb{R}^3$ of 3-tuples of real numbers used as a model for the space continuum, in particular the dense order of the set of real numbers, from which the impossibility of immediate successiveness (adjacency) between its elements is derived: between any two real numbers there is always an infinitude of other different real numbers. This simple and well-known numerical fact imposes certain restrictions on how space intervals can be expanded and deformed, which poses new difficulties for physical theories that make use of such expansions and deformations of space.

Keywords: space expansion, space deformation, length contraction, set densely ordered, immediate successiveness, adjacency, real numbers, real intervals, Principle of Formal Dependence, self-creation, self-destruction.

25.1 Introduction

The deficient use of formal language, even ordinary language, in contemporary physical theories has already been discussed in previous articles in this series, especially in the first two of them. Indeed, the inappropriate (because inaccurate) use of expressions such as adjacent points, contiguous points, point-to-point, etc. are not uncommon in the primary literature of physics, not to mention its secondary literature. The inappropriate use of certain terms and expressions has serious consequences, providing authors with a distorted and erroneous view of some foundational aspects of physical sciences. In T. Maudlin words [234, p. xiv]:

> Unfortunately, physics has become infected with very low standards of clarity and precision on foundational questions, and physicists have become accustomed (and even encouraged) ti just "shut up and calculate," to consciously refrain from asking for a clear understanding of the ontological import of their theories.

The same is true of certain mathematical and geometrical concepts that are essential to the construction of physical theories. These conceptual errors prevent them from seeing the real problems posed by the infinitist objects they use in their mathematical modeling of the physical world. And the problems are so serious that they call into question the physical theories themselves. All this apart from the formal inconsistency of the Actual Infinity Hypothesis.

The deficient use of language, and even of logic, in physical theories has already been denounced by other authors, and in a very solvent way. This is, the case, for example, of the article *'A Bang into Nowhere'* by C. Antonopoulos, from which I include the following quotes on the expansion and deformation of space, somewhat related to the content of this article [48]:

> The growing (or expansion) of the balloon *presupposes* space. But the growing of Space cannot similarly presuppose space because, presumably, Space itself is being created *by* such growing (or expansion) and is presently just as large as it has grown, and not more [48, p. 49].

> If Space is curved, and therefore if Space has a shape, then *Space* occupies some parts of Space but not some *other* parts of Space. Or, somewhat differently, Space can be found only in *some* points of Space but not in *all* the points of Space [48, p. 55].

Over the last century we have become accustomed to strangeness in scientific theories (strangeness that is often confused with difficulty), to the point of evaluating the strangeness content of scientific theories as positive, as if this strangeness added some kind of positive and attractive value to them. Some theories presume to be strange, and therefore difficult to understand. I wonder what will happen when it is discovered that this strangeness actually results from the inconsistency of some of the foundational hypotheses of such theories, as is the case of the Actual Infinity Hypothesis that underlies the mathematical language of contemporary physics, a language that physics never questions. The lack of formal rigor in the case of the (primary and secondary) literature on the expansion and deformation of space can reach levels that, according to Antonopoulos, are already very close to the ridiculous.

25.2 Points have neither extension nor shape

As noted in Chapter 7 the firsts Pythagorean believed in the existence of indivisible geometrical points with an extension δ greater than zero [231, pp. 11-16]. But the Pythagorean themselves later discovered the existence of incommensurable lengths, and with them the extension of points disappeared forever. In effect, since then we suppose that points have no extension (and instants have no duration). And by means of a little transfinite arithmetic, it is possible to prove in formal terms that this is the case. Recall the following theorem, proved in Chapter 17:

> Theorem 32 of Abstract Points: In the spacetime continuum, points have neither size nor shape, and instants have not duration.

Unfortunately, Pythagoreans thinkers did not have the occurrence of integer division and the subsequent possibility of a discrete arithmetic. On the contrary, the era of the continuous space and time was inaugurated. Confirmed, moreover, by our sensory perception of the physical world as a continuous world. Think, for example, of the continuous perception of motion. The invention of cinematography, more than twenty-five centuries later, came too late and our science is completely based on the continuum of the real numbers that models spacetime. The problem is that that spacetime continuum is inconsistent as was proved in Theorem 8 of the Inconsistent Continuum (page 23). And this will change all, because no scientific theory can make use of an inconsistency as one of its fundamentals.

25.3 Experiments and theories

Educated as a naturalist, I am a strong advocate of experimental science, although experiments are also subject to error, almost always related to precision. It is experimental results (including all kinds of observations) that should guide the development of natural sciences. In this regard, it seems appropriate to recall the following points:

1. Experimental results may be compatible with more than one theory. For example, the first experimental confirmation of time dilation resulted from the Ives-Stilwell experiment, which was compatible with Einstein's theory of special relativity, and with H. E. Ives' own theory (absolute space and time) [169, 172, 173, 351, 350]. This last detail is one that is almost always forgotten.

2. Experimental results are also compatible with the apparent, not real, nature of certain physical phenomena: we can experimentally

measure the refractive deformation of a rod partially immersed in water and confirm Snell Law over and over again, but the rod is only apparently deformed, not really deformed.

3. The relativistic deformations (special relativity) of material objects, or of some mechanisms (e.g. clocks), do not depend on the nature of the deformed objects (or on the type of mechanisms in the case of mechanisms): they only depend on the relative velocity at which they are observed; i.e. they do not depend at all on the deformed object itself but on the way they are observed from outside the object. But this is also the case for refractive deformations. And as with refractive deformations, relativistic deformations could also be only apparent.

4. In some cases, such as the relativistic space contraction, or the relativistic time dilation, the experimental confirmations must be double and symmetric, but almost never are: if from a reference frame A it is observed the contraction of a ruler in another reference frame B, then from the reference frame B the same contraction of the same type of ruler must be observed in the reference frame A. And the same applies to inertial time dilation and inertial phase difference in synchronization.

5. Experimental confirmation of a theory can also occur if it results from interpreting an essentially discrete physical world (including space and time) in terms of continuum-based mathematics: the relativistic Lorentz factor coincides with the conversion factor between the discrete and the continuum versions of Pythagoras Theorem.

6. In any case, we should avoid that some scientific theories, such as the theory of special relativity (SR), end up usurping the role of the fundamental laws of logic, as is already happening: if this or that statement is not compatible with SR, then that statement is false. This is equivalent to equating SR with the fundamental laws of logic.

All of which suggests a humble and not arrogant attitude in the defense of scientific theories, however experimentally proven they may be.

25.4 Expanding and contracting the space continuum

With regard to the deformations of space, and in spite of the arrogance with which contemporary orthodoxy is defended, the reader should remember that things are far from being resolved: there is a deep division among its proponents (see previous article on the physical/geometrical nature of space):

25.4 Expanding and contracting the space continuum

1. Those who believe that space is unreal, perhaps the majority of authors: space is just an illusion; space is a fiction: space is an empty conception; etc.

2. And those who defend its reality (like the second Einstein from 1916) [185, 186]): space is endowed with physical qualities [186, p. 68]; spacetime is also a form of matter [371, p. 180], etc.

And, obviously, the division is very significant from the point of view of the space deformations: can a non-real object be deformed and the deformation be empirically measured?

Contemporary physical theories are essentially mathematical theories, modelizations in the form of equations of different types (linear, nonlinear, differential, in partial derivatives etc.) developed always with the same language of infinitist mathematics, the mathematics based on the Hypothesis of the Actual Infinite subsumed in theAxiom of Infinity. But one thing is the model and another the physical world. It seems appropriate to recall here again the following words of P. Dirac written more than 60 years ago [81, p. viii] (cited in [46, p. 11]:

> Mathematics is only a tool and one should learn to hold the physical ideas in one's mind without reference to the mathematical form.

But the most relevant aspect of the expansion (contraction) of the space modeled by the set $\mathbb{R}^3$ of all 3-tuples (x, y, z) of real numbers is the existence of certain restrictions imposed by the dense order of the real numbers, and consequently by the impossibility of adjacency (immediate successiveness) between the elements of that set, as well as by the null extension of the points of the continuum R^3 (Theorem 33, page 208). Among those restrictions, we have to consider the following:

1. Points cannot be deformed because they have no shape to deform. Therefore a space of points cannot be deformed by deforming its points.

2. Points have no extension:

    a) Therefore a space of points cannot contract by the contraction of its points: points have no extension to contract.

    b) Neither can the space be stretched by stretching its points: those points would cease to be inextense, and therefore would cease to be points.

3. A space of points cannot be deformed by sliding its points: they would leave gaps of indeterminable extension and space would no longer be a continuum of points.

4. Furthermore, collisions between points would occur, which is im-

possible because the collided points would have to be contiguous, which is not possible because contiguous points are impossible in a densely ordered set of points.

5. The deformation of a space of points cannot occur either by destruction of points or by creation of new points:

   a) In the first case, gaps of indeterminable extension ($0 \times 2^{\aleph_o}$ is a mathematical indeterminacy) would be created and the continuum would cease to be a continuum.

   b) In the second case, there would have to be gaps of indeterminable extension for the new points, which is impossible in a continuum.

6. A deformed space would be indistinguishable from an undeformed space: in any direction of the space, deformed or not, there would always be the same number of inextensive, formless, non-deformable and non-adjacent points: always $2^{\aleph_o}$ of such points.

Consequently, it seems impossible to deform a space composed of points without size, without shape, without contiguity, and of which the same number would exist in any region of the universe, whatever its size, from a Planck volume to the entire three-dimensional universe.

25.5 The relativistic contraction of space

In the year 1889 G. F. FitzGerald [133], and in 1892 H. A. Lorentz [226], proposed independently a real length contraction of moving objects in the direction of absolute motion through the ether in order to explain the negative results of the Michelson-Morley experiment (in the case of FitzGerald it would be an expansion in the orthogonal direction [218]). According to both authors, the contraction was caused by changes in the intermolecular forces of the moving objects as a consequence of their interactions with the ether, even though there was no empirical evidence for these changes.

Some years later, in 1905, Einstein writes on the different lengths of a rigid rod when measured at rest and in relative uniform motion [96, p. 95]. This relativistic contraction is immediately deduced from the Lorentz Transformation [208, pp. 70-71]. It is the well-known FitzGerald-Lorentz contraction, according to which a straight line AB in its proper reference frame is observed contracted by a factor f when its length is measured in relative motion:

$$f = \gamma^{-1} \cos \alpha_o = \sqrt{1-k^2} \, \cos \alpha_o \tag{1}$$

where γ is the Lorentz's factor; $v = kc$, $0 < k < 1$, is the relative velocity;

25.5 The relativistic contraction of space

and α_o the angle that the straight line AB makes in its proper reference frame with the direction of the relative motion. And that is all.

As can be seen, nothing is said about the way in which the line is contracted; the restrictions and possibilities outlined in the previous section are not mentioned. If we change the line AB for a physical object, for example a metal ruler R, maintaining the conditions of the relative velocity motion, we would see the ruler R with all its marks and numbers contracted in the direction of v by the same above relativistic factor f.

The problem with the rod R is that it can be observed with a multitude of different relative velocities, and therefore the measurement of its length will be different for each of these measurements. This opens the question about the real or apparent nature of the FitzGerald-Lorentz contraction, because it is difficult to assume that the same physical object can simultaneously have so many different lengths. Today there are supporters of both alternatives. Those who defend the reality of contraction would have to explain:

1. How is it possible that all physical objects, whatever their composition and internal structure, contract in exactly the same way exclusively dependent of the relative velocity at which they are observed?

2. What is the physical cause of the contraction? the contraction of atoms? the contraction of interatomic distances?

3. How can the same physical object have different sizes at the same time?

4. Can there be as many simultaneous superimposed physical realities as there are different velocities at which their objects and events can be simultaneously observed?

5. FitzGerald-Lorentz contraction is incompatible with some well established physical laws and empirical data, for instance [208, pp. 64-355]:

    - It is impossible for an elastic cord free of forces to be enlarged only in some of its parts.
    - The laws of mechanics governing the deformation of rigid materials as glass cannot be violated.
    - It is impossible to deform continuously a metal object without any force act upon it.
    - It is impossible to make disappear the rotation of a physical object.

- It is impossible for the hydrostatic pressure not to be the same in all directions.
- All optical isotropic media behave as anisotropic when observed in relative motion.
- The speed of a photon through a transparent medium should be the same in all reference frames, but it is not.
- Optical isotropy cannot be observed in relative motion.
- Optical isotropic material only exist at rest.
- The relativistic bipolar anisotropy does not exist in real media at rest.
- The laws driving the kinetics of oscillating reaction cannot be violated.
- Two identical clocks identically accelerated should not behave in different ways while they are accelerated.
- Extreme anisotropy is unknown in optical crystallography.
- It is impossible that a simple reflection of a photon on a mirror changes the speed of the reflected (re-emitted) photon by hundred of thousands km/s.
- The second law of the reflection of light can be violated in certain relativistic circumstances.
- Snell Law is also violated in certain relativistic conditions.
- The total internal reflection of light can also occurs under anomalous relativistic circumstances.
- It is impossible for a rigid body to enlarge indefinitely.
- It is impossible that an explosion occurs and does not occur.
- It is impossible that a standard ideal pendulum swings faster in one direction than in the other.
- It is impossible to accelerate a physical object without a force acts on it.
- It is impossible to accelerate photons if they always moves with the same universal speed.
- The impossibility to describe motion with respect to external references does not imply that motion does not exist. It necessarily existed without observers who could describe it for the first few million years of the universe's history.
- The universality of the laws of logic is incompatible with the lack

of relativistic simultaneity.
- etc. (see [208])

Those who defend the apparent nature of the FitzGerald-Lorentz contraction have also a some questions to answer:

1. If time dilation and phase difference in synchronization (local simultaneity) are also deduced from the same Lorentz Transformation as the FitzGerald-Lorentz contraction, are they also apparent?
2. If the three relativistic effects can be experimentally confirmed and deduced from the same Lorentz Transformation, why the FitzGerald-Lorentz contraction would be apparent and the other two real effects? Where does special relativity (SR) establish that some inertial spacetime deformations are real and others are apparent?
3. If the three effects are only apparent, would not SR be explaining a reality that is observed to be deformed because of relative motion but is not actually deformed?
4. If that were the case, would not SR be a theory of apparent deformations of little interpretive value of the real physical world?

25.6 The expansion of intergalactic space

As is well known, the experimental verification of the expansion of the universe predates the Big-Bang theory, which was precisely inspired by that expansion. What we also know at least since 1998 is that this expansion is accelerated. The theory that tries to explain the facts concludes that it is space itself that expands, as in the original times of the great cosmic inflation, but in a more moderate way: at a speed of about 70 Km/s per megaparsel. Moreover, not all space expands, the space inside the galaxies does not: the objects of a galaxy maintain their relative distances. Naturally, the general theory of relativity is an essential part of the theoretical framework that explains the expansion of space, which naturally excludes any possibility of absolute motion, and therefore of absolute space and absolute time.

But as we have already seen in this series of articles, on the very existence of space as a physical entity in its own right, there is a deep division of opinions. For some authors (the majority) there is no such physical entity:

> It is only an illusion. A good way of simplifying the concept of general relativity to the public (see Chapter 24).

For others, space exists as a physical entity:

> A form of matter [371, p. 180]. A complex medium full of spontaneous activity [370, p. 1]. Modes of the dynamic ex-

istence of matter [342, p. 146], etc. (see Chapter 24).

Consequently, and on the issue of the expansion of space, we would have to highlight the following:

1. For the supporters of the non-existence of physical space:
    1. The universe evolves *as if* there were something (which in reality does not exist) that expands according to certain equations. This attitude is compatible with the consideration of any other object that does not exist but has the appropriate properties to explain whatever one wishes to explain, and to assume that, indeed, that is the explanation one is looking for. These objects are rejected by formal logic as objects defined on purpose (*ad hoc*) to achieve the desired end.
    2. The problem with imaginary objects is that they do not physically exist, and trying to explain the evolution of what does physically exist with objects that do not physically exist is not physically acceptable; or at least it is not a complete physical explanation.
    3. If we assume that physical objects in the universe interact only with other physical objects in the universe, it is not possible to explain the evolution of the universe by the interaction of real physical objects with imaginary non-real objects. It seems reasonable to require that reality must be explained only by the properties of real objects.
    4. Even though it is imaginary, this unreal space would have the ability to create itself, which goes against Theorem 20 of the Formal Dependence (page 43):

        Theorem 20 (of Formal Dependence).-*No concept defines itself; no statement proves itself; no physical object is the cause of itself; and no cause is the cause of itself.*

2. For the supporters of the existence of physical space who assume the physical space is properly modeled by the continuum R^3:
    1. The points of the continuum R^3 must represent something like the physical points of physical space. A physical point would have to be a physical object (not a quality of a physical object). But how can something that has no extension be a real physical object?
    2. Any region of physical space has the same number ($2^{\aleph_0}$) of inextensive physical points before and after any expansion. Under these conditions, how is the expansion of physical space possible if neither the number of its points nor the size of its points

increases?

3. The expansion of physical space requires its capacity for self-creation, which goes against Theorem 20 of the Formal Dependence (page 43).

25.7 The gravitational deformation of space

Although there are alternatives, the hegemonic theory of gravity in contemporary physics is the one based on the general theory of relativity. Gravity would not be a force but the consequence of a geometrical deformation of space: physical objects, including photons, would move along geodesics that can be straight lines or lines curved by the nearby presence of objects with the appropriate mass.

To those who defend the unreal nature of physical space it makes sense to pose a couple of questions:

1. How is it possible to deform something that does not exist?
2. How is it possible for physical objects to move along non-existent geodesics? What quality of the object determines the next (non-existent) point to which the object is to move?

Those who defend the real nature of physical space can be asked the same questions they were asked in the case of the expansion of space, and they will have to draw the same conclusions: their theory is incomplete and will remain incomplete because of the impossibility to describe how these deformations occur in fact. They can only quantify them.

In any case, it should be noted that with respect to gravitational deformations other much simpler theories could be developed based on preinertia, and in which it is not necessary to deform any space, be it a real space or an imaginary space. Since preinertia is a universal property of all physical objects (including photons), there must exist in all of them something common that is the cause of that preinertia. The simplest explanation could be that that something is mass (an undefinable primitive concept), the same that originates the inertial mass and gravitational mass of all material objects. Thus, although much remains to be discussed on mass, the simplest explanation of the above three facts would be that all material objects have a property capable of modifying the properties of space (gravitational fields), of offering a certain resistance to their own changes of motion (inertial mass), and of inheriting (in vector terms) the motion of the body from which it starts its own independent motion. This latter property manifests itself even in supposedly massless objects such as photons. There is a possibility that photons have a rest mass of the order of 10^{-64}

Kg, which could be called quantum mass m_q [208, p. 235]:

$$m_q = \sqrt{\frac{G\hbar^3 R_\infty^4}{c^5}} = \hbar\, t_p R_\infty^2 = 6.238883052 \times 10^{-64} Kg \qquad (2)$$

where t_p is the Planck time and R_∞ is the universal Rydberg constant, which is specific to each chemical element and varies slightly with its mass.

26. On time deformations

This chapter is also an article of the series Towards a Discrete Cosmology [213].

Abstract.-The continuum $\mathbb{R}^+$ modeling physical time, real or not, poses the same problems in the case of time as in the case of physical space discussed in Chapter 25. In the case of time, moreover, the continuous model of time makes impossible the solution of one of the oldest physical problems that remains unsolved: the problem of change (discussed in Chapter 18). The main objective of this chapter is the discussion of the real or apparent nature of relativistic time deformations. Here, it is proved that, despite their experimental confirmations, relativistic time dilation and local simultaneity (phase difference in synchronicity) could be apparent, not real, in the same way that all refractive deformations are apparent, no matter how many experiments confirm Snell's Law. This Chapter also demonstrates the impossibility of describing relativistic time dilation in terms of the successive instants that supposedly define the passage of time. The discrete and real nature of time, already demonstrated in previous chapters of book, not only indicates a way to solve the problem of change, but could also explain all relativistic deformations of space and time.

Keywords: time dilation, difference in phase synchronization, set densely ordered, immediate successiveness, adjacency, real numbers, real intervals, twin robots paradox, impossible pendulums.

26.1 Introduction

We will never know what would have happened if the pre-Socratics, who initially considered that the points of space had a non-zero extension before discovering the incommensurability of the side of a square with its diagonal, had also discovered integer division and discrete arithmetic (Chapter 7). Or that the discrete view of the physical world (including space, time and motion) introduced in the tenth century by the Islamic school of thought Kalam [176, p. 62-68] would have been accepted and developed by the modern science that emerged in the sixteenth century. But neither possibility occurred. What did occur was the appearance of the irrational numbers, and eventually the set $\mathbb{R}$ of the real numbers and the actual infinite, numerable and non-numerable. In my opinion, with Cantor, science entered a phase of infinitist bewitchment in which it still remains, and from which I believe it will be very difficult to emerge. Like the Pied Piper of Hamelin, the actual infinite has been guiding the steps of mathematics and physics for the last century and a half.

In Chapter 3 of this book, the reader was given a very simple proof of the inconsistency of actual infinity, and thus of the spacetime continuum (Theorem 8 of the Inconsistent Continuum, page 23). The inconsistency of actual infinity should change everything, although it will do so by creating many difficulties. In the meantime, this chapter examines some of the consequences of using the set $\mathbb{R}$ of real numbers as a guide and model. Some of the aspects discussed in the previous chapter with respect to space deformations are discussed here with respect to relativistic time deformations. The discussion leads to results that are unacceptable from both a logical and a physical point of view. The solution might be to assume that these deformations are not real but apparent. Or, alternatively, to assume that space and time are both discrete, discontinuous.

26.2 Instants have neither duration nor contiguity

Whether real or not, physical time in contemporary physics is represented (modeled) by the continuum $\mathbb{R}^+$ of positive real numbers. And it is within that model that the Theorem 32 was proved. According to this theorem the instants of time have no duration, nor do the points of space have extension. Neither do they have immediate successiveness (adjacency, contiguity): between any two instants there is always an infinite number of different instants, successively increasing, but without any of them being the immediate successor of another; as it happens with the natural numbers, where any of them, n, does have an immediate successor $n+1$ (Peano's Axiom of the Successor [272, p. 1]), and so that between n and $n+1$ there are no other natural numbers.

That topological feature of $\mathbb{R}^+$ (the lack of points adjacency) raises numerous problems related to the way in which space and time deformations could be carried out. In the previous Chapter it was shown that such deformations cannot occur:

1. By deforming the points (instants)

2. By changing the size of points (instants).

3. By adding or removing individual points (instants).

They are, therefore, formally impossible. In the case of relativistic time dilation and relativistic local simultaneity we know only the global factors:

$$\gamma = (1 - k^2)^{-1/2} \qquad \text{(dilation of time)} \qquad (1)$$

$$\frac{\gamma L_o k}{c} \qquad \text{(phase difference in synchronization)} \qquad (2)$$

that say nothing about how the deformations occur ($v = kc$ is the relative velocity ($0 < k < 1$), and L_o is the proper separation between two events in the direction of the relative velocity v).

26.3 The model R^+ of time and the problem of change

The problem of change was discussed in Chapter 18, where the Theorem of Change was proved:

> Theorem 38 of Change: Canonical changes are instantaneous and then impossible in the spacetime continuum.

It was also shown in Chapter 18 that immediate successiveness (adjacency) is a necessary condition to consistently explain any canonical change of any kind. Obviously, adjacency is possible in discrete time, with minimal indivisible and adjacent units of time (qutits), so that the problem of change could find a solution in the framework of discrete space and time, as proposed in this book. And remember that, however forgotten the problem of change may be, until it is solved we will not have a sufficient explanation of the physical world, precisely because the physical world is in continuous change.

(The following three sections are taken from [208])

26.4 The Ives-Stiwell experiment

The first suggestion that the ticking of clocks could change with movement was published in 1887 [360]. And perhaps the first suggestion of how this alteration could be verified experimentally came from a debate between A. Einstein and W. Ritz, but at that time the experiment was considered unfeasible [174]. Some twenty years later, in 1938, the experiment was performed by H. E. Ives and G. R. Stilwell [174, 168, 167, 170, 171], a short modern review can be found in [110].

Ives-Stilwell used a Dempster tube of canal rays and the Transversal Doppler shift of the radiation emitted by the moving particles. The spectrographic analysis produced results consistent with Larmor-Lorentz predictions. Or in other words, compatible with Ives's theory of absolute space and time and with Einstein's theory of relativity. As is well known, after the Ilves-Stilwell experiment, many different experimental verifications of (the relativistic) dilation of time have been carried out. But here we are interested in the first one for the following reasons:

1. It so happens that, invariably, whenever the Ives-Stilwell experiment is cited, it is done to indicate that it was the first experimental verification of Einstein's theory. But it is (almost) never noticed that it was also an experimental test of Ives's theory of absolute space and time. Einstein knew the Ives-Stilwell experiment and its results, although he never cited them (just as he never cited the Sagnac effect or the Michelson-Gales experiment) [350, p. 44, 84-85]. By the way, with that same theory, Ives was also able to explain the anomaly of Mercury's orbit, previously explained by Einstein in 1916, but in the case of Ives with different methods of classical mechanics that do not assume the curvature of space-time.

2. We are then in front of experimental results that are compatible with two very different theories (and surely with others): Einstein's theory of relativity and Ives' theory of absolute space and time. A significant detail forgotten by the most fervent believers of Einsteinian relativism.

3. As with most time dilation experiments, the Ives-Stilwell experiment takes place within the strong gravitational field of the Earth. This detail should be taken into account when it is stated that these experiments demonstrate the inertial (exclusively due to rectilinear and uniform motion) dilation of time. In any case we have to remember again that a verification of the inertial dilation of time requires two things: to verify from an inertial reference frame A that the clocks of another inertial reference frame B, both in non-zero relative motion, run slower than A clocks. And at the same time, check from the inertial frame of reference B that the clocks of A go slower than those of B. If this is not the case, the proof is incomplete.

   This double requirement is at the base of Herbert Dingle's critique (see Chapter 52) which, in my opinion, has not been satisfactorily answered. In short: if there is only a unique objective reality, then clock A ticks and does not tick slower than clock B; and clock B ticks and does not tick slower than clock A. And if there is no single objective reality, there would have to be as many simultaneous and superimposed realities as there are relative ways of observing the clocks A and B.

26.5 Relativistic dilation of time

On the other hand, it is appropriate to emphasize again that the relativistic dilation of time has to be universal: the same for all kinds of clocks. Mechanical, electrical, electromagnetic, electronic, atomic, biological clocks... all of them would undergo the same increase in the duration of the periodic events by which they measure time. And this raises an unavoidable question (although not very frequent in relativistic catechisms) related to that universality: how is it possible that so many different objects (in their chemical composition, in their internal structure, in their periodic mechanisms) all suffer the same increase in the duration of the periodic events by which they measure time when they are observed in relative uniform motion? And let us not forget that for most physicists time is only an illusion, a fiction to express certain relationships between physical objects.

As a comparative reference, let us consider any number of rigid rods of the same dimensions, immersed partially in water under the same conditions: all of them will show the same deformation regardless of their chemical composition and internal structure (glass, plastic, copper, aluminum, ceramic, wood etc.). That under the same conditions exactly the same deformation occurs in such a great variety of materials suggests that, in fact, these observed deformations are only apparent.

And these time dilations being exclusively related to the relative velocity at which the clocks are observed, we would have to admit, moreover, that if these time dilations were real, all clocks of all imaginable types would have to function simultaneously in as many different ways as there are different relative velocities at which they can be observed. Or, alternatively, they should start running at a certain rate when they begin to be observed with a new relative velocity; and stop running at a certain rate when they cease to be observed with a certain relative velocity (uniform velocities in all cases). In addition, and in some way, the respective clocks must receive the information that they begin (or end) to be observed at a certain relative velocity. Which is still a rather twisted conclusion and alien to the simplicity of all known natural laws.

Taking into account the enormous diversity of all existing and imaginable clocks, the answer to the question on how they can all modify their respective ticks in exactly the same way exclusively dependent on the uniform relative velocity at which they are observed may have to be sought at a more basic level than the level of physical laws, or simply from the perspective of appearances (as in the case of refractive deformations). And from the purely descriptive point of view: how can a densely ordered succession of instants be dilated if its instants cannot be dilated and their number cannot increase either: it will always be $2^{\aleph_o}$, both in one second and in the whole history of the universe?

26.5 Relativistic dilation of time

The following argument is reminiscent of the famous twin paradox, although it contains new elements that allow different conclusions to be drawn. Indeed, consider three *identical* robotic observers, the robot A, the robot B and the robot C, in their proper frame RF_o, and assume they are programmed to carry out the following tasks:

a) At the instant t_{oo} of RF_o, the robots A and B accelerate in exactly the same conditions until each of them reach the same uniform speed u with respect to RF_o. The only difference is that A and B are accelerated in opposite senses of the same direction parallel to X_o.

b) Once reached the speed u, A and B end their corresponding acceleration, and each of them moves for one hour according to their respective identical clocks, at the same uniform speed u in the same direction but in opposite senses with respect to the robot C.

c) Therefore, for one hour each robot remains in an inertial reference frame from whose perspective the other robot moves at a uniform velocity v.

d) During that hour, and only during that hour, A and B register in an appropriate physical support (the same in both cases) the successive ticks of their corresponding clocks, one tick per second, each tick recorded as a short beep.

e) After the programmed hour, A and B accelerate and then decelerated in such a way that at the instant t_{o1} of RF_o they recover their initial condition of being at rest in RF_o together with the robot C.

Once at rest in RF_o, the recordings of A and B are compared (Figure 26.1). And when comparing the recordings of the beeps produced by A's clock and B's clock, one of the following alternatives will occur:

(a) The recordings do not match: in one of them the time interval between any two successive beeps is greater than in the corresponding beeps of the other.

(b) Both recordings match: the beeps are separated by the same time intervals between any two successive beeps.

What cannot happen is that the beeps recorded by A are at the same time less temporally spaced and more temporally spaced than those recorded by B simply because this possibility goes against the Second Law of Logic. By the way, this possibility is at the core of the problem posed by H. Dingle cited above.

We must, therefore, consider only the above two alternatives. But before, recall that according to SR, for the robot A, the clock of B runs

26.6 Relativistic local simultaneity

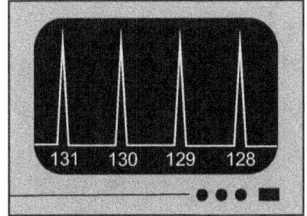 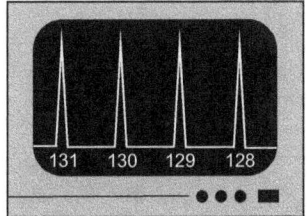

Figure 26.1 – Recordings of the twin robots: one beep per second during one hour of their respective proper time.

slow than its proper clock, and vice versa. Or in other words: robot A's clock is and is not further ahead than robot B's clock; and robot B's clock is and is not further ahead than robot A's clock. And recall that both clocks have been registering their respective beeps while they were moving at a uniform velocity v relative to each other.

In the case of the alternative (a), the symmetry of SR would not hold, and not all inertial reference frame would be equivalent, which goes against the Principle of Relativity. In the case of the alternative (b) the dilation of time observed from different inertial reference frames would only be apparent, as apparent as the refractive deformation of a rod partially submerged in water.

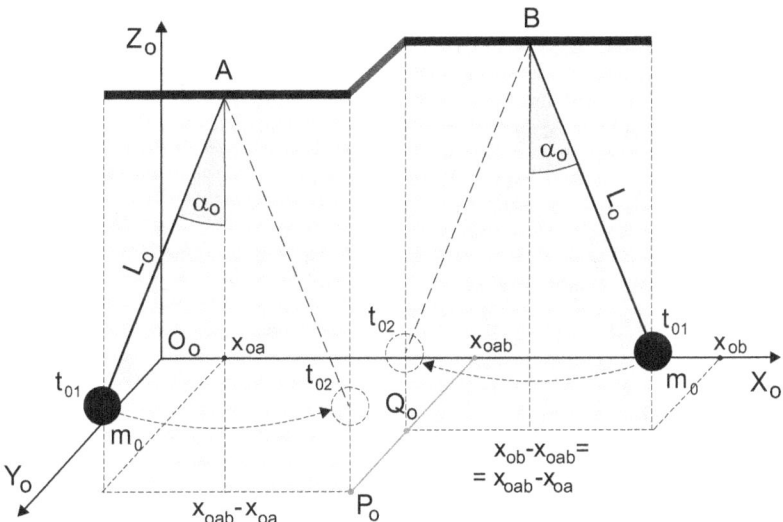

Figure 26.2 – Two identical pendulums A and B in their proper reference frame RF_o just at the instant t_{o1} when they begin to swing.

26.6 Relativistic local simultaneity

The scenario of the argument developed in this section is represented in Figure 26.2. In that scenario, two identical (and ideal) pendulums oscillate in the same conditions in planes parallel to the plane $X_o Z_o$ of their proper reference frame RF_o. And they oscillate as follows:

1. At the instant t_{o1} of RF_o both pendulums begin to oscillate simultaneously and in the same conditions defined by the angle α_o.

2. The ball of the pendulum A begins to oscillate from its initial position whose coordinate on the axis X_o is x_{oa}. This first oscillation of A is to the right (the increasing direction of X_o).

3. The ball of the pendulum B begins to oscillate from its initial position whose coordinate on the axis X_o is x_{ob}. This first oscillation of B is to the left (the decreasing direction of X_o).

4. At the instant t_{o2} of RF_o both balls reach the end of their first oscillation, which in each case occurs at a different point P_o and Q_o, although both points have the same coordinate x_{oab} on the X_o axis.

5. Both pendulums have the same amplitude of oscillation:

$$x_{oab} - x_{oa} = x_{oab} - x_{ob} \tag{3}$$

6. At the instant t_{o2} both pendulums start their second oscillation: the pendulum A to the left, and the pendulum B to the right, until reaching again their initial positions, and then they begin their third oscillation analogous to the first one.

7. According to the laws of mechanics, swings to the right take the same time for both pendulums as swings to the left.

8. The oscillations of both pendulums are repeated a certain number of times with a certain frequency.

We will now examine the oscillations of the two pendulums A and B from the perspective of the inertial reference frame RF_v which, as always in this book, coincides at a certain instant with the reference frame RF_o (the proper reference frame of both pendulums).

From the perspective of RF_v, the frame RF_o moves with speed $v = kc$, $(0 < k < 1)$, parallel to the X_v axis, in the direction of its increasing values. It is important to note that both RF_o and RF_v are two inertial reference frames to which special relativity can be applied, even though the movements of the pendulums are not uniform. The argument that follows is just a consequence of the Lorentz Transformation, whatever the particular implications of SR and GR may have.

Since in RF_o the two identical pendulums A and B start to oscillate simultaneously at the instant t_{o1} and in the same conditions (angle α_o), in the reference frame RF_v (Figure 26.3), and according to LT, they also oscillate in the same conditions, now defined by the angle α_v, which is related to the proper angle α_o according to:

$$\tan \alpha_v = \gamma^{-1} \tan \alpha_o \tag{4}$$

but they do not begin to swing simultaneously: A starts a time δt_{v1} before B, because when in RF_o both pendulums start to oscillate they are separated in the direction of the relative motion by a proper distance d_o given by (Figure 26.2):

$$d_o = x_{ob} - x_{oa} \tag{5}$$
$$= 2L_o \sin \alpha_o \tag{6}$$

where L_o is the proper length of each of the two pendulums. Consequently, the pendulum A begins a time δt_{v1} before the pendulum B, which is deduced from the Lorentz Transformation and is given by:

$$\delta t_{v1} = \frac{k(x_{ob} - x_{oa})}{c\sqrt{1 - k^2}} \tag{7}$$

Accordingly, the observers of RF_v will have to describe the oscillations of the pendulums A and B as follows:

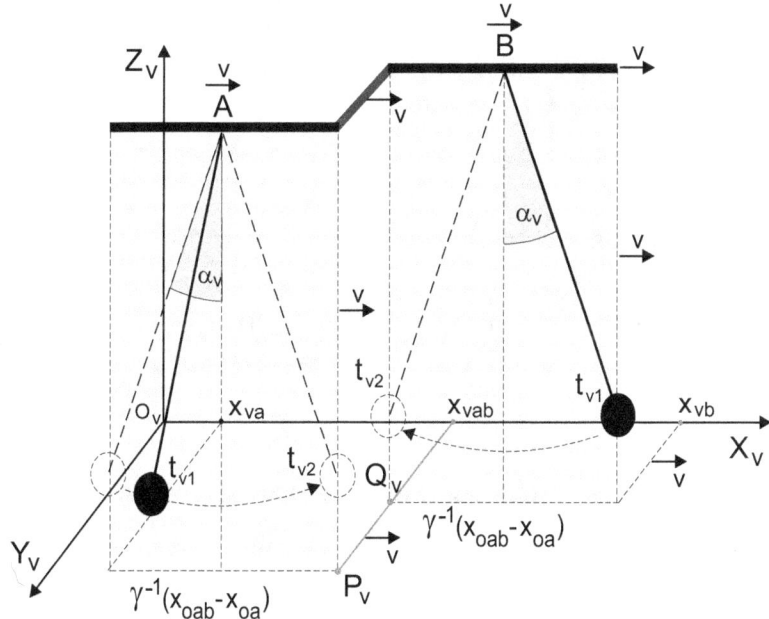

Figure 26.3 – The pendulums A and B from the perspective of the reference frame RF_v.

1. The pendulum A starts its first oscillation a time δt_{v1} before the pendulum B.
2. When the pendulum B starts its first oscillation, the pendulum A

has already covered part of its first oscillation.

3. Both pendulums end their first oscillation at the same instant t_{v2} and at two points P_v and Q_v with the same coordinate x_{vab}, because the distance in the direction of the relative motion between the points P_o and Q_o of RF_o where the pendulums A and B respectively end their first oscillation is zero.

4. Both pendulums have the same amplitude of oscillation:

$$\sqrt{1-k^2}(x_{oab} - x_{oa}) = \sqrt{1-k^2}(x_{oab} - x_{ob}) \tag{8}$$

5. Both pendulums start their second oscillation at the same instant t_{v2}.

6. The pendulum A finishes its second oscillation a time $\delta t_{v2} = \delta t_{v1}$ before the pendulum B.

7. When the pendulum A moves from left to right, it moves slower than the pendulum B.

8. When the pendulum A moves from right to left, it moves faster than the pendulum B.

9. The length of both pendulums first increase and then decrease along each complete oscillation because only in the vertical position $L_v = L_o$, at any other angle $L_v < L_o$ because its component on the X_v axis is not zero and then it is contracted by the factor γ^{-1}.

Thus, according to the observers in RF_v, i.e. according to the Lorentz Transformation:

1. Both pendulums swing slower from left to right than from right to left

This consequence deduced from the Lorentz Transformation is contradictory to all the theoretical and empirical knowledge about the oscillation of pendulums.

26.7 Consequences of the relativistic time deformations

In the case of time dilation, within the framework of H. Ives' theory of absolute space and time, the contradiction that occurs in the case of relativistic time dilation does not occur, when the dilation is considered real instead of apparent. Therefore, in the case of H. Ives it could be a real dilation, or the consequence of a possible anisotropy in the speed of light.

In the case of special relativity, and in accordance with the previous two sections, relativistic time dilation and local relativistic simultaneity imply some consequences incompatible with some of the fundamentals of logic and mechanics:

1. Given any two identical clocks, one of them is and is not further ahead than the other (robots argument).

2. A freely swinging pendulum swings faster in one direction than in the opposite direction without any force other than gravity acts on the pendulum mass (pendulums argument).

The solution of both incompatibilities could be that the relativistic deformations of time are only apparent. Therefore, the special theory of relativity is either inconsistent or describes deformations that are not real but apparent, as is the case with refractive deformations.

26.8 Time in CALM

Introducing CALMs as a model to explain the physical world does not solve any of the limitations imposed on human knowledge by the (potentially) infinite regress of definitions, arguments and causes. Limitations that are unavoidable for human knowledge, but which are rarely echoed in written or spoken human knowledge; as they do, with some pomp, of other much narrower and more debatable limitations occasioned by the debatable self-reference [201]. In Chapter 5 of this series, the inevitability of the limitations imposed by the Aristotelian infinite regress of arguments was demonstrated, and in this series of articles it has been extended to definitions and causes:

> Any recursive sequence S of proofs, definitions or causes in which there is a last element to be proved (defined, caused) and each element has an immediate predecessor that proves (defines or causes) it, is uncompletable.

Changing the reference model (the infinite continuum for the finite discreteness) to explain the world does not mean overcoming these limitations. We still need to use primitive concepts. We should not forget this inevitable limitation of human knowledge.

Whether continuous or discrete, the concept of time is surely one of those basic indefinable, primitive concepts. What can be affirmed is that as opposed to the infinitist time continuum modeled by $\mathbb{R}^+$, time in a CALM is always finite and discrete, with minimal indivisible units that here we are calling qutits. The qutits also present adjacency (contiguity, immediate successiveness): each qutit is immediately followed by another qutit without any other qutit existing between them, in the same way that every natural number n is followed by another natural number $n+1$, without any natural number existing between n and $n+1$.

Recall that a CALM could exist in a sequence of alternating states: the perceptible *permanence mode*, in which the state of its qusits (discrete space units) does not change; and the imperceptible *interacting mode*, which is simultaneous with the permanence mode, although it remains in an imperceptible background (such as computer applications running in the background) while changing the state of each qusit for the next qutit, according to the CALM's rules.

Unlike instants (which have no duration), qutits do have a duration greater than zero, all of them the same. Thus, in a CALM time is not a densely ordered sequence of instants without duration. In a CALM time is a strictly ordered (with immediate successiveness) and finite sequence of immediately successive qutits. And a time interval will always be defined by a finite number of successive and adjacent qutits. In a CALM time is therefore absolute. And since space is also absolute (article 12 of this series), only absolute motion is possible in CALMs. Although in the physical world (whether or not modeled by a CALM) preinertia makes it impossible to detect and measure absolute motion. Only relative motion resulting from different absolute motions of physical objects through the real physical space can be observed and measured. Except, perhaps, in the conditions analyzed in the Santiago del Collado experiment.

27. The Shame of Physics

Abstract.-This paper develops an argument about the nature of motion based on preinertia, a universal property of all physical objects that physics has not yet discovered. The argument is closely related to the historical Clarke-Leibniz epistolary debate, which is now considered settled in favor of Leibniz's relativistic thesis, the same thesis that modern physics has adopted and enforced in an absolutist manner since the beginning of the twentieth century. The brief preinertial argument developed in this article proves that the debate may indeed be settled, but in the opposite direction. This really justifies the title of the article, not because of the probably wrong solution imposed (science is trial and error), but because of the totalitarian way of imposing it. And, of course, for not having yet discovered preinertia, surely the most universal property of all physical objects.

Keywords: preinertia, absolute motion, relative motion, physical space, gravitational waves, special relativity.

27.1 Introduction

As is well known, the enigma of parallels was called *the shame of mathematics* in the 18th century, among others by the "prince of mathematics" C. F. Gauss (quoted in [264, p. 9]). It may have been an exaggeration, for as E. Beltrami proved in the following century, non-Euclidean geometries are formally consistent, implying that Euclid's fifth postulate cannot be deduced from the other four [25, 26]. So it was not an embarrassment, but an impossibility. What would have been possible was to change Euclid's postulates (including a productive definition of a straight line) and prove the famous postulate on the new formal basis [196, 200].

It is appropriate to ask whether there is anything similar in physics that could be called "the shame of physics". In my opinion the answer is yes. Like geometry with its unsolved problems of parallels, physics has its own similar problem with the relative or absolute nature of motion, also discussed for centuries and only seemingly resolved, as the reader of this article will see. The novelty of the argument developed here is the use of a new concept (which should be classical because of its overwhelming empirical evidence): preinertia, a universal property of all physical objects by virtue of which each of them inherits the velocity VECTOR of the object on which it is at rest when it is set in motion (the absolute/relative nature of this inherited velocity will be discussed below).

It can be formally demonstrated that even photons are preinertial [208, pp. 337-356], although the demonstration is not necessary given the overwhelming empirical evidence for preinertia, for example, every time one of our objects falls to the ground; or every time we land on the same place we jumped (an argument used by some classical Greeks to defend the immobility of the Earth); or every time we start walking; or every time we throw a stone, a ball, an arrow, a rocket... In all these cases, a physical object is set in motion from another physical object (the Earth) moving at 370 km/s (1332000 km/h) in the direction of galactic coordinates (264.4±0.3, 48.4±0.5). If an object set in motion from the Earth did not inherit this motion from the Earth, it would shoot out in the opposite direction at about 1332000 Km/h, which obviously never happens. And the reason it never has and never will is preinertia, the shame of physics, because physics has not yet discovered preinertia, the most universal property of all physical objects, as will be seen here.

On the other hand, it is remarkable that physics has always used preinertia implicitly (I would say that more than implicitly it has always been used unconsciously) in almost all its arguments and theories. Let us recall, for example, the following words of Galileo, published in 1632 [125, p. 213]:

> But the diurnal revolution, by its own natural motion, is imparted to the globe, and consequently to all its parts, and, so far as it is impressed by Nature, it is ineradicable to them. Thus the stone at the top of the tower has as its primary instinct to revolve around the center of its whole in twenty-four hours, and exercises this natural talent eternally, whatever state it is in.

This primary instinct is, of course, the universal property I have just called preinertia, a universal physical property of all physical objects that physics should begin to consider, make its existence explicit, and use it in the discussions about the nature of motion, and even in the discussions about the nature of objects with the ability to move.

The main argument developed in the fourth section of this chapter is an example of how the concept of preinertia can be used, and at the same time a proof of its enormous importance. As you will see, it is an unexpected argument with strong historical resonances ranging from the Clarke-Leibniz controversy to the relativistic spacetime continuum of our own day. Besides being unexpected, it is, in my opinion, far from irrelevant for the future of physics: preinertia proves that absolute motion is undetectable and that motion is absolute, being the relative motions, the only observable motions, obvious consequences of the different unobservable absolute motions.

Apart from this introduction, the chapter consists of four sections. The second section introduces the concept of preinertia and briefly discusses some of its more immediate implications, such as the impossibility of detecting absolute motion; or the relation that preinertia might have to inertial mass and gravitational mass. The third section is a short argument on rotations, which serves as a preamble to the fourth section, which develops the main argument of the paper on the absolute or relative nature of motion. The conclusion of the argument is confirmed by another, even shorter, argument on the real nature of physical space. The fifth section briefly discusses the consequences of this conclusion for the real or apparent nature of relativistic inertial deformations of space and time. Obviously, if the argument developed in the fourth section is consistent, these deformations can only be apparent.

27.2 Preinertia: the vectorial inheritance of motion

Preinertia is a universal property with much more empirical evidence than inertia. The problem is that its existence and importance for understanding the nature of motion, and even of the physical objects, had not occurred to us. It can be defined in the following terms:

Definition 20 (of Preinertia) *Capacity of a physical object to inherit the velocity vector of the proper reference frame in which it is set in motion.*

The penultimate section of this chapter (section 27.4) proves that the inherited velocity can only be absolute in nature. And although it can be formally demonstrated from the Lorentz Transformation that even photons are preinertial [208, pp. 337-356], it seems appropriate, in view of their enormous empirical evidence, to suggest their inclusion in the statement of the Principle of Inertia:

Principle 4 (of Inertia) *All physical objects are preinertial and maintain their state of motion as long as no external agent acts upon them.*

The universal reality of preinertia can also be demonstrated experimentally. A forthcoming work on the experiment being carried out in

Santiago del Collado (Avila, Spain) will confirm this is the case. Preinertia is also consistent with the conservation principles of physics, to which it may be closely related.

On the other hand, it is logical to think that preinertia makes it impossible to detect the absolute motion of a reference frame by setting in motion its own objects. And the reason could not be clearer: the objects used in the detection attempt, including photons, would have the same (inherited) component in their motion as the motion to be determined. Or to put it in an elementary example: it is impossible to detect the velocity of a train by dropping an object on the floor of the train: the object inherits the velocity vector of the train when it is dropped (preinertia), maintains it while falling (inertia), and will always fall in the same place, regardless of the velocity of the train. Preinertia is also consistent with the conservation principles of physics, to which it may be closely related. The formal proof that follows is also very simple (taken from [208, p. 328-330]):

Assume, just for a moment! that there exist an absolute reference frame RF_a (perhaps made of the indivisible quantum space units qusits) through which physical objects can move in absolute terms. Let RF_o be a reference frame at rest in RF_a, and let A be any physical (point) object at rest in RF_o, where it is placed in the position (x_1, y_1, t_o) of RF_o (for simplicity, we dispense with the z-coordinate). Let A be set in motion at t_1, $(t_o < t_1)$ with a uniform velocity $\vec{u}$ so that at the instant t_2 it is placed in the position of coordinates (x_2, y_2, t_2).

Consider now that RF_o moves in RF_a with an absolute and uniform velocity $\vec{v}$, and let A be set in motion under the same above conditions when RF_o was at rest in RF_a. Thanks to preinertia, A *inherits* the absolute velocity vector $\vec{v}$ of RF_o with respect to RF_a, and thanks to the Principle of Inertia, A *maintains* $\vec{v}$ along its own motion with respect to RF_o. Let O_o be the origin of coordinates of RF_o. This point O_o moves with respect to RF_a at a velocity $\vec{v}$, while A moves with respect to RF_a at a velocity:

$$\vec{w} = \vec{u} + \vec{v} \tag{1}$$

The object A (that moves with respect to RF_a at the velocity $\vec{w}$ given by (1)) will move with respect to O_o (that moves with respect to RF_a at the velocity $\vec{v}$) at a velocity $\vec{u'}$ given by:

$$\vec{u'} = \vec{w} - \vec{v} \tag{2}$$
$$= \vec{u} + \vec{v} - \vec{v} \tag{3}$$
$$= \vec{u} \tag{4}$$

which is the same velocity as if RF_o were at rest with respect to RF_a.

27.2 Preinertia: the vectorial inheritance of motion

In consequence, the coordinates of A in RF_o at t_2 will be the same as in the first case when RF_o was at rest in RF_a. So, the coordinates of A at t_2 will also be (x_2, y_2, t_2), and they cannot be used to detect the absolute motion of RF_o.

Since RF_o is any reference frame, A any physical object initially at rest in RF_o, and $\vec{u}$ any uniform velocity, we must conclude that the absolute motion of a reference frame is undetectable by setting into motion any physical object (or objects) of that reference frame.

The impossibility to detect any possible absolute motion due to preinertia is already a sufficient reason to consider this universal property of all physical objects. And there is still the most important reason, which will be discussed in section 27.4 of this chapter.

Another important aspect of preinertia is its possible relationship to inertial mass and gravitational mass. Remember that:

1. Every material object offers resistance to change its state of motion (inertial mass).

2. Every material object alters the properties of physical space (gravitational mass).

3. Every material object is sensitive to the gravitational fields created by other objects (gravitational mass).

4. Every physical object inherits the velocity vector of the reference frame in which it is set in motion (preinertia).

And the inevitable question is: do these four properties of material objects have the same common cause? The possible affirmative answer could be a fundamental mass, from which inertial mass, gravitational mass, and preinertia are derived. Note, however, that photons, which are supposedly massless, are also preinertial. But photons, as such particles, have only spin 1. Other particles have a different spin (1/2) and yet all have the same preinertia. So it does not seem reasonable to think that spin is the fundamental cause of preinertia. On the other hand, it is also worth considering that there are extremely small universal masses, such as the mass I have called QUANTUM MASS or RYDBERG MASS. [208, p. 235]:

$$m_q = \sqrt{\frac{G\hbar^3 R_\infty^4}{c^5}} = \hbar\, t_p R_\infty^2 = 6.845023 \times 10^{-64} Kg$$

where t_p is the Planck time and R_∞ is the universal Rydberg constant, which is specific to each chemical element and varies slightly with its mass. It then seems reasonable to propose some fundamental mass for photons, such as the quantum mass mentioned above, as the cause of their preinertia, and we would have the same cause as all other preinertial objects. Moreover, if photons had mass, one could analyze the possibility that their gravitational interaction with very massive objects bends their trajectories, rather than those very massive objects bend the physical space itself, as general relativity proposes, a bending much more bulky than that of photon trajectories.

In any case, it seems reasonable to propose that all, absolutely all, physical objects capable of motion have the same property (fundamental mass?) responsible for resisting changes in their state of motion; for modifying the state of motion of other objects at a distance by changing the properties of the surrounding space; and for inheriting in vector terms (!) the state of motion of the object on which they were at rest when they were set in motion. Obviously, it would be a key property in the evolution of the universe.

27.3 An elementary preamble on rotations

The Newton's Bucket experiment [260, p. 131-132] [259, p. 80-81] is, in my opinion, one of the most important real experiments in the history of physics, both for its results (which are not sufficiently appreciated in our relativistic days) and for the controversies that it gave rise to, especially the one between Clarke (absolutist) and Leibniz (relativist) [233, p. 34-46][234, p. 67-86]. Two hundred years later, E. Mach revived the controversy by proposing that the water in Newton's bucket was in fact rotating WITH RESPECT TO the sphere of fixed stars (SFS) in the place of AROUND an internal axis of rotation [233, p. 83-84][234, p. 45].

Consider the daily rotation of the Earth around its internal axis of rotation. As a result of this rotation, from the Earth, the Sun is observed to rotate around the Earth daily. But this daily rotation of the Sun is not only apparent, it is impossible: for the same reason as in the case of the Earth, the Sun would also have to rotate around each of the bodies that orbit around it and revolve around an internal axis of rotation (Mercury, Venus, Mars, etc.). Therefore, each point of the Sun would have to describe a large number of different circular trajectories around different centers of rotation at the same time, which is physically and geometrically impossible. But the appearance, as such appearance, is real. Therefore the rotation of the Earth that produces it can only be a real rotation.

Consider again the Earth rotating around its internal axis of rotation, the axis Ax. Each point of the Earth describes a circle around a unique point, its center of rotation on the axis Ax. Consequently, and since a point cannot rotate around two or more centers of rotation at the same time, the rotation of the Earth can only be referred to its internal axis of rotation Ax. Thus:

1. The rotation of the Earth is real, as evidenced by the apparent rotations it causes in other celestial bodies as the Sun.
2. The rotation of the Earth can only be referred to its own axis of rotation.
3. Therefore, the rotation of the Earth can only be an absolute rotation, i.e. an absolute motion.

The same conclusion, and for the same reasons, must apply to the billions of celestial bodies that rotate around an internal axis of rotation. The vast majority of these celestial objects have motions defined by multiple components, one of which is an absolute rotation around an internal axis. The question is how can a motion, one of whose components is an absolute motion, be relative? The following section points to a very simple answer.

27.4 A preinertial argument on the nature of motion

As we will see, preinertia reopens the classic Clarke-Leibniz epistolary debate about the absolute/relative nature of motion, albeit in very different non-theological terms. Indeed: Suppose that at a given instant t_o a cosmic object A is uniformly moving with respect to any other cosmic object X with a given relative velocity vector $\vec{v}_{xa}$. According to the principle of inertia, all objects at rest on A move with the same relative velocity vector $\vec{v}_{xa}$ with respect to X, and thanks to preinertia they inherit this relative velocity vector when they are set in motion from A itself.

Suppose then that at the precise instant t_o one of these objects at rest on A, say B, is set in motion with a rectilinear and uniform velocity vector $\vec{v}_{ab}$ with respect to A. And let us also suppose that all motions are relative, that absolute motion does not exist, as is assumed in the hegemonic relativistic stream of contemporary physics, for which absolute motion is meaningless [74, p. 341], is anathema. The preinertia of B implies that B inherits the relative velocity vector $\vec{v}_{xa}$ of A with respect to X as a component of its own velocity vector $\vec{u}_b = v_{xa} + v_{ab}$, so that B moves with respect to X with a velocity vector $\vec{u}_b$.

Now then, A has billions of relative velocity vectors (most of them variable with time due to different cosmic incidents) with respect to the billions of different objects in the universe (photons, neutrines, electrons, planetesimals, planetoides, planets, stars etc.). Although (in principle) it could be sufficient to inherit only one of them, for example the relative velocity vector $\vec{v}_{xa}$, there is no physical or logical reason to inherit one of them and not any of the others. But things are much more complicated because we also have to take into account:

1. The continuous variation of all the billions of relative motions due to continuous interactions of all kinds (collisions, accelerations, decelerations, explosions, etc.) that produce changes in the billions of the relative velocity vectors of the corresponding objects.

2. In addition to the storage system of the information corresponding to all these billions of relative velocity vectors, and taking into account that the vast majority of these objects are not quantum entangled, there would have to exist in each object a mechanism of emission and reception of the information corresponding to all those billions of changes in the relative velocity vectors of the different cosmic objects, and in addition there would have to exist a way to propagate that information through distances of billions of light years.

3. The object that is set in motion could in turn be the object from which other objects are set in motion.

Under these conditions, each object O set in motion with the ability to set other objects in motion (including, for example, electrons that could emit photons) would have to inherit the information of all the billions of relative velocity vectors of the object A from which it was set in motion, in order to transmit them in turn to the new objects O' that could be set in motion from the object O. Otherwise, the object O would not have the necessary information to update the successive changes in the relative velocities of the billions of objects whose relative velocity with respect to the first object A would have changed in the past and not yet been updated in A when O was set in motion from A.

Consequently, every natural object (including living organisms!) would have to have a system for storing information about all the relative velocity vectors of all the billions of other physical objects moving with respect to it, as well as a system for transmitting and receiving all those inevitable changes in relative velocity. In addition, and as indicated above, there would also have to be a way to propagate throughout the billions of light-years of all space the information of all the changes in velocity that the billions of cosmic objects may undergo as a result of all kinds of cosmic interactions. Obviously, none of this seems to exist, nor does it seem reasonable to assume that it could exist in any of the objects of the universe that can be set in motion from another moving object on which they were at rest. This is simply absurd, and there is an extremely simple alternative:

> The unique velocity vector inherited in preinertia is the absolute velocity vector (THROUGH the real physical space) of the object from which any other object at rest (in the first

object) is set in motion.

The only real motions would be the absolute motions THROUGH absolute space, as Newton defended [259]. The different absolute motions of the different objects being the cause of their different relative motions, which are the only motions we can detect for the time being, just because of preinertia.

The above conclusion about motion is confirmed by another completely independent argument concerning the physical reality of space (which is still denied by many contemporary physicists). Indeed, the empirical detection of gravitational waves proves the physical reality of space, the existence of a space matter, since what does not exist cannot vibrate, nor can it transmit its own vibrations, nor can it modify the size of other physical objects as the arms of the interferometers that detect those space vibrations. The only objects with empirically detectable physical properties are real physical objects; fictional objects have no empirically detectable physical properties. Consequently, and once the vibrations of space (gravitational waves) have been empirically detected [1, 23, 40, 49, 50, 62, 100, 119, 146, 154, 250, 261, 299, 333, 372], it must be admitted that space is a real physical object with real and empirically detectable physical properties. This real space is the unique common space for all real physical objects (except space itself). Consequently, motion THROUGH a real and unique common space can only be considered as absolute motion. Therefore, the entire argument of this section proves the following theorem.

Theorem 41 (of Absolute Motion) *The universality of preinertia and the reality of absolute physical space prove that all motion through that absolute space is absolute motion.*

On the other hand, everything would be much simpler if that were the case. And it is worth remembering that the physical world bears in its essence the signature of simplicity, as we are reminded by Ockham's Razor and the following words of Galileo, with which I concluded the previous argument about preinertia and the nature of motion [125, p. 183-184]:

> Now, if in order to achieve the same effect in a precise way, it is just as important that the Earth alone should move, stopping all the rest of the Universe, as it is that the whole Universe should move with a single movement, who would want to believe that Nature (which, according to common agreement, does not do by the intervention of many things what it can do by means of a few) has chosen to make an immense number of very large bodies move, with inestimable velocity, in order to achieve what can be obtained by the moderate

movement of a single body around its own center?

27.5 Consequences on the theory of special relativity

The above Theorem of Absolute Motion would be confirming the apparent, not real, nature of the inertial deformations of spacetime deduced from special relativity (actually from the Lorentz Transformation). Indeed, the spacetime deformations of special relativity could be only apparent, as apparent as the refractive deformations: no matter how many times we experimentally confirm Snell's Law, the rod partially submerged in water is not really bent. In the case of the relativistic FitzGerald-Lorentz contraction, a good part of relativists think that it is not real, but apparent, because an object cannot have different sizes at the same time; nor can an elastic band at rest be more stretched in some parts than in others if it is free of external forces; see the elastic band argument in Chapter 15 (page 165), and many more in [208].

Now then, if one of the consequences of the Lorentz Transformation is apparent, are the other consequences also apparent? If the answer is no, what part of the theory of special relativity determines which of these deformations are apparent and which are not? and why should some be apparent and others real? Moreover, let us not forget that the experimental confirmations of special relativity must be:

1. SYMMETRIC: If from an inertial reference frame RF_A a spacetime deformation is observed in another inertial reference frame RF_B, then *at the same time* and from the reference frame RF_B the same deformation must be observed in the reference frame RF_A. A symmetry that, as far as I know, has never been confirmed.

2. UNIVERSAL: All objects contract in the direction of relative motion in exactly the same way, regardless of their composition and internal structure: wood, paper, steel, elastic bands, glass, etc., all contract in the same way and without any external force explaining the contraction. And the ticking of clocks also expands in the same way in all imaginable types of clocks: mechanical, electrical, electronic, biological, etc., without any cause that explains the change in the corresponding mechanisms that cause their respective periodic events (tic-tac) used to measure time. Of course, modern clocks that display time on large alphanumeric displays call into question inertial time dilation and relativistic local simultaneity, unless those displays simultaneously display as many different times as different relative velocities at which they can be observed [208].

3. ACAUSAL: The relativistic spacetime deformations have no specific physical cause that produces them. The only cause of their ex-

istence would be the relative velocity at which the corresponding objects and events are observed. The problem is that we can observe and measure deformed objects that are not really deformed but apparently deformed, a deformation that also depends on the way in which these apparently deformed objects are observed, in this case partially submerged in water.[1]

All this points to the fact that the relativistic inertial deformations of space and time, as mentioned above, could only be apparent. Not to mention the more than possible inconsistency of the infinitist spacetime continuum, if any of the more than 40 proofs of the inconsistency of the Hypothesis of the Actual Infinity in [212, pdf] is correct. Let me end by recalling one of those iconic images of a mad, genius-looking scientist proudly posing in front of a blackboard full of mathematical signs. Doesn't the reader think that science has too much ego, too much author mania? And what will happen if this infinitist mathematics turns out to be wrong? The reader can get an idea of this possibility by taking less than five minutes to read the two final appendices to this article.

---

[1] Are we really so idiot?

28. A Special Relativity Inconsistency

Abstract.-After proving that rotations can only be absolute motions, the relativistic assumption that all uniform motions are relative is used here to prove the existence of two inertial reference frames A and B in which the relative uniform velocity of A with respect to B is real, while the reciprocal relative uniform velocity of B with respect to A is only apparent, as apparent as the daily rotation of the Sun around the Earth. An asymmetry due to the fact that only A underwent a brief acceleration that changed its previous uniform relative velocity with respect to B, so that the corresponding reciprocal change in the uniform relative velocity of B with respect to A, as observed from A, has no physical cause. It is therefore an apparent change in velocity, from which only an apparent uniform velocity can result. This symmetry breaking violates the principle of relativity.

Keywords: rotation, relative motion, absolute motion, change in velocity, causal change, non-causal change, absolute space, preinertia, special relativity, Principle of Relativity, inconsistency of the non-causal relativism.

Note.-As usual, in this chapter the reciprocal velocity of the relative uniform velocity of an object A with respect to another object B will be the relative uniform velocity of B with respect to A.

28.1 The memory of a historic debate

For the reasons that will be explained in the following section, and in spite of G.W. Leibniz and E. Mach, it seems difficult to reject Newton's conclusion about the absolute nature of the rotation of the water in the bucket of his famous experiment [260, p. 131-132] [259, p. 80-81]. In the case of Leibniz, even though two of his logical principles (the Principle of Sufficient Reason and the Principle of Identity of Indiscernibles) were involved, his argument was more theological than physical and logical, as mentioned on page 127 of this book:

Since (according to Leibniz) two different and indiscernible things cannot exist, and since in Newton's absolute space things could be located in several different and indiscernible ways, Leibniz argued that God would have had to choose one of these indiscernible ways, without any reason to choose one of them in preference to the others, which for Leibniz is not proper to God. Therefore, absolute space cannot exist.

In Mach's case, his argument was neither physical nor logical, but arbitrary: the water in Newton's bucket does not rotate absolutely, but WITH RESPECT to the dark background of fixed stars [233, p. 34-46][234, p. 67-86]. I have emphasized the words "with respect to" because they were the initial reason for my critique of those arguments, the critique which is presented, along with its conclusions, in this chapter.

The expression "with respect to" is the least appropriate to refer to a rotation. Objects rotate AROUND something, for example AROUND an internal axis, as is the case with the daily rotation of the Earth. In these cases, the word "around" expresses that each point of the rotating object (for example around an internal axis) always describes the same circular trajectory whose only center is always the same point on the axis of rotation. This uniqueness of the trajectories of rotating objects will be one of the critical elements in the analysis of rotations introduced in the previous chapter and recalled in this one. This uniqueness is also, I suspect, the source of so much confusion and historical debate.

28.2 Rotations are always absolute motions

(This section repeats an argument given in Chapter 12)

To avoid duplication of explanations and arguments, only pure circular rotations will be analyzed, although for reasons that will be seen in the arguments, the uniqueness of other types of closed trajectories (e.g. elliptic) allows them to be immediately included in the arguments developed for pure circular rotations, which will henceforth be referred to simply as rotations.

Consider the daily rotation of the Earth around its internal axis of rotation. As a result of this rotation, the Sun is observed from the Earth to rotate daily around the Earth. But this daily rotation of the Sun is not only apparent, it is impossible: for the same reason as in the case of the Earth, the Sun would also have to rotate around each of the bodies that orbit it and revolve around an internal axis of rotation (Mercury, Venus, Mars, etc.). Therefore, each point of the Sun would have to describe at the same time a large number of different circular trajectories around different centers of rotation, which is physically and geometrically impossible. Well then, even though it is impossible, the apparent rotation of the Sun around the Earth, as such an APPARENT rotation, has a real cause: the daily rotation of the Earth around its internal axis of rotation. Therefore, the daily rotation of the Earth around its axis of rotation is a REAL ROTATION.

Again, consider the Earth rotating around its internal axis of rotation. Each point of the Earth describes daily a circle around a unique point, its center of rotation on the axis of rotation. Consequently, and since a point cannot rotate around two or more centers of rotation at the same time, the rotation of the Earth can only be referred to its internal axis of rotation. Thus:

1. The daily rotation of the Earth is real, as evidenced by the apparent rotations it causes in other celestial bodies as the Sun.

2. In its daily rotation, each point of the Earth moves always along the same circular trajectory whose center is always the same point on its axis of rotation.

3. So, and due to the uniqueness of the trajectories described by its points around the internal axis of rotation, the daily rotation of the Earth can only be referred to its own axis of rotation. Then, the daily rotation of the Earth cannot be relative to any other object different from its axis of rotation.

4. Therefore, and being a real rotation that can only be referred to its internal axis of rotation, the daily rotation of the Earth can only be absolute, i.e. it can only be an absolute motion.

The same conclusion, and for the same reasons, must apply to the billions of celestial objects that rotate around an internal axis of rotation. The vast majority of these celestial objects have complex motions defined by multiple components, one of which is an absolute rotation around an internal axis. The question is how can a motion, one of whose components is an absolute motion, be relative? The following sections points to a very simple answer.

28.3 Real and apparent velocity changes

Let us consider a space probe P that is teleguided from the Earth and moves free of gravitational interactions at a certain uniform velocity $\vec{v}_{ep}$ with respect to the Earth. This relative uniform velocity of P with respect to the Earth implies that P also moves relative to the rest of the celestial bodies. So P will be moving with billions of different relative uniform velocities, one for each of the billions of celestial objects uniformly moving (COUMs for short). Suppose that at a certain instant the space probe P is accelerated from Earth by starting its small engine for a few seconds, so that its new uniform velocity relative to Earth doubles.

Once P reaches its new uniform velocity with respect to the Earth, its billions of uniform relative velocities with respect to each of the billions of COUMs in the universe will have changed, regardless of their distance from P. But the real change in the velocity of P occurs only in P, because the cause of the velocity change (the propulsion caused by its small engine, activated for a few seconds from the Earth) is only in P. This change in the velocity of P changes the relative uniform velocity of P with respect to the Earth and with respect to each of the COUMs. And since these new uniform velocities of P are relative to each of the billions of COUMs, it will be observed from P that it is the reciprocal relative uniform velocity of each of these billions of COUMs with respect to P that has changed to a new uniform relative velocity.

Now then, while there is a physical cause that explains the changes in each of the relative uniform velocities of P with respect to each of the billions of COUMs, there is no physical cause for the change in the reciprocal relative uniform velocity of each of those billions of COUMs with respect to the P, as observed from P. And in a consistent universe no physical change in a physical object can occur without a physical cause producing the change. In consequence, all those billions of changes in the relative uniform velocity of each of the COUMs with respect to P can only be apparent, unreal. And since the resulting velocity from an apparent change of velocity can only be an apparent velocity, the new uniform velocity of each COUM with respect to P observed in P can only be apparent, as apparent as the velocity of the daily rotation of the Sun around the Earth, as observed from the Earth.

The above argument of the space probe P can be applied to any one of the billions of COUMs, including the smaller ones (planetoids, comets, meteorites, dust particles, etc.), that undergoes a change in its uniform velocity due to any physical cause, e.g. due to a gravitational interaction. In fact, consider any one object X among the billions of COUMs that at any given moment undergoes a change in its relative uniform velocity with respect to another much more massive COUM Y due, for example, to a slight gravitational interaction between X and Y.

Once X has reached its new uniform relative velocity with respect to Y, all the relative uniform velocities of X with respect to each of the rest of the billions of COUMs will also have changed. And since they are relative uniform velocities, the reciprocal relative uniform velocities of each of these billions of COUMs with respect to X will also have changed, as is observed from X. But if these changes in the relative uniform velocities of each of the billions of COUMs with respect to X

were real changes, they would be physical changes without physical causes producing them. So they can only be apparent changes in velocity, and the resulting new relative uniform velocity of each COUM with respect to X can only be an apparent velocity, as apparent as the velocity the daily rotation of all celestial bodies around the Earth, as observed from the Earth.

From the above argument follows an inescapable conclusion: if there are only relative velocities in the universe, if absolute velocities are meaningless [74, p. 341], as special relativity claims, then billions of changes in relative uniform velocities are continuously occurring in the universe such that these changes are real (by virtue of their physical causality) in certain inertial reference frames, and only apparent, unreal, (by virtue of their physical non-causality) in other inertial reference frames. In consequence, there exist in the universe millions of COUMs, as the above X, whose uniform relative velocity with respect to any other COUM is real while the reciprocal relative uniform velocity of this COUM with respect to the first one is only apparent. Or in other words, according to special relativity, in the universe it is possible to consider two inertial reference frames A and B such that the uniform velocity of A with respect to B is real, while the reciprocal uniform velocity of B with respect to A is only apparent. A conclusion that breaks the symmetry of the Principle of Relativity. It is a special relativity inconsistency. As we will see in the next section, this inconsistency has a very simple non-relativistic solution.

28.4 Inconsistency of the non-Causal Relativism

It seems reasonable to propose that the uniform velocity vector of any object X in the present universe is the result of the particular history of its successive velocity changes, changes caused by its successive physical interactions with other physical objects in the universe. The problem we encountered in the previous section is that while there are physical causes that explain the changes in the relative uniform velocity of X with respect to the rest of COUMs, there is no physical cause that explains the reciprocal change in the relative uniform velocity of the rest of COUMs with respect to X when observed from X and interpreted within the framework of special relativity. Consequently, and according to this theory, billions of physical causal changes of relative uniform velocities must be continually occurring in the universe together with their reciprocal non-causal changes of relative uniform velocities. This is the inconsistency of the non-causal relativism.

Evidently, the above inconsistency of the non-causal relativism breaks the symmetry established by the Principle of Relativity, because while the uniform relative velocities resulting from causal (real) changes in the velocity of a given physical object are real, those resulting from the corresponding reciprocal non-causal (non-real) changes can only be apparent. An asymmetry that will occur whenever a physical object undergoes a causal (real) change of its uniform velocity that the billions of celestial objects uniformly moving (COUMs) relative to that object do not undergo.

As Galileo would say, humans do not have the sensors to perceive uniform rectilinear motion as we have to perceive heat or pressure [126, p. 529]. Nor can we perceive space as such a physical object, which was, and continues to be, only a useful fiction for a good part of modern physicists (XX and XXI centuries). Although it is to be expected that the physical reality of physical space will eventually be accepted, once some of its physical properties have been empirically detected and measured (gravitational waves [214]). But as real as it is, physical space is completely transparent to all physical objects, which can move through it without encountering any resistance that could be used in the detection of that motion. And as if these problems were not enough, preinertia makes it impossible to detect the possible absolute motion of an object THROUGH space, hence the repeated failures of Michelson-Morley type experiments [208]. For the time being, we are left with only logic to try to solve the above inconsistency of non-causal relativism and its corresponding symmetry breaking of the Principle of Relativity.

A very simple solution to this inconsistency would be to consider that all physical objects move THROUGH the same physical space with different absolute velocity vectors, which would result in the observed different relative velocities between such physical objects. Moreover, if space had a structure, its structural elements could serve as reference elements to describe the absolute motion of any object. This possibility is not so far-fetched if, as it seems, infinity and infinite divisibility are inconsistent [212, 213] while the universe is consistent. Under these conditions, there would exist minimal and indivisible units of space (qusits) and time (qutits) which, once discovered, could serve to define an absolute reference frame in order to describe all absolute motions of all physical objects THROUGH the real physical space.

The problem with this simple solution is that it has no place in modern relativistic mechanics: space is relative (and fictitious to most modern physicists), time is relative (and fictitious to most modern physicists), and then motion can only be relative (though real in this case!). Absolute motion is therefore meaningless, even though there are absolute rotations (as proved above). Under these conditions, the proposed solution to the inconsistency of non-causal relativism is already condemned to contempt and ostracism. It will not even be considered by relativistic officialism, because, as is well known, special relativity is *sufficiently confirmed by experience*. In the following section we will recall some aspects of these empirical confirmations of the special relativity that are very rarely found in the special relativity literature.

28.5 On the empirical confirmation of special relativity

(This section partially repeats some of the contents of the Chapters 15 and 27)

To begin this new discussion, it is worth remembering that an experiment can confirm more than one theory. This is what happened, for example, with the first experimental confirmation of time dilation in the Ives-Stilwell experiment performed in 1938 by H. E. Ives and G. R. Stilwell [174, 168, 167, 170, 171] (a short modern review can be found in [110]). This experiment confirmed H.E. Ives' own theory of absolute space and time, a theory completely different from special relativity (Ives is often regarded as the most important opponent of Einstein's theory of relativity of his time). With this theory, Ives was also able to explain the anomaly of Mercury's orbit, which had been explained by Einstein in 1916, but in Ives' case with different methods of classical mechanics, which do not assume the curvature of spacetime. [169, 172, 173, 351].

On the other hand, the observed relativistic inertial deformations of space and time could also be the consequence of explaining a discrete physical world with indiscrete mathematics based on the infinitist spacetime continuum, simply because the factor that converts between both versions (discrete and continuous) of the Pythagorean Theorem (a key theorem in determining distances) is precisely the relativistic Lorentz factor γ [208]:

$$\gamma = \frac{1}{\sqrt{1-k^2}}; \ k = v/c; \ 0 < k < 1 \tag{1}$$

where v is the relative velocity at which the relativistic deformations are observed. As is well known γ intervenes in the definition of all these inertial deformations of space, time and mass in special relativity.

Moreover, the deformations of space and time deduced from special relativity could be only apparent, as apparent as the refractive deformations: no matter how many times we experimentally confirm Snell's Law, the rod partially submerged in water is not really bent. In the case of the relativistic FitzGerald-Lorentz contraction, a good many relativists think that it is not real, but apparent, because an object cannot have different sizes at the same time; nor can an elastic band at rest be more stretched in some parts than in others if it is free of external forces (see the elastic band argument in page 165 of this book, and much more in [208]). Now, if one of the consequences of the Lorentz Transformation is apparent, are the other consequences also apparent? If the answer is no, what part of special relativity determines which of these deformations are apparent and which are not? And why should some be apparent and others real?

Finally, let us recall that the experimental confirmations of special relativity must be:

1. SYMMETRIC.-If from an inertial reference frame A it is observed that in all physical objects of another inertial reference frame B there is a contraction of length in the direction of the relative motion between the two frames (FitzGerald-Lorentz contraction), then at the same time from the reference frame B it must also be observed that in all objects of A there is exactly the same contraction of length in the same direction of the relative motion between the two frames. The same is true for the inertial dilation of time: if from the reference frame A it is observed that all clocks of the reference frame B run slower than the proper clocks of A, then at the same time from the reference frame B it must be observed that all clocks of A run slower than the proper clocks of B. And whatever the events are, two simultaneous events in A that occur at a non-zero distance in the direction of relative motion will be observed as non-simultaneous in B. And two simultaneous events in B separated by a non-zero distance in the direction of relative motion will be observed as non-simultaneous in A.

2. UNIVERSAL.-Provided they are observed at the same relative velocity $v = kc$, $(0 < k < 1)$, all objects will contract in the direction of relative motion by the same factor $\sqrt{1-k^2}$, regardless of their size, composition, and internal structure. A steel cube and a foam rubber cube will undergo the same amount of contraction in the direction of relative motion. Time dilation and phase difference in synchronization are also universal: the same for all kinds of clocks: mechanical, electrical, electronic, atomic, chemical, biological, etc.

3. ACAUSAL.-The sole and exclusive cause of all relativistic inertial spacetime deformations is the relative velocity at which the involved physical objects and events are observed. There is no physical agent involved in these deformations. Only the relative velocity at which they are observed.

This obviously *relativizes* the validity of the experimental confirmations of special relativity.

29. Two fallacies in modern physics

Abstract.-This short paper proves the falsity of two popular and important statements of modern physics.

Keywords: comprehensible universe, conscious observers.

29.1 On spooky actions and double slits

A year after publishing (together with B. Podolsky and N. Rosen) the famous paper in which its authors questioned the completeness of quantum mechanics, A. Einstein published an essay entitled PHYSIK UND REALITÄT [92], in which he continued to insist on the problems accumulating in the foundations of physics. Among them were the problems related to quantum non-locality. From this second essay comes a statement that would eventually become one of the most popular quotes of its author [99, p. 315]:

> Man kann sagen: Das ewig Unbegreifliche an der Welt ist ihre Begreiflichkeit. Dass dieSetzung einer realen Aussenwelt ohne jene Begreiflichkeit sinnlos wäre, ist eine der grossen Erkenntnisse Immanuel Kants.

> One can say: The eternal incomprehensible of the world is its comprehensibility. That the setting of a real external world would be senseless without that comprehensibility is one of the great insights of Immanuel Kant.

Also summarized and highly cited in the forms:

> The eternal mystery of the world is its comprehensibility... The fact that it is comprehensible is a miracle.

> The most incomprehensible thing about the universe is that it is comprehensible.

About 50,000 pages on the web currently echo Einstein's statement, in one or another of its various versions. It also appears very frequently in the primary and secondary literature of physics, so that it has been and continues to be widely read and commented on by tens of thousands of researchers, professors, and students around the world.

Let us also recall another widely reported statement, in this case from the literature on the diffraction of photons and other particles through a double slit, and then from the some interpretations of quantum mechanics (for instance, von-Neumann-Wigner interpretation, or the relational interpretation, also being compatible with the widely assumed Copenhagen interpretation). The following four are appropriate examples of this statement from four physicists (emphasis is mine):

> M. Planck: I consider matter to be a derivative of CONSCIOUSNESS [277].
>
> A. Goswami: No object exists in spacetime without a CONSCIOUS observer observing it [132, p. 60].
>
> A.D. Linde: The universe and the observer exist as a duo (quoted in [319, p. 210]).
>
> C. Rovelli: ...the INCORRECT notion that generates the unease with quantum mechanics is the notion of OBSERVER INDEPENDENT state of a system [301, p. 1]

29.2 Two false assertions in modern physics

To conclude that something is, or is not, comprehensible can only be done by a rational mind after correctly applying the methods of the formal and experimental sciences. But on Earth, rational minds are produced (or modulated) by human brains. So here on Earth, it takes a human being to come to a rational conclusion about whether something is comprehensible or not. The problem is that human beings took more than 13700 million years to appear in this universe, and not as a random and sudden event, but as a result of the evolution of the universe itself. An evolution that, in the case of the Earth, has been partially recorded in its rocks since they existed at least 4.4 billion years ago. This record is only compatible with a universe that always evolves in the same direction of increasing its entropy[1] and under the control of the same consistent set of physical laws, i.e. with an evolving universe that is comprehensible in rational terms. Consequently, and contrary to Einstein's opinion, we can affirm:

> It would be incomprehensible if a universe that produces rational beings were not comprehensible.

With respect to the necessary presence of conscious observers to explain the existence of quantum objects and of physical reality itself, something similar could be said: the universe, at whatever level it is considered, had to exist very long before the appearance of CONSCIOUS

---

[1] Compatible with open and informed systems evolving to complexity, as is the case of living beings [192, 193].

observers (human observers, others we do not know and cannot do science with them). An existence, therefore, objective and independent of the presence of conscious observers. Consequently, and as in the case of Einstein's quote, the fact that human observers exist in our universe implies the opposite of what many quantum physicists claim. So we can write:

> The existence of human observers confirms the existence of a consistent physical reality independent of human observers.

Recall that quantum objects interact and produce macroscopic objects, which in turn interact with other macroscopic objects. All of these interactions have been produced incessantly in the universe, and in the same direction of increasing its entropy, an increase that is compatible with the appearance of open systems such as living beings [192, 193]. The vast majority of these interactions have occurred without the presence of living beings, conscious or unconscious. It does not seem necessary that the universe has to be observed to evolve consistently over billions of years in the same direction of increasing its entropy while at the same time producing complex open systems that can evolve toward rationality and consciousness, as is the case.

29.3 Discussion

Considering the simplicity of the above arguments, which demonstrate the falsity of their corresponding initial assertions, one may wonder how it is possible that these initial assertions have been repeated and accepted in academic circles for more than a century without the slightest critical attitude, as if biology and geology were not scientific disciplines. I have an answer, which I will present in a forthcoming short article, although the reader can elaborate his own.

30. Discrete conclusions

30.1 Introduction

This chapter presents some of the conclusions that can be drawn from the formal results obtained in the previous chapters. All of them point to a real physical space in a consistent universe. Which, taking into account the inconsistency of actual infinity, demonstrated in various ways in this book and especially in [212], implies that physical space, besides being a real physical object, must be finite and discrete, with indivisible minimal units. Of course, everything remains to be done in the construction of a physical theory of discrete space in our observable universe. In this book we have only confirmed the desirability of beginning this construction.

30.2 Some classical questions to start with

The application of the scientific method always begins with asking the right question(s) about the problem to be solved. With these questions, we unpack the problem we are dealing with into its many details, which helps us to understand the true nature of the problem and may even help us to find solutions. In our case, the problem we are dealing with here is nothing less than the nature of physical space. Of course, we do not intend to solve it, but rather to criticize the inconsistent position of contemporary physics on this issue and, at the same time, to propose a radical change of perspective both in the approach to the problem and in the alternatives for its possible solution.

As we shall see, a major problem in understanding the problem posed by the reality of physical space is the uniform motion of material objects through it (Principle of Inertia, Chapter 19). In this sense, recall that Galileo defended the idea that uniform linear motion could not be sensed: we do not have sensors for uniform motion as we do, for example, for heat or sound. Galileo was therefore convinced that uniform motion can only be perceived when a change in position with respect to some external object is observed [126, p. 529].

This theoretical conviction, together with the empirical data from his famous inclined plane experiments, led Galileo, and then Newton, to the Principle of Inertia, the first law of mechanics. Although it is an inductive principle, the induction must be experimental because, as just noted, we have no sensory experience of uniform linear motion. Moreover, the principle of inertia is far from being intuitive (although we do have empirical knowledge of sliding objects and rolling balls...), especially in the case of objects in uniform rectilinear motion, as can be tested by classical authors such as Aristotle [16, Books 3]. For all these reasons, it can be considered one of the great achievements of mankind. But physical explanations should not stop at the Principle

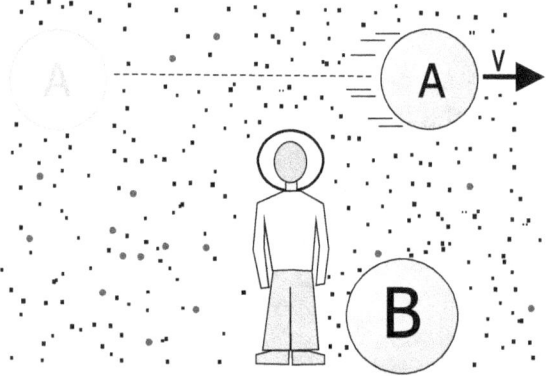

Figure 30.1 – The balls A and B follow the Principle of Inertia.

of Inertia, where they actually do. As we will see, it motivates some of the most important questions about motion and physical space that we can imagine. As an example, consider two identical balls A and B at rest in the inertial reference frame RF_o. At a given instant, A is pushed and set in motion so that from that instant on it moves with a uniform velocity v with respect to B, which remains at rest in RF_o (Figure 30.1). The following pertinent questions are not so easy to answer:

1. What determines and controls the linear trajectory of A, its *successive* positions along the *successive* instants?

2. How does A remember that it was pushed? Where is the imprint of this action?

3. What, if anything, has changed in its internal structure as a result of being set in motion?

4. What distinguishes a ball that has been pushed from one that has not?

5. If space and the ball are interpenetrating, how can it be explained that neither is (at least apparently) affected?

30.2 Some classical questions to start with

6. Is space somehow affected by a moving ball?
7. Knowing that A has been pushed and B has not been pushed, is it the same to say that A moves with respect to B as to say that B moves with respect to A?
8. Is moving A the same as moving the rest of the universe?
9. Is there an absolute, describable reality?
10. If there is no absolute describable reality, are there as many realities as there are relative forms of observing them? To observe what?
11. Could the universe, as such an object, be described from outside the universe?
12. Are we living beings with the capacity to reason, but not to observe reality?
13. etc.

As Feynman said, we know how objects move, but not why they move (why they move in a straight line) [116, p. 18]. The main goal of physics should be to discuss the possibilities of extending physics by asking questions similar to those above. Remember that every application of the scientific method begins with a significant question.

The above set of questions, very significant with respect to the intimate nature of material objects and their inertial motion through physical space, can be extended with others equally significant with respect to the nature of physical space:

1. How can the vacuum (which by definition has no substance) have measurable physical properties such as magnetic permeability or electrical permittivity?
2. A significant number of contemporary physicists argue that the universe originated from a "fluctuation of nothing". But doesn't that imply that nothing is not nothing, but something with the ability to fluctuate?
3. How is it possible that something that does not exist, such as space, can expand, deform, and vibrate? If the vibrating medium does not exist, can its vibration exist?
4. Can physical reality be explained by the behavior of objects that are not real?
5. Is not Theorem 3 of the Actual Infinity (page 22) a proof of the inconsistency of the actual infinity? And consequently, is not the infinitist spacetime continuum inconsistent? (Theorem 8 of the Inconsistent Continuum, page 23).

6. Can a theory based on an inconsistent axiom be consistent?
7. Should contemporary physicists not pay more attention to the infinitist foundations of their mathematical language?
8. Is there anything more fundamental in physics than change?
9. And since the problem of change has not yet been solved, how is it possible that physics, the science of change, has forgotten the problem of change?
10. Zeno dichotomies (still unsolved) pose an essential problem for the physics of motion. How could physics have forgotten them?
11. How is it possible that contemporary physics is unable to explain how a simple change of position occurs?
12. On the other hand, and accepting the inevitability of hegemonic currents of thought in physics, must they be so hostile to dissent?
13. What does it take for contemporary physics to consider ideas outside its mainstream?

To this scenario of essential questions, which do not even arise in contemporary physics, we must add an equally fundamental problem, although easier to solve: the correct use of ordinary language. In fact, in the primary and secondary literature of physics, it is normal to find the use of the same word for two different concepts, and the same two different words for each of these two concepts. This is the case, for example, with the pairs of words [202, 194]:

1. Order and organization
2. Information and entropy.
3. Vacuum and nothingness.
4. Point and space.
5. Instant and time.

And above, all the incorrect use of certain notions, incorrect because they contradict the infinitist foundation of the mathematical language of physics (see Chapter 22), for instance the notion of point:

1. "the space around a point"
2. "a tiny ball or point"
3. "point-like charge"
4. "mass point"
5. "mass concentrated in a point"
6. "infinitesimal point"
7. "the motion of points with mass and charge"
8. "propagation proceeds point by point"

9. "propagate through the contiguous points"
10. "creating changes at adjacent points"
11. "what happens to the field at adjacent points
12. etc.

which show the use of qualities that points do not have: extension and immediate successiveness, or adjacency (points do not touch each other).

On the other hand, physics should prepare itself to dispense with the concept of infinity in all its models and theories of the physical world. It is rare to find a physics book (especially those related to the physics of space and time) in which the word infinity does not appear dozens of times, for example more than 240 times in [310], 170 times in [128], more than 140 times in [303], more than 70 times in [188]. And in phrases like:

... escape to infinity.
... at infinite distance.
... repelled to infinity.
... extends to infinity.
... continues to infinity.
... an infinitude of positions.
... infinitely many points.
... etc.

which are always used imprecisely, without ever specifying what kind of infinity they are referring to: actual or potential infinity; numerable or non-numerable. As if they were all the same.

30.3 Real or fictitious?

As already indicated in Chapter 23, the opinion of modern physicists (20th and 21st centuries) on space has always been divided, although the majority opinion has always been that space is not real, it is only a useful fiction. Let us recall some of the quotations reproduced on page 281 of this book:

> Spacetime is not a fabric, it is not material. Space is just an illusion, time is just an illusion therefore spacetime is just an illusion and a good way of simplifying the concept of general relativity to the public.

This has also been the opinion of many relevant authors in the history of science and thought (particularly empiricists): G. Leibniz, D. Hume, C. Huygens, E. Mach, H. Poincaré, E. Borel, L. Wittgenstein etc. And of the vast majority of contemporary physicists. For example [325, p. 266]:

> ... space and time, like society, are in the end also empty conceptions. They have meaning only to the extent that they stand for the complexity of the relationships between the things that happen in the world.

At the same time, practically all contemporary physicists defend the idea that space can extend, deform, vibrate and be the transmitting medium of its own vibrations. And in this book we have always wondered how it is possible that something that does not exist can extend, deform, vibrate and be a medium transmitting transverse waves.

But this picture must begin to change because physical properties of space have been EMPIRICALLY DETECTED. And this empirical detection is only possible if a real physical object with these properties exists. Indeed, in Chapter 23 (page 281) of this book a definition of the concept of physical space was proposed, a definition that has only been possible once gravitational waves have been detected, which proves the reality of space as a physical object provided with certain empirically detectable physical properties. In contrast, the following definitions correspond to definitions of the abstract concept of space, most of which are circular. Probably because the ABSTRACT concept of space is a primitive concept. Among these definitions we find (italics are mine):

> DICCIONARIO DE LA LENGUA ESPAÑOLA: *Extension* containing all existing matter [105].

> DICCIONARIO MARÍA MOLINER: *Magnitude* in which all the bodies that exist at the same time are contained and in which these bodies and the separation between them are measured [247].

> BRITANNICA ONLINE: A boundless, three-dimensional *extent* in which objects and events occur and have relative position and direction.

> CAMBRIDGE DICTIONARY ONLINE: The *area* around everything that exists, continuing in all directions.

> GRAN ENCICLOPEDIA LAROUSSE: Indefinite *extension*, unbounded medium containing all finite extensions [353].

> MERRIAM-WEBSTER ONLINE: (substantival) A boundless three-dimensional *extent* in which objects and events occur and have relative position and direction. (relational) The only thing real are the spatial relations between physical objects.

> DICCIONARIO DE CIENCIA Y TECNOLOGÍA: Term used to refer to real three-dimensional Euclidean *space* [230].

OXFORD DICTIONARY OF PHILOSOPHY: An objective *thing* comprised of points or regions at which, or in which, things are located [39].

OXFORD DICTIONARY OF PHYSICS: a *property* of the universe that enables physical phenomena to be *extended* into three mutually perpendicular directions [335].

COLLINS WEB-LINKED DICTIONARY OF MATHEMATICS: A primitive concepts that in Newtonian mechanics is assumed to be a Euclidean space with a set of Cartesian coordinates ... [45]

HARPER COLLINS DICTIONARY OF MATHEMATICS: A *set of points* endowed with a structure that is usually defined by specifying a set of axioms to be satisfied by the points.

And especially Newton's definition, which expresses very well the difficulty of defining a primitive concept [259, p. 77]:

I do not define time, space, place and motion, as being well known to all.

As was done in Chapter 23, and from a finite and discrete perspective (the alternative is inconsistent), it is possible to define space as a physical object endowed with certain empirically detectable and measurable physical properties, as its capacity to vibrate, and to be the transmitting medium of its own vibrations, at a speed of 299972458 m/s. But before recalling that definition it is necessary to recall some of the results that have been formally demonstrated in the previous pages of this book.

30.4 Continuous or discrete?

If it is impossible to exaggerate the importance of space in physics [233, p. 25], it is also impossible to exaggerate the importance of the continuous or discrete nature of space. To decide on this alternative we have formal results, duly proved, that are only compatible with the discrete alternative. They all follow from the Theorem 1 of the Actual Infinity (Page 17):

The actual infinity is inconsistent.

Among many others, the following results can be deduced from the above theorem:

- The actual infinite division of any finite real interval is inconsistent (page 185).

- In the Euclidean space $\mathbb{R}^3$ every line with two endpoints has a finite length (page 185).

- In the spacetime continuum, the distance between any two points and the time elapsed between any two instant is always finite (page 187).

- The universe evolves under the control of a unique set of invariant and consistent physical laws (page 42).

- The laws of physics apply to all regions of space and time (page 43).

- No physical object or phenomenon can be fully explained without a first cause that cannot be explained in terms of other causes (page 49).

- In a consistent reality only a finite number of universe could exist (page 25).

- A consistent universe cannot contains an actual infinite number of physical objects (page 25).

- The mass and the energy of the observable universe cannot be actually infinite (page 26).

- There is an indivisible minimum of space (time) of which all space (time) intervals are an integer multiple (page 209).

- The laws of physics do not apply in spaces smaller than the indivisible unit of space nor in times smaller than the indivisible unit of time, both being of non-zero extension (duration) (page 209).

- Every space interval (or time interval) is finite and can only be divided into an integer number of adjacent qusits (qutits) (page 210).

- Canonical changes are instantaneous and then impossible in the spacetime continuum (page 214).

And taking into account the physical reality of space (see Chapter 23), we can add:

1. Physical space is a real physical object with certain physical properties that can be tested and measured in experimental terms.

2. Physical space can only consist of discrete units which, as such units, are of a non-null extension, indivisible and contiguous in all directions.

3. Time is a discrete magnitude whose discrete units, as such units, have to be indivisible, contiguous and of a non-null extension.

all of them essential to initiate a discussion on the nature of space (and time) in a consistent universe. Taking all of them into account, the following definition of physical space is proposed:

> Space is a real physical object formed by a finite number

of indivisible and contiguous units of a non-null extension (qusits) that contains, and possibly generates, all the material objects of the universe, to which it offers no resistance to their motions and makes possible their mutual interactions.

30.5 Special relativity is not compatible with discreteness

According to the Principle of Relativity, the laws of physics are the same in all inertial reference frames. And according to the Theorem of the Consistent Universe 42, the universe evolves under the control of a unique set of invariant and formally consistent laws. This includes all the universal constants involved in such laws. So, being l_p and t_p two universal constants, they should be universal constants in all inertial reference frames, which poses the problem of their respective relativistic contraction and dilation. Or in other words, the problem of the relativity of the intervals of space and time below which physical laws do not apply, will now depend on the relative velocity at which the corresponding events are observed. This problem has already been dealt with by some authors, although they have not had a great impact [11, 175, 5, 4, 162].

Even less impact have had until now my two suggestions about length contraction and time dilation predicted by the special theory of relativity and confirmed experimentally with different observations and experiments:

1. Relativistic inertial length contraction, inertial time dilatation, and inertial phase difference in synchronization could be only apparent, as apparent is the deformation of a rod partially introduced in water; deformation that can also be observed and experimentally measured. But the deformation of the rod is not real, it is only apparent.

2. Relativistic inertial length contraction, inertial time dilatation, and inertial phase difference in synchronization could be the consequence of explaining a discrete reality with indiscreet (continuum-based) mathematics. It also happens that the relativistic Lorentz factor coincides with the factor that converts between the continuous and discrete versions of Pythagoras Theorem (Chapter 16, 187), which has a capital importance in the calculation of distances and other invariants of Euclidean geometry.

It could be argued that the Lorentz Transformation does not hold for lengths and times respectively less than l_p and t_p. So, let L_o be the length of a macroscopic rule parallel to X_o in its proper inertial reference frame RF_o. If it were an integer multiple of the Planck length, we

would have:
$$L_o = n_o l_p; \quad n_o \in \mathbb{N} \qquad (1)$$

Let RF_v be another inertial reference frame that coincides with RF_o at a certain instant and from whose perspective RF_o moves with a uniform velocity $v = kc$, $(0 < k < 1)$, in the direction of the increasing axis X_v. In accordance with the Lorentz Transformation, the moving rule will be observed with a length L_v such that:
$$L_v = \gamma^{-1} L_o \qquad (2)$$

where $\gamma = 1/(\sqrt{1-k^2}$ is Lorentz factor. If l_p is also a universal constant in RF_v, and L_v also an integer multiple of the Planck length, we will have:
$$L_v = n_v l_p; \quad n_v \in \mathbb{N} \qquad (3)$$

In consequence, it must hold:
$$n_v l_p = \gamma^{-1} L_o = \gamma^{-1} n_o l_p \qquad (4)$$
$$n_v l_p = \gamma^{-1} n_o l_p \qquad (5)$$
$$n_v = \sqrt{1-k^2}\, n_o \qquad (6)$$

which is impossible because $\sqrt{1-k^2}$ is not a natural number. The same argument applied to any proper interval of time $t_o > t_p$ leads to:
$$k_v = \sqrt{1-k^2}\, k_o; \quad k_v, k_o \in \mathbb{N} \qquad (7)$$

which for the same above reason is also impossible. We have to conclude that the theory of special relativity is not compatible with a discrete space and a discrete time. Therefore, the special theory of relativity requires that one of the following two alternatives be satisfied:

1. The laws of physics hold for any time interval and length interval respectively less than t_p, and l_p.

2. The speed of light is undefined for any time interval and length interval respectively different from nt_p, and nl_p, for any natural number n.

30.6 A discreet model to start with

In order to start building a new foundational basis for the concept of physical space, we will now recall the space in CALMs (Cellular Automata Like Models) introduced in Chapter 18. In these theoretical objects:

30.6 A discreet model to start with

1. Space is made of indivisible and contiguous elements we call qusits (cells in the jargon of cellular automata).
2. Time is a sequence of indivisible and contiguous elements we call qutits.
3. Being contiguous means that between any two of contiguous qusits (qutits) no other qusit (qutit) exists (immediate successiveness or adjacency).
4. Every region of a CALM has a finite number of qusits.
5. Every interval of time in a CALM has a finite number of qutits.
6. The state if each qusit is defined by a set of variables.
7. During at least one qutit, each qusit exhibits the same state, while in the background, the interactions between the states of the different qusits are performed, which will determine their corresponding states to be exhibited in the next qutit.
8. The changes of state are driven by the laws of the CALM.
9. Some groups qusits can be temporarily or permanently linked: they are the objects of the CALM.
10. From the initial configuration of the CALM and its rules, a wide variety of objects emerge.
11. Some CALM objects are intertwined by new laws that emerge from the evolution of the automaton.

31. The pending revolution in physics

31.1 It is impossible to exaggerate the importance of ...

As has been repeated several times in the pages of this book, I agree with T. Maudlin that it is impossible to exaggerate the importance of space in physics [233, p. 25], despite the fact that for many physicists space does not really exist. It also seems impossible to me to exaggerate the importance of the actual infinity in space, in time, and in all of physics. And yet contemporary physicists are not interested in analyzing the consistency of the actual infinity, which is the only infinity of the infinitist mathematics they use to build their theories. Although, again, it is impossible to exaggerate the controversial nature of the actual/potential infinity throughout its entire history of more than 25 centuries. Indeed, from Parmenides and Zeno of Elea to G. Cantor, it was impossible to prove whether the actual infinity was consistent or not, with a historical division of opinion on the matter, so that in the end its consistency had to be established by an axiom, the Axiom of Infinity assumed by contemporary mathematics:

$$\exists N((0 \in N) \wedge (\forall x \in N,\ s(x) \in N)) \tag{1}$$

that reads: there exist a set N [symbols: $\exists N$] such that 0 belongs to N [symbols: $0 \in N$] and for all element x in N [symbols: $\wedge\ \forall x \in N$] the successor of x, denoted by $s(x)$, also belongs to N [symbols: $s(x) \in N$]. Making use of Dedekind's definition of infinite set, it can be easily proved that the infinity subsumed in the Axiom of Infinity can only be the actual infinity (Theorem 6, page 21). Or put in less formal and more Aristotelian terms: the Axiom of Infinity states that the list of natural numbers in their natural order of precedence exists as a COMPLETE list, although the existence of a last natural number to complete this list is impossible.

I have the impression that most physicists think and use infinity in terms of the potential infinity instead of the actual infinity, which underlies the mathematics of their theories. So much so that in ordinary language they often use expressions that contradict the actual infinity, such as adjacent points and the like. It is somewhat ironic, on the other hand, that it was set theory, which includes the Axiom of Infinity among its foundations, that finally provided the formal tools to demonstrate the inconsistency of the Hypothesis of the Actual Infinity, and to be able to do so in more than forty different ways [212]. But again, it is impossible to exaggerate the difficulty of changing such a hegemonic stream of thought as that of the actual infinity in contemporary mathematics and physics. And it is also impossible to exaggerate the consequences of the inconsistency of the actual infinity for contemporary mathematics and physics. In page 22 of this book you can examine a very brief proof of that inconsistency

31.2 A revolution in three words

As might be expected, the observations and measurements made in physics throughout its history have always been finitist: finite sets of physical objects have always been analyzed (even the estimated total number of elementary particles in the universe is considered finite), and all measurements have always had a finite number of decimal places (an accuracy of 30 decimal places is considered an extraordinary success). Nothing infinite has ever been observed, and no object has ever been divided into an infinite number of parts. Thus, there is no empirical evidence for infinity. Experimental physics has always been, and I believe always will be, finitist.

This is not the case with physical theories, since almost all of them use infinitist concepts such as the spacetime continuum. This is the case, among many others, with the theories of relativity. Even the physical theory that comes closest to the discrete world, quantum mechanics, is expressed in infinitist mathematical terms (for instance in Hilbert spaces of infinite dimensions). Thus, even if experimental physics is safe from the consequences that may follow from the inconsistency of the actual infinity, most physical theories will be seriously affected by this inconsistency. Their theoretical models and interpretations of the physical world will necessarily have to be changed if, as everything seems to indicate, the actual infinity is an inconsistent notion.

From the inconsistency of the actual infinity some fundamental theoretical results for physics follow almost immediately. Let us recall some of those that have been proved in other chapters of the book:

Theorem 7 of the Inconsistent Dense Order (page 22): Densely ordered sets are inconsistent.

31.2 A revolution in three words

Theorem 8 of the Inconsistent Continuum (page 23): The spacetime continuum is inconsistent.

Theorem 9 of the Discrete Sets (page 23): All discrete sets are finite.

Theorem 10 of the Strictly Ordered Sets (page 23): Every strictly ordered set is discrete.

Theorem 9 of the Finite Number of Universes (page 25): In a consistent reality only a finite number of universe could exist.

Theorem 11 of the Finite Universe (page 25): A consistent universe cannot contains an actual infinite number of physical objects.

Corollary 12 of the Finite Mass-Energy (page 26): The mass and the energy of the observable universe cannot be actually infinite.

Corollary 18 of Discrete Values (page 182): The number of all possible values of a variable magnitude is finite, and all of them can be arranged in a discrete set with a minimum and a maximum value.

Theorem 31 of the Finite Distances and Durations (page 187): In the spacetime continuum, the distance between any two points and the time elapsed between any two instant are always finite.

Theorem 18 of the Consistent Universe (page 42): The universe evolves under the control of a unique set of invariant and consistent physical laws.

Theorem 21 of the Reference Frames (page 44): The laws of physics are the same in all discrete reference frames.

Corollary 17 of the First Cause (page 49): A first unexplained cause is inevitable in every natural process.

Theorem 35 of Indivisible Units (page 209): There is an indivisible minimum of space (qusit) and time (qutit) of which all space (time) intervals are an integer multiple.

Corollary 24 of Finite Space and Time (page 210): Every space interval (or time interval) is finite and can only be divided into an integer number of adjacent qusits (qutits).

Corollary 23 of Discrete Threshold (page 209): The laws of physics do not apply in spaces smaller than the indivisible unit of space nor in times smaller than the indivisible unit of time, both being of non-zero extension (duration).

Theorem 38 of Change (page 214): Canonical changes are instantaneous and then impossible in the spacetime continuum.

The content of this last section of this chapter of the book is based on three results, one theoretical and the other two empirical. The theoretical result is the inconsistency of the actual infinity, of which a formal proof and some formal consequences of interest to physics have just been recalled.

The first empirical result is preinertia, the property of all physical objects to inherit the velocity vector of the proper reference frame in which they are set in motion. It is empirical because it is detected every time, for example, an object falls to the ground; or every time we jump and fall in the same place where we started the jump; or because we do not notice that we are actually moving at more than one million three hundred thousand kilometers per hour (approximately 1321200 Km/h, i.e. 367 Km/s in the direction of galactic coordinates (264.4, 48.4)). Physicists have always implicitly used preinertia (since the time of Galileo) without explicitly declaring its existence and therefore without using it in their arguments and models of the physical world (did they realize that they were not using it?). The universal preinertia of all physical objects (including photons) is the reason why absolute motion cannot be detected, which, together with the non-perception of the absolute physical space THROUGH which all physical objects move, has led us to the hegemonic strict physical relativism of contemporary physics.

The second empirical result is the experimental detection of gravitational waves, i.e. the detection of the vibrations of physical space and of the ability of space itself to transmit those vibrations at 299792.458 km/s. The physical-formal consequence of this empirical detection was discussed in Chapter 23. It is an extraordinarily significant consequence: physical space is a real physical object with empirically detectable and measurable physical properties (see the just mentioned Chapter 23). And being real and consistent (Theorem 18, page 42), it must be finite and discrete: formed by minimal indivisible units (qusits of Planck volume $4.22167 - 10^{-105}$ m^3 ?). The same conclusion of finiteness and discreteness would have to be applied to time, although time would not be a physical object but a universal magnitude measuring in minimal indivisible units (qutits, Planck time $5.39124 - 10^{-44}$ seconds ?) the stability of the states of physical objects.

The universe would consist of a finite set of discrete objects evolving in a finite and discrete space over an equally finite and discrete time. The discreteness of space and time would join the already assumed and proven discreteness of all other physical entities: space, time, matter, mass, energy and charges (Theorem 14). The relative motions of objects, the only observable ones, would be a consequence of the different absolute motions through absolute space, the only real motions. The relative motions would be only apparent, although they would be the only ones we can observe and measure (except may be in the case of the Santiago del Collado experiment [208, p. 371-378]). We have, then, a pending revolution in physics. A revolution that could be summarized in three words: finite, discontinuous and absolute, instead of respectively infinite, continuous and relative. This discrete revolution will have to look for new models to explain the world. CALMs could be a good starting point.

Appendix A:
Hilbert machine

(Appendix taken from [212])

A.1 Hilbert Hotel

In the next discussion we will make use of a supermachine inspired by the emblematic Hilbert Hotel [156, p. 730] [22, p. 42-50] [66, p. 237-239]. But before beginning, let us relate some of the prodigious, and suspicious, abilities of the illustrious Hotel.

Figure A.1 – The power of the ellipsis: An infinitist way of making money.

Its director, for example, has discovered a fantastic way of getting rich: he demands one euro to R_1 (the guest of the room 1); R_1 recovers his euro by demanding one euro to R_2 (the guest of the room 2); R_2 recovers his euro by demanding one euro to R_3 (the guest of the room 3); and so on. Finally all guests recover his euro, because there is not a last guest losing his money. Our crafty director then demands a second euro to R_1 which recovers again his euro by demanding one euro to R_2, which recovers again his euro by demanding one euro to R_3, and so on and on. Thousands of euros coming from the (infinitist) nothingness to the pocket of the fortunate director.

Hilbert Hotel is even capable of violating the laws of thermodynamics by making it possible the functioning of a perpetuum mobile: in fact we would only have to power the appropriate machine with the calories obtained from the successive rooms of the prodigious hotel in the same way its director gets the euros.

Incredible as it may seem, infinitists justify all those absurd pathologies, and many others, in behalf of the *peculiarities* of the actual infinity. They prefer to assume any pathological behavior of the world before examining the consistency of the pathogene. In the next discussion, however, we will come to a contradiction that cannot be easily justified by the picturesque nature of the actual infinity.

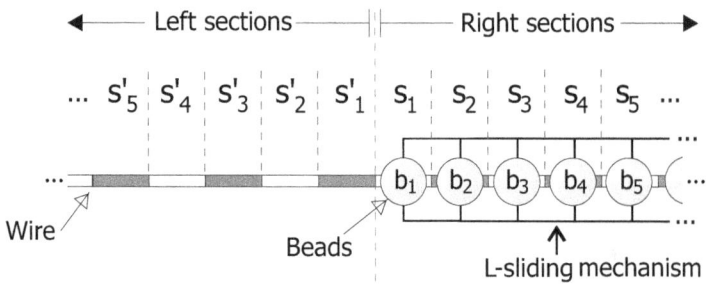

Figure A.2 – Hilbert machine just before performing the first L-sliding.

A.2 Hilbert machine

In the following conceptual discussion we will make use of a theoretical device, inspired by the emblematic Hilbert Hotel, that will be referred to as *Hilbert machine* and denoted by H_ω, composed of the following elements (see Figure A.2):

a) An infinite horizontal wire divided into two infinite parts, the left side and the right side:

  1) The right side in turn is divided into an ω-ordered sequence of disjoint and adjacent sections $\langle S_i \rangle$ of equal length indexed from left to right as $S_1, S_2, S_3, \ldots$. They will be referred to as right sections.

  2) The left side is also divided into an ω-ordered sequence of disjoint and adjacent sections $\langle S'_i \rangle$ of equal length, the same length as the right sections, and indexed now from right to left as $\ldots, S'_3, S'_2, S'_1$; being S'_1 adjacent to S_1. They will be referred to as left sections.

b) An ω-ordered sequence of indexed beads $\langle b_n \rangle$ strung on the wire, so that they can slide on the wire as the beads of an abacus, being the center of each bead b_i initially placed on the center of the right section S_i.

c) All beads are mechanically linked by a sliding mechanism that slides simultaneously all beads the same distance along the wire.

A.2 Hilbert machine

d) The sliding mechanism is adjusted in such a way that it slides simultaneously each bead exactly ONE, AND ONLY ONE, SECTION to the left (L-sliding).

Obviously, Hilbert machine H_ω is a theoretical artifact, and its functioning is a simple thought experiment that illustrates a formal argument to test ω-order, the type of order of the well-ordered set $\mathbb{N}$ of the natural numbers, whose ordinal number is ω, the least transfinite ordinal [56, p. 160, Theorem §15 A]. This is not, therefore, a discussion on the physical restrictions and consequences of performing a particular sequence of physical actions.

Since the sections $\langle S'_i \rangle$ of the left side of the wire are ω-ordered, each section S'_n has an immediate successor section S'_{n+1} just on its left (ω-successiveness). In accord with the Hypothesis of the Actual Infinity all those infinitely many left sections exist as a complete totality in spite of the fact that there is not a last section completing the sequence. The same applies to the right sections $\langle S_i \rangle$.

I will assume H_ω always works according to the following:

Restriction 1 (of Hilbert Machine) *An L-sliding will be carried out if, and only if, after being performed all beads remain strung on the wire. Otherwise, the corresponding L-sliding will be undone so that every bead recovers its previous position on the wire and then the machine stops.*

Note this restriction could also be applied to the supertask of counting the successive natural numbers: stop counting if the number n just counted is not the immediate successor of the previously counted number $n - 1$; or, alternatively, count the next number only it is the immediate successor of the number just counted. As we sill see next, it is possible to demonstrate in at least two different ways (by Modus Tollens and by induction) that for each natural number v it is possible to perform the first v L-slidings.

P9 MODUS TOLLENS. Let us begin by proving that for each $v \in \mathbb{N}$ the first v L-slidings can be carried out according to Restriction 1. Assume this assertion is not true. There will be a natural number $n \leq v$ such that it is impossible to perform the nth L-sliding according to Restriction 1. But this is impossible because whatsoever be the left section occupied by b_1 just before performing the nth L-sliding, there always be a left section contiguous to that section, otherwise b_1 would be in the impossible last left section of an ω-ordered sequence. So, b_1 can L-slide to that contiguous left section, and every bead $b_{i,i>1}$ can move to the section previously occupied by b_{i-1}. Therefore, the nth L-sliding can be carried out according to Restriction 1. Consequently our assumption is not true, and for each $v \in \mathbb{N}$ it is possible to carry out the first v L-slidings according to Restriction 1. □

P10 INDUCTION. The following inductive argument leads to the same conclusion as the previous Modus Tollens P9. It is clear that the first L-sliding can be performed: b_1 slides to S'_1 and every $b_{i;i>1}$ to the section previously occupied by b_{i-1}. Suppose that, for any natural number n, the first n L-slidings can be carried out. Since each L-sliding moves each bead one section to the left, all beads will have been moved n sections to the left, so that b_1 will be in the left section S'_n, because S'_n is n sections to the left of S_1, the section initially occupied by b_1. And since S'_n has an adjacent left section S'_{n+1} (ω-successiveness), b_1 can slide to S'_{n+1} and each $b_{i;i>1}$ to the section previously occupied by b_{i-1}. So, if for any n the first n L-slidings can be carried out, the first $n+1$ L-slidings can also be carried out. And since the first L-sliding can be carried out, we inductively conclude that for any $v \in \mathbb{N}$ the first v L-slidings can be carried out. □

A.3 Hilbert machine contradiction

From now on, to carry out an L-sliding means to carry out it according to Restriction 1. That said, assume that while the successive L-slidings can be carried out, they are carried out. It is immediate to prove the following two contradictory theorems (Hilbert contradiction):

Theorem 42 (All in Wire) *Once performed all possible L-slidings all beads remain strung on the wire.*

Proof: It is an immediate consequence of Restriction 1: if an L-sliding removes a bead from the wire, that L-sliding would be undone and the machine stops with every bead strung on the wire in the section occupied just before that L-sliding. In addition, since an L-sliding simultaneously moves each bead one section, and only one section, to the left, and the first bead to the left of all beads is b_1, it had to be b_1, and only b_1, the unique bead that came out of the wire by one L-sliding. Otherwise, if the first n beads were simultaneously removed from the wire by one L-sliding, then each bead $b_{i>1}$ would have been moved more than one section to the left by one L-sliding, which is impossible according to the functioning of the machine (item d, page 365). In consequence, and being b_1 the unique bead removed from the wire, b_2 would have to be in the impossible last section of an ω-ordered sequence $\langle S'_i \rangle$ of sections. So, once all possible L-slidings have been done, all beads remain strung on the wire. □

Theorem 43 (None in wire) *Once performed all possible L-slidings no bead remains strung on the wire.*

Proof: Let b_v be any bead and assume that once performed all possible L-slidings it is strung on the right section S_k. It must be $k < v$ because all L-slidings are towards the left, the direction towards which the indexes of the right sections $\langle S_i \rangle$ decrease. Since b_v was initially placed on S_v only a finite number $v - k$ of L-slidings would have been performed, and then it would not have been possible to perform the the first $v - k + 1$ L-slidings, which goes against P9 and P10, because $v - k + 1$ is a natural number. A similar reasoning can be applied if b_v were finally strung on a left section S'_n, being now the number of performed L-slidings exactly $v + n - 1$ and then it would not have been possible to perform the first $v + n$ L-slidings, which also goes against P9 and P10, because $v + n$ is also a natural number. Thus, since b_v is any bead, if all possible L-slidings have been performed, then no bead remains strung on the wire. Note this is not a question of indeterminacy but of impossibility: the set of possible sections any bead b_v could be finally occupying is the empty set. □

It is remarkable the fact that in the above demonstration of Hilbert's contradiction it has only been assumed that, under the Hypothesis of the Actual Infinity, all possible L-slidings have been performed. The reader can easily prove a corollary of the Theorem 43: all beads stop being inserted in the wire at the same instant, an instant at which L-slidings are no longer performed.

A.4 Discussion

Let us compare the functioning of the above Hilbert machine H_ω with the functioning of a finite version of the machine (symbolically H_n). This finite machine has a finite number n of both right and left sections (Figure A.3). A finite sequence of n beads are initially strung on the right side of the wire, the center of each bead b_i placed on the center of the right section S_i. It is immediate to prove that H_n can only perform n L-slidings because not having a left section S'_{n+1}, Restriction 1 will stop the machine with each left section S'_i occupied by the bead b_{n-i+1} and all right sections empty, and this is all. No contradiction is derived from the functioning of H_n. Thus for any natural number n, the corresponding machine H_n is a consistent theoretical artifact. Only the infinite Hilbert machine H_ω is inconsistent.

What the above Hilbert contradiction proves is not the inconsistent functioning of a supermachine. What it proves is the inconsistency of ω-order itself because of ω-successiveness. Perhaps we should not be surprised by this conclusion. After all, an ω-ordered sequence is one which is both complete (as the actual infinity requires) and incompletable (there is not a last element that completes the sequence). On

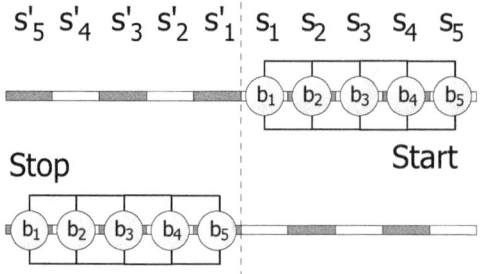

Figure A.3 – A finite machine of five sections.

the other hand, and as Cantor proved [56, p. 160, Theorem §15 A], ω-order is an inevitable consequence of assuming the existence of infinite sets as complete totalities. An existence axiomatically stated in our days by the Axiom of Infinity, in all axiomatic set theories including its most popular versions as ZFC [338, 336]. It is, therefore, that axiom the ultimate cause of contradiction of theorems 42 and 43.

Appendix B:
A disturbing supertask

Chapter taken from [212]

B.1 Introduction

This chapter examines the consistency of ω-order by means of a supertask that works as a sort of trap for the assumed existence of ω-ordered collections, which are simultaneously complete (as is required by the Actual infinity) and incompletable (because no last element completes them). Cantor himself proved [56, P. 160, Teorema §15 A], that ω-order is a formal consequence of assuming the existence of denumerable sets as complete totalities. Although it is hardly recognized, to be ω-ordered means to be both complete and incompletable. In fact, the Axiom of Infinity states the existence of complete denumerable totalities, the most simple of which are ω-ordered, i.e. with a first element and such that each element has an immediate successor, and then an immediate predecessor, except the first one. Consequently, there is not a last element that completes ω-ordered totalities.

To be complete and incompletable is a modest eccentricity in the highly eccentric infinite paradise of our days, but its simplicity is just an advantage if we are interested in examining the formal consistency of ω-order. In addition, ω is the first transfinite ordinal, the one on which all successive transfinite ordinals are built up. This magnifies the interest of its formal analysis, because if the basis of the construction is inconsistent, all constructions built on that basis will also be inconsistent. The short discussion that follows is based on a supertask conceived to put into question just the ability of being complete and incompletable that characterizes ω-order and all ordinals of the second class second kind defined, according to Cantor terminology.

B.2 The last disk

Consider a hollow cylinder C and an ω-ordered collection of identical disks $\langle d_i \rangle$ such that each disk d_i fits exactly within the cylinder (Figure B.1). Let a_1 be the action of placing the disc d_1 completely inside the cylinder C, and let $a_{i>1}$ be the action of replacing the disk d_{i-1} inside the cylinder by its immediate successor the disk d_i, which is accomplished by placing d_i completely within the cylinder. Consider the ω-ordered sequence of actions $\langle a_i \rangle$ and assume that each action a_i is carried out at the instant t_i, being t_i an element of the ω-ordered strictly increasing and convergent sequence of instants $\langle t_i \rangle$ in the finite real interval (t_a, t_b) such that t_b is the limit of $\langle t_i \rangle$. Let S_ω be the supertask of performing the ω-ordered sequence of actions $\langle a_i \rangle$

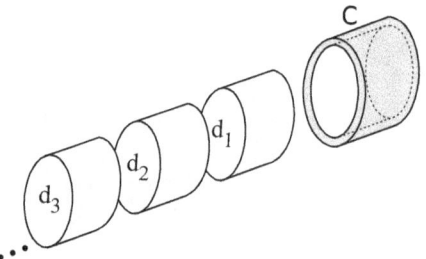

Figure B.1 – The hollow cylinder C and the ω-ordered collection of discs $\langle d_i \rangle$.

Let us impose to S_ω the following:

Restriction 2 (of the Last Disk) *Each action a_i of $\langle a_i \rangle$ will be carried out if, and only if, it leaves the cylinder completely occupied by the disc d_i.*

It is immediate to prove that all actions $\langle a_i \rangle$ observe restriction 2: in fact it is clear that a_1 observes restriction 2 because it leaves the cylinder completely occupied by the disk d_1. Assume the first n actions observe Restriction 2. It is quite clear that a_{n+1} also observes Restriction 2: it leaves the cylinder completely occupied by the disk d_{n+1} because, by definition, it consists just in placing d_{n+1} completely inside the cylinder. Consequently all actions $\langle a_i \rangle$ observe restriction 2. Consider now the

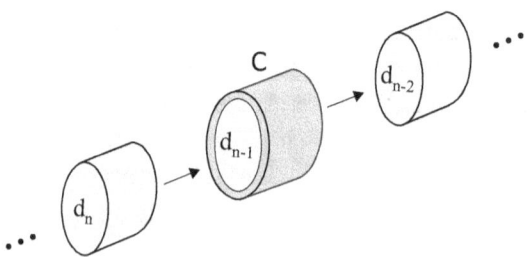

Figure B.2 – The action a_n of supertask S_ω about to be carried out.

one to one correspondence f between $\langle t_i \rangle$ and $\langle a_i \rangle$ defined by $f(t_i) = a_i$. Since t_b is the limit of $\langle t_i \rangle$ (the first instant after all instants of $\langle t_i \rangle$), at t_b all actions $\langle a_i \rangle$ will have been carried out, and supertask S_ω will have been completed.

B.3 Discussion

With respect to the possibilities of being occupied by the disks $\langle d_i \rangle$, the cylinder C can exhibit one, and only one, of the following three alternative states:

1. Empty, occupied by no disk.
2. Partially or completely occupied by one disk.
3. Partially or completely occupied by two disks.

According to the way the successive actions $\langle a_i \rangle$ are carried out, the third state is impossible because each action $a_{i,i>1}$ consists in removing from the cylinder C the disk d_{i-1} by introducing the disk d_i completely inside C. So, once performed the infinitely many actions $\langle a_i \rangle$ of the supertask S_ω, the cylinder C can only be either empty or (partially or completely) occupied by one disk of the collection $\langle d_i \rangle$.

P11 At instant t_b the cylinder C cannot be occupied by a disk d_v, whatsoever it be, because in such a case only a finite number v of disks would have been introduced inside the cylinder and the supertask S_ω would not have been completed. In consequence, at t_b once the supertask S_ω has been completed, C must be empty. $\square$

The problem is: how C becomes empty if none of the performed actions leaves it empty? There are two alternatives regarding the completion of the ω-ordered sequence of actions $\langle a_i \rangle$ of the supertask S_ω:

a) The completion is an additional ($\omega+1$)-th action.

b) The completion is not an additional ($\omega+1$)-th action. It simply consists in performing each one of the infinitely many actions $\langle a_i \rangle$, and only them.

Let us examine the first alternative. The supposed ($\omega+1$)-th action can only occurs at t_b because, being t_b the limit of $\langle t_i \rangle$, for any instant t prior to t_b there is an instant t_v of $\langle t_i \rangle$ such that $t < t_v$ and there still remain infinitely many actions $a_v, a_{v+1}, a_{v+2} \ldots$ of $\langle a_i \rangle$ to be carried out. Whatever be the instant we consider, if it is prior to t_b, there will remain infinitely many actions to be carried out and only a finite number of them will have been carried out.

Therefore, the assumed ($\omega + 1$)-th action must occur at the precise instant t_b. In consequence, at t_b the cylinder has to be occupied by a disk, otherwise, if the cylinder were empty at t_b, the supposed (ω+1)-th action, which occur at t_b and consists just in leaving the cylinder empty, would not be the cause of leaving the cylinder empty as it is assumed to be, because it is already empty. We will have, therefore, a disk d_v inside the cylinder at t_b. And, for the reasons given in P11, this is impossible if S_ω has been completed: the disk d_v within the cylinder would be proving that only a finite number v of actions would have been carried out. Thus, the first alternative is impossible.

We will examine, then, the second alternative. According to it, the cylinder becomes empty as a consequence of having completed the countably many actions a_1, a_2, a_3,... and only them. Thus, either the successive actions have an accumulative effect capable of leaving finally the cylinder empty, or the completion has a sort of sudden final effect on the cylinder as a consequence of which it results empty. We can rule out this last possibility for exactly the same reasons we have ruled out the above (ω+1)-th additional action: that additional action would have to take place at t_b, and then at t_b there would be a disk d_v inside the cylinder proving that at t_b only a finite number v of actions would have been performed. The only possibility is, therefore, that the cylinder C becomes empty as a consequence of a certain accumulative effect of the successively performed actions.

In defense of this alternative of the accumulative effect, Benacerraf's infinitist followers would argue as follows: Let v_i be the volume inside the cylinder which is not occupied by the disk d_i once d_i is placed inside the cylinder by the action a_i, i.e. let v_i be the empty volume inside C once d_i has been placed in C. According to the above definition of $\langle a_i \rangle$ we will have:

$$v_i = 0, \ \forall a_i \in \langle a_i \rangle \qquad (1)$$

Let us then define the series $\langle s_i \rangle$ as:

$$s_i = v_1 + v_2 + \cdots + v_i, \ \forall i \in \mathbb{N} \qquad (2)$$

The ith term s_i of this series represents, therefore, the empty volume inside the cylinder once performed the firsts i actions of $\langle a_i \rangle$. Evidently we will have:

$$s_i = 0, \ \forall i \in \mathbb{N} \qquad (3)$$

$\langle s_i \rangle$ is therefore a series of constant terms. So, we will have for the final empty space inside de cylinder C:

$$\sum_{1}^{\infty} s_i = 0 + 0 + 0 \cdots = 0 \times \aleph_0 \qquad (4)$$

which is indeterminable, and then we cannot say nothing on the final

B.3 Discussion

empty space in the cylinder C. Consequently, we can neither say that it is empty nor that it is not empty at the instant t_b.

To this argument, I oppose the following one. Let v_i be the volume inside the cylinder which is not occupied by the disk d_i once d_i is placed inside the cylinder by the action a_i, i.e. let v_i be the empty volume inside C once d_i has been placed in C. According to the above definition of $\langle a_i \rangle$ we will have:

$$v_i = 0, \ \forall a_i \in \langle a_i \rangle \tag{5}$$

If the empty volume inside C were the result of an accumulative effect, a certain empty volume greater than zero would have to have been created in C at some instant t before t_b, otherwise that empty space would not be created accumulatively, but all at once. Now then, it is impossible that at the instant t the cylinder has accumulated an empty volume greater than 0, because being t_b the limit of the sequence $\langle t_i \rangle$, we will have:

$$\exists\, t_n \in (t_a, t_b) : t < t_n < t_b;\ n \in \mathbb{N} \tag{6}$$

According to (5), and being n a finite natural number, we will have:

$$v_1 + v_2 + \cdots + v_n = n \times 0 = 0 \tag{7}$$

Consequently, the assumed empty space at any instant t before t_b is impossible. So once completed the ω-ordered sequence of actions $\langle a_i \rangle$, the cylinder C cannot be empty of discs as a consequence of an accumulative effect of the successively performed actions a_i. Therefore, the completion of the ω-ordered sequence of actions $\langle a_i \rangle$ does not leave the cylinder empty. It must therefore be concluded that supertask S_ω leads to a contradiction: the completion of $\langle a_i \rangle$ leaves and does not leave the cylinder empty of disks.

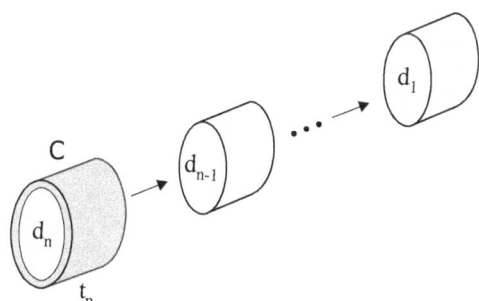

Figure B.3 – The finite version S_n of S_ω with a finite number n of discs.

I will consider now the finite version S_n of S_ω for each natural number n (Figure B.3). For this let n be any natural number and $\langle d_i \rangle_{i=1,2...n}$ the finite collection of the first n disks of $\langle d_i \rangle$. As in the case of S_ω, let a_1 be the first action of placing the disc d_1 inside the cylinder C, and let $a_{i, 1 < i \leq n}$ be the action of replacing, at the instant t_i, the disk

d_{i-1} inside C with its immediate successor the disk d_i. Let S_n be the task of performing the finite sequence of all actions $\langle a_i \rangle_{1,2,...n}$. It is immediate to prove that at t_n all these actions will have been performed and the cylinder will finally contain the last disk d_n placed within it. No contradiction arises here. And this holds for every natural number:

S_n is consistent for every natural number n

Only S_ω is inconsistent. But the only difference between S_ω and $S_n, \forall n \in \mathbb{N}$ is just the ω-order of S_ω. The contradiction with S_ω can only derive from this type of infinite ordering, and then from the Axiom of Infinity, of which it is a formal consequence. Thus, the argument above is not on the impossibility of a particular supertask, but on the inconsistency of ω-order. Being complete and incompletable could be, after all, a formal inconsistency rather than an eccentricity of the first transfinite ordinal.

Appendix C:

Proving Unproved Euclidean Propositions

An extended and updated version (29/09/2021) of the published paper [196].

Abstract.-This paper introduces a new foundational basis for Euclidean geometry that includes productive definitions of concepts so far primitive, or formally unproductive, allowing to prove a significant number of axiomatic statements, unproved theorems and hidden postulates, among them the strong form of Euclid's First Postulate, Euclid's Second Postulate, Hilbert's Axioms I.5, II.1, II.2, II.3, II.4 and IV.6, Euclid's Postulate 4, Posidonius-Geminus' Axiom, Proclus' Axiom, Cataldi's Axiom, Tacquet's Axiom 11, Khayyām's Axiom, Playfair's Axiom, Euclid's Postulate 5. The proposed foundation is more formally detailed and productive than other classical and modern alternatives, and at least as accessible as any of them.

Note: Only the foundational base of the new Euclidean geometry is included, the complete development can be said in [200]. As this is an appendix to an already extensive book, I have not followed the same design and formatting criteria as those used in my book [200], in which the reader always has the proofs and their corresponding figures in view (on the same page or pages). Given the formal relevance of their results, the proofs have been broken down into tiny steps of a line of text, or little more, at the end of which the formal reason for what is stated in that step of the proof is always indicated in parentheses.

C.1 Introduction

After more than two millennia of discussions on Euclid's original geometry and at a time in which such discussions have been practically abandoned, this article introduces a new foundational basis for Euclidean geometry that includes productive definitions of concepts so far formally unproductive, as sidedness, betweenness, straight line, straightness, angle, or plane, among others (all of them properly legitimized by axioms or by formal proofs). The result is an enriched Euclidean geometry in which it is possible to prove some theorems that were proved to be unprovable on other Euclidean geometry bases. It will be introduced in the next sections. Conventions and general fundamentals are the objectives of Section C.2. Section C.3 introduces the new foundational basis: 29 definitions, 10 axioms and 44 corollaries (8 of the 10 axioms and most of the 45 corollaries are implicit (hidden) postulates in other Euclidean geometries).

C.2 Conventions and general fundamentals

The nth axiom, corollary, definition, postulate, and theorem will be referred to, respectively, as [Ax. n], [Cr. n], [Df. n], [Ps. n], [Th. n]. The same letters, for instance AB or BA, will be used to denote a line of endpoints A and B [Df. 21], as well as its length [Df. 29], and the distance between A and B [Df. 35] if AB is a straight line [Df. 31]. Unless otherwise indicated, different letters will denote different points, including endpoints. When convenient, lines will also be denoted by lower case Latin letters, whether or not indexed. Symbols as $0, +, -, =, \neq, \leq$, etc. will be used conventionally. The expressions 'point in a line,' and 'point of a line' will be used as synonyms. The same goes for 'line in a plane' and 'line of a plane'. Closed lines [Df. 22] will be referred to as such closed lines, or by specific names, as circle [Df. 39]. As in classical Euclidean geometry [280, p. 8], [150, p. 153], in Euclidean geometry a straight line is a particular type of line. So, and in contrast with modern English, in Euclidean geometry 'line' and 'straight line' are not synonyms. Asterisked expressions as 'for instance*', 'for example*', 'assume*' etc., will always indicate that only one of the possible alternatives in a proof will be considered and proved, because the other alternatives can be proved in the same way. Proofs begin with the symbol ▷ and end with the symbol □. In figures, the symbol ∟ (in any orientation) will represent a right angle. And an arc of a circle between two straight lines will represent an angle. The biconditional logical connective will be shortened by the term 'iff'. It goes without saying that the three fundamental laws of logic and the basic rules of inference are assumed. And, unless otherwise indicated, the word 'number' will always mean natural number.

C.2 Conventions and general fundamentals

The following four definitions and three postulates are not exclusive to geometry, they have a general use in all sciences. For that reason they have been separated from the very fundamentals of geometry and named with letters in the place of numbers.

The following four definitions and three postulates are not exclusive to geometry, they have a general use in all sciences. For that reason they have been separated from the very fundamentals of geometry and named with letters in the place of numbers.

Definition A *A quantity to which a real number can be assigned is said a numerical quantity. Numerical quantities that can be symbolically represented and operated with one another according to the procedures and laws of algebra, are said operable values.*

Definition B *An operable value is said to vary in a continuous way iff for any two different operable values of the corresponding variation, the variation contains any operable value greater than the less and less than the greater of those two operable values.*

Definition C *Metric properties and metric transformations: properties (transformations) to which operable values that vary in a continuous way are univocally assigned: to each quantity of the property (transformation) a unique and exclusive operable value, even zero, is assigned.*

Definition D *To define an object is to give the properties that unequivocally identifies the object. Objects with the same definition are said of the same class. To draw objects is to make any descriptive representation of them by means of graphics or texts, or by both of them, without the drawing modifies neither their established properties nor their established relations with other objects, if any.*

Postulate A *Of any two operable values, either they are equal to each other, or one of them is greater than the other, and the other is less than the one. Symbolic representations of equal operable values, or of equal objects, are interchangeable in any expression where they appear.*

Postulate B *To be less than, equal to, or greater than, are transitive relations of operable values that are preserved when adding to, subtracting from, multiplying or dividing by the same operable value, the operable values so related. Metric properties (transformations) are algebraically operable through their corresponding operable values.*

Postulate C *Belonging to, and not belonging to, are mutually exclusive relations. Belonging to is a reflexive and transitive relation.*

Contrarily to, for instance, fuzzy set theory or non-Boolean logics, this Euclidean geometry assumes [Ps. C], according to which it is not possible for an object to partially belong and partially not to belong to another object.

C.3 Foundational basis of Euclidean geometry

FUNDAMENTALS ON LINES

Definition 21 (Endedness) *Endedness.- A point at which a line ends is said endpoint. If such a point belongs to the line, the line is said closed at that end; if not, the line is said open at that end. Two endpoints of a line, whether or not in the line, define two opposite directions in that line, each from an endpoint, said initial, to the other, said final.*

Definition 22 (Segment) *Collinearity.-Of the points that belong to a line is said they are points of the line, or points that are on the line; and the line is said to pass through them. A line whose points belong, all of them, to a given line is said a segment of the given line. Two points of a line are said different iff they are the endpoints of a segment of the line. Two lines are said different, iff one of them has at least one point that is not in the other. Different points and segments of the sane line are said collinear; points and segments that do not belong to the same line are said non-collinear.*

Note.-The expression line passing through one or more points may be simplified to line through one or more points.

Definition 23 *Commonness.-Points and segments belonging to different lines are said common to them, otherwise they are said non-common to them. Non-collinear lines with at least one common segment are said locally collinear. Lines without common segments but with at least one common point are said intersecting lines, and their common points are also said intersection points. Intersecting lines are said to cut or to intersect one another at their intersection points.*

Definition 24 *Adjacency.-Lines whose unique common point is a common endpoint are said adjacent at that common endpoint iff no point of any of them is a non-common endpoint of any of the others. Lines containing all points of a given line, and only them, are said to make the given line.*

Definition 25 *Sidedness.-Adjacent lines containing all points of a given line, and only them, whose common endpoint is a given point of the given line and whose non-common endpoints are the endpoints of the given line, if any, are said sides of the given point in the given line.*

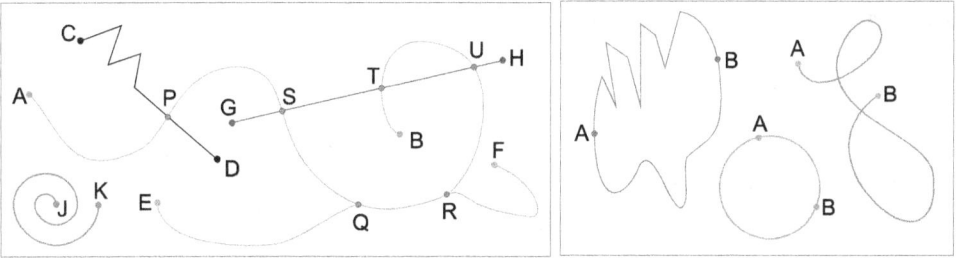

Figure C.1 – Left: A, B: endpoints of AB; C, D: endpoints of CD etc. AB, EF: locally collinear lines. AP, PS: lines (segments) adjacent at P. AP, PB: sides of P in AB. QR: common segment of AB and EF. S is between A and Q; between P and R etc. Right: self-closed lines.

Definition 26 *Betweenness.-A point is said to be between two given points of a line, iff it is a point of that line and each of the given points is in a different side of the point in that line.*

Definition 27 *Self-closed line: a line in which each pair of its points are the common endpoints of two of its segments, said complementary, whose points contain all points of the line, and whose only common points are their common endpoints. Lines with self-closed segments are said self-intersecting. Self-closed and self-intersecting lines are also called figures.*

Definition 28 *Uniformity.-Lines whose segments have the same definition as the whole line are said uniform. Two or more uniform lines are said mutually uniform iff any segment of any of them has the same definition as any segment of any of the others.*

Definition 29 *Metricity.-Length (area) is an exclusive metric property of lines (figures) of which arbitrary units can be defined. Lengths (areas) are said equal iff their corresponding operable values are equal. Lines (figures) with a finite length (area) are said finite. If the sides of a point of a line have the same length, the point is said to bisect the line.*

Axiom 3 *Point, line and surface are primitive concepts of which any number, and in any arrangement, can be considered and drawn.*

Axiom 4 *A line has at least two points, at least one point between any two of its points, and at most two endpoints, whether or not in the line.*

Axiom 5 *Two adjacent lines make a line, and a point of a line can be common to any number of any other different lines, either collinear, or non-collinear, or locally collinear.*

Axiom 6 *Being not a figure, each point of a line, except endpoints, has just two sides in that line, whose lengths are greater than zero and sum the length of the whole line.*

Unless otherwise indicated, from now on figures will be given particular names, for example circle, and will always be referred to by those particular names. The rest of the lines will be closed at their endpoints, if any.

Corollary 25 *The number of points of a line is greater than any given number.*

▷ It is an immediate consequence of [Axs. 3, 4]. □

Corollary 26 *Each side of a point, except endpoints, of a line is a segment of the line and both sides make the line.*

▷ Except endpoints, a point P of a line l [Ax.3, Cr. 25]

has two, and only two, sides in l [Ax. 6],

which are two lines adjacent at P [Df. 25]

containing all points of l, and only them [Df. 25].

So, each side is a segment of the line [Dfs. 25, 22],

and both sides make the line l [Ax. 5, Df. 24]. □

Corollary 27 *Any point of a line is in one, and only in one, of the two sides of any other point, except endpoints, of the line.*

▷ Except endpoints, a point P of a line l [Ax.3]

has two, and only two, sides in l [Ax. 6].

Any other point of l [Cr. 25]

will be in one of such sides [Cr. 26],

and only in one of them, otherwise both sides would not be adjacent at P [Df. 24],

which is impossible [Dfs. 25, 24]. □

Corollary 28 *A point is in a line with two endpoints iff, being not an endpoint of the line, it is between the endpoints of the line.*

▷ If a point P is between the two endpoints of a line AB [Axs. 3, 4, 6, Df. 26],

it is in AB [Df. 26].

If a point P is in a line AB and is not an endpoint of AB [Cr. 25], it has just two sides in AB [Ax. 6],

whose respective non-common endpoints are the endpoints A and B of AB [Dfs. 25, 24].

So, P is between both endpoints A and B [Df. 26]. □

Note.-Unless otherwise indicated, from now on a point P of a line AB will be a point of AB between A and B.

Corollary 29 *Any two points of a line are the endpoints of a segment of the line. And the line has a number of segments and a number of points between any two of its points greater than any given number.*

▷ Let P and Q be any two points of a line l different from its endpoints, if any [Ax.3, Cr. 25].

Q has two sides in l [Ax. 6],

which are two lines l_1 and l_2 adjacent at Q [Df. 25]

that contains all points of l and only them [Cr. 26].

So, in one, and only in one, of such lines, for instance in l_1, will be P [Cr. 27].

In turn, P has two sides in that side l_1 of Q [Df. 25, Ax. 6],

the side PQ in which it is Q and the side in which it is not Q [Cr. 27].

PQ is a line [Df. 25]

all of whose points belong to l_1 [Df. 25]

and therefore to l [Ps. C].

Hence, PQ is a segment of l [Df. 22].

Being P and Q any two of its points, l has a number of segments and a number of points between any two of its points greater than any given number [Crs. 25, 28]. □

Corollary 30 *A segment of a segment of a line, it is also a segment of that line.*

▷ Let RS be a segment of a segment PQ of a line l [Ax.3, Cr. 29].

PQ is a line whose points belong to l [Df. 22].

RS is a line whose points belong to PQ [Df. 22],

and then to l [Ps. C].

So, RS is a segment of l [Df. 22]. □

Corollary 31 *If a point is between two given points of a given line, it is also between the given points in any other line of which the given line is a segment.*

▷ Let R be a point of a segment PQ of a line l' [Ax.3, Cr. 29],

which is a segment of another line l [Cr. 29].

Since PQ is a segment of l', it is also a segment of l [Cr. 30].

So, R is a point of a segment PQ of l [Df. 22],

and then a point of l [Df. 22]

between P and Q [Cr. 28]. □

Corollary 32 (A variant of Hilbert's Axiom II.2) *At least one of any three points of a line is between the other two.*

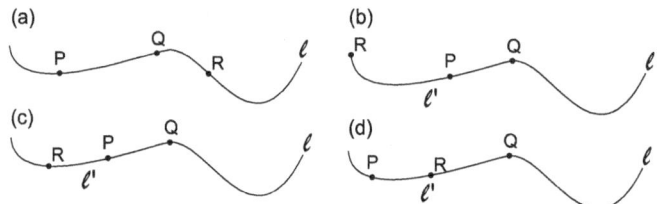

Figure C.2 – Corollary 32

▷ Let P, Q and R be any three points of any line l [Ax.3, Cr. 25].

At least one of them, for example* Q, will not be an endpoint of l [Ax. 4].

P can only be in one of the two sides of Q in l [Cr. 27].

R can only be in one of the two sides of Q in l [Cr. 27].

So, either P and R are in different sides of Q in l, or they are in the same side of Q in l. If P and R are in different sides of Q in l (Fig. C.2 (a)), then Q is between P and R in l [Df. 26].

If not, P and R are in the same side of Q in l, which is a segment l' of l [Cr. 26],

one of whose endpoints is Q [Df. 25].

If R is an endpoint of l' (Fig. C.2 (b)), P can only be between the endpoints Q and R of l' [Cr. 28],

and then between Q and R in l [Cr. 31].

If R is not an endpoint of l', it has two sides in l' [Ax. 6]:

the side RQ in which it is Q, and the side in which it is not Q [Cr. 27].

If P is in RQ (Fig. C.2 (c)), P is between R and Q in l' [Cr. 28], and then between R and Q in l [Cr. 31].

If P is in the side of R in l' in which it is not Q (Fig. C.2 (d)), then P and Q are in different sides of R in l', and R is between P and Q in l' [Df. 26]

and then between P and Q in l [Cr. 31].

So, in all possible cases [Ax. 6, Cr. 27]

at least one of the three points is between the other two in l. □

Corollary 33 (Hilbert's Axioms II.3, II.1) *One, and only one, of any three points of a line is between the other two.*

▷ Let P, Q and R be any three points of any line l [Ax.3, Cr. 25].

At least one of them, for example* Q, will be between the other two, P and R, in l [Cr. 32],

in which case Q is a point of PR [Cr. 28].

So, Q has two sides in PR [Ax. 6],

which are two lines, QP and QR, adjacent at Q [Df. 25].

P cannot be between Q and R, otherwise it would be in QR [Cr. 28],

QP would be a segment of QR [Cr. 29],

all points QP [Cr. 25]

would be points of QR [Df. 22],

and QP and QR would not be adjacent at Q [Df. 24],

which is impossible [Df. 25].

For the same reasons R cannot be between P and Q either. Therefore, one [Cr. 32],

and only one, of any three points of a line is between the other two. □

Corollary 34 (a variant of Hilbert's Axiom II.4) *Of any four points of a line, two of them are between the other two.*

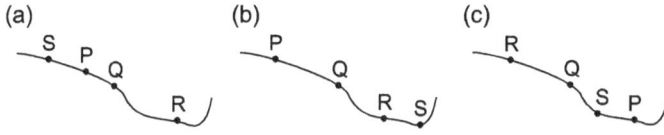

Figure C.3 – Corollary 34.

▷ Let P, Q, R and S be any four points of a line l [Ax.3, Cr. 25].

Consider any three of them, for instance* P, Q and R. One, and only one, of them, for instance* Q, will be between the other two, P and R [Cr. 33], and Q will be in PR [Cr. 28].

Of the other three points P, R and S, one, and only one, of them will be between the other two [Cr. 33]:

if P is between S and R (Fig. C.3 (a)), it is in SR [Cr. 28],

so that PR is a segment of SR [Cr. 29],

Therefore Q, which is in PR, is also in SR [Cr. 30].

So, Q and P are between R and S [Cr. 28].

For the same reasons, if R is between P and S (Fig. C.3 (b)) then Q and R are between P and S; and if S is between P and R (Fig. C.3 (c)), then Q and S are between P and R. So, in all possible cases [Ax. 6, Cr. 27] two of the four points are between the other two. □

Corollary 35 *Two segments can only be either collinear or non-collinear. And if a segment of a given line is non-collinear with another segment of another given line, then both given lines are also non-collinear.*

▷ Since belonging to is a reflexive relation [Ps. C]

and segments are lines [Df. 22],

any two segments l_1 and l_2 [Ax. 3]

belong to a line, even if the line is the own segment itself [Df. 22].

So, l_1 and l_2 will be either collinear, or non-collinear, or collinear and non-collinear. If they were collinear and non-collinear they would be segments that belong to the same line l [Df. 22],

and segments that do not belong to the same line l [Df. 22],

which is impossible [Ps. C].

So, l_1 and l_2 can only be either collinear or non-collinear. Let now l'_1 be a segment of a line l_1, and l'_2 another segment of a line l_2 [Cr. 29],

such that l'_1 and l'_2 are non-collinear [Df. 22].

If l_1 and l_2 were collinear, they would be segments of the same line l [Df. 22],

and being their respective segments l'_1 and l'_2 also segments of l [Cr. 30],

l'_1 and l'_2 would also be collinear [Df. 22],

which is not the case. Hence, l_1 and l_2 must also be non-collinear. □

Corollary 36 *If two points of a line have a given property, and all points between any two points with the given property have also the given property, then the line has a unique segment whose points are all points of*

C.3 Foundational basis of Euclidean geometry

the line with the given property.

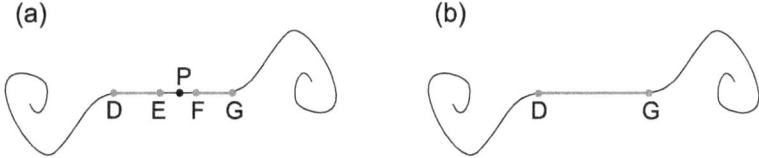

Figure C.4 – Corollary 36.

▷ Let A and B be two points [Ax.3, Cr. 29]

with a given property (gp-points for short) of a line l such that all points of l between any two of its gp-points are also gp-points. So, l has a number of gp-points greater than any given number [Cr. 29].

Let a segment whose points are gp-points, except at most its endpoints, be referred to as gp-segment. Any gp-point C of l is at least in the gp-segment AC of l [Crs. 29, 28].

So, all gp-points of l are in gp-segments. If all gp-points of l were not in a unique gp-segment, they would be in at least two gp-segments DE and FG of l [Cr. 29],

so that, being* E and F between D and G [Cr. 34],

DG is not a gp-segment. If so, there will be at least one point P between D and G that is not a gp-point. P has two sides in DG, namely PD and PG [Ax. 6, Df. 25].

E must be in the side PD of P in DG in which it is D, otherwise it would be in the side PG of P in DG in which it is not D [Cr. 27],

P would be between D and E [Df. 26],

it would be a point of DE [Cr. 28],

and being gp-points all points of DE, except at most D and E [Cr. 28],

P would be between any gp-point of DP and any gp-point of PE [Ax. 4],

and P would be a gp-point, which is not the assumed case. So, DE is a segment of the side PD of P in DG [Crs. 29].

For the same reasons, FG is a segment of the other side PG of P in DG. Hence, P is between any gp-point of DE and any gp-point of FG [Df. 25].

It is then impossible for P not to be a gp-point, and for DG not to be a gp-segment. And l has a unique gp-segment DG. □

Corollary 37 *The length of a finite line is greater than the length of each of the sides of any of its points, except endpoints, and it is greater than* zero. *The length of each side is equal to the length of the whole line minus the length of the other side. And the length of a segment of the line is less than the length of the whole line if at least one endpoint of the segment is not an endpoint of the line.*

▷ Let P be a point of a finite line AB [Df. 29, Axs. 3, 4].

Assume the length AP is not less than the length AB. It will be $AP \geq AB$ [Ps. A],

and being $AB = AP + PB$ [Ax. 6],

it would hold $AP \geq AP + PB$ [Ps. A].

Hence, $0 \geq PB$ [Ps. B],

which is impossible [Ax. 6].

So, it must be $AP < AB$ [Ps. A].

And for the same reasons $PB < AB$. Therefore, and being $0 < PB$ [Ax. 6],

it holds $0 < AB$ [Ps. B].

So, the length of any line is greater than zero. And from $AP + PB = AB$ [Ax. 6],

it follows immediately $AP = AB - PB$; $PB = AB - AP$ [Ps. B].

Let now Q be any point of AB different from P [Crs. 25].

It will be in one, and only in one, of the sides of P in AB [Cr. 27],

for instance* in AP. It has just been proved that $AP < AB$. If Q were the endpoint A of AP we would have $QP = AP$ [Ps. A].

If not, and for the same reasons above, it will be $QP < AP$. So, we can write $QP \leq AP$, and then $QP < AB$ [Pss. B, A].

Therefore, the length of a segment of AB is less than AB if at least one if its endpoints P is not an endpoint of AB. □

Fundamentals on straight lines

Definition 30 *Extensible lines.-To produce (extend) a given line by a given length is to define a line, said production (extension) of the given line, so that the production is adjacent to the given line, has the given length, and the production and the produced line are lines of the same class as the given line. Lines that can be extended from each endpoint and by any given length are called extensible lines.*

Definition 31 *Straight lines: extensible and mutually uniform lines that can neither be locally collinear nor have non-common points between common points.*

C.3 Foundational basis of Euclidean geometry

Definition 32 *Straightness.-Three or more points are said to be in straight line with one another iff they are in the same straight line, whether or not produced. A point is said in straight line with a given straight line iff it is in straight line with at least two points of the given straight line, whether or not produced. Only the straight segments of the same straight line, whether or not produced, are said to be in straight line with one another. Otherwise it is said that they are not in a straight line.*

Axiom 7 *Any two points can be the endpoints of a straight line, and only both points are necessary to draw the straight line.*

Corollary 38 *A segment of a straight line is also a straight line.*

▷ It is an immediate consequence of [Ax. 7, Dfs. 31, 28]. □

Corollary 39 (Strong form of Euclid's First Postulate) *Any two points can be the endpoints of one, and only of one, straight line.*

▷ Assume two different straight lines l_1 and l_2 have the same endpoints A and B. At least one of them will have a point which is not in the other [Df. 22].

And they would have at least one non-common point between the two common points A and B, which is impossible [Df. 31].

So, any two points can be the endpoints of one [Ax. 7],

and only of one, straight line. □

Note.-Unless otherwise indicated, hereafter, to join two points will mean to consider and draw the unique straight line whose endpoints are both points.

Corollary 40 (Strong form of Euclid's Second Postulate) *There is one, and only one, way to produce a given straight line by any given length and from any of its endpoints, being the produced line a straight line; and the given straight line and its production, adjacent straight lines in straight line with each other.*

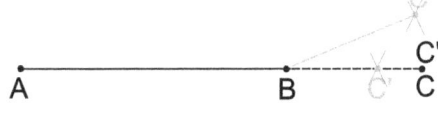

Figure C.5 – Corollary 40.

▷ Let AB be any straight line [Ax. 3, Cr. 39].

AB can be produced from any of its endpoints, for example* from B, by any given length [Dfs. 31, 30] to a point C, so that BC and AC are straight lines [Dfs. 31, 30, D],

and AB and BC are adjacent segments [Dfs. 31, 30].

Assume AB can be produced from B by the same given length to another point C'. The straight lines AC, AC' [Dfs. 31, 30]

would have a common segment AB [Cr. 29];

they would be collinear since they cannot be locally collinear [Dfs. 31, 23, Cr. 35];

and BC and BC' would be two segments of the same line l [Cr. 29],

both adjacent at B to AB [Ax. 7, Df. 30],

and so with a common endpoint B. And being C and C' different points of the same line l, one of them, for example* C', would be between B and the other in l [Cr. 33],

and we would have $BC' < BC$ [Cr. 37],

which is not the case. So, C' can only be the point C. And being BC a straight line [Dfs. 31, 30, D],

it is the unique straight line joining B and C [Cr. 39].

So, there is a unique way of producing a straight line by a given length from any of its endpoints. And AB and BC are the unique straight lines joining respectively A with B and B with C [Cr. 39],

and being A, B and C points of the straight line AC [Dfs. 32, 31, 28],

the straight lines AB and BC are segments of the same straight line AC [Dfs. 22, 24].

Therefore, the straight lines AB and BC are in straight line with each other [Df. 32]. □

Corollary 41 *Through any two points, any number of collinear straight lines of different lengths can be drawn.*

▷ It is an immediate consequence of [Df. 22, Crs. 39, 40]. □

Corollary 42 *Two straight lines with two common points belong to the same straight line.*

▷ Let AB and CD be two straight lines with two common points P and Q [Cr. 41].

Consider one of them, for instance* AB. Every point R of AB is in straight line with two points, P and Q, of CD [Df. 32].

Therefore, every points R of AB belongs to CD, whether or not produced [Df. 32].

In consequence, AB is a segment of CD, whether or not produced [Df. 22, Cr. 40].

Hence, AB and CD belong to the same straight line: CD or a production of CD [Cr. 40]. □

Corollary 43 *Being in a straight line is a transitive relation of straight lines.*

▷ Suppose that a straight line AB is in a straight line with another straight line CD, which in turn is in a straight line with another straight line EF. AB and CD belong to a straight line r_1. CD and EF belong to a straight line r_2 [Df. 32].

Since CD belongs to r_1 and r_2, the straight lines r_1 and r_2 have two common points C and D, so they belong to the same straight line r_3 [Cr. 42].

Consequently, A, B, E and F belong to r_3, and AB and EF are segments of r_3 [Cr. 29].

So, they are in straight line with each other [Ps. C, Df. 32]. □

Corollary 44 *A point is in straight line with a given straight line, iff its is in straight line with any two points of the given straight line.*

▷ Let l be any straight line [Cr. 39].

A point P in straight line with l is in straight line with at leas two points Q and R of l, produced or not [Df. 32, Cr. 40].

So, P, Q and R belongs to l, produced or not [Df. 32, Cr. 40].

And being a point of l, P belongs to the same straight line l as any couple of points of l; and P is in straight line with them [Df. 32].

Alternatively, if P is in straight line with any two points of l, then it is in straight line with l [Df. 32]. □

Corollary 45 *Any point between the endpoints of a given straight line can be common to any number of intersecting straight lines not in straight line with the given straight line, and that point is the only common point of those straight lines and the given straight line, even arbitrarily producing them and the given straight line.*

▷ Any point P between the endpoints of a straight line AB [Ax. 3, Cr. 39]

can be common to any number n of non-collinear straight lines [Ax. 5], which being non-collinear are not in straight line with the given straight line AB [Dfs. 32, 22].

Assume there is a second common point Q of AB and of any one of those n intersecting straight lines l, whether or not producing AB and l [Cr. 40].

Both straight lines would belong to the same straight line [Cr. 42],

which is not the case, because they are non-collinear [Df. 23].

Therefore, P is the only intersection points of AB and each of those n intersecting straight lines, even arbitrarily producing AB and any of the n intersecting straight lines. □

Corollary 46 *There is a number of points greater than any given number that are not in straight line with any two given points, or with a given straight line.*

▷ Let A and B be any two points [Ax. 3].

Join A and B [Cr. 39],

and let PC be a straight line non-collinear with AB that intersects AB at P [Cr. 45].

P is the only common point of both straight lines even arbitrarily produced [Cr. 45].

So, PC has a number of points greater than any given number [Cr. 25]

none of which, except P, is in straight line with A and B because none of them belong to AB, produced or not [Df. 32].

On the other hand, if AB is any straight line, it has just been proved there is a number greater than any given number of points that are not in straight line with the points A and B. So, there is a number greater than any given number of points that are not in straight line with AB [Cr. 44]. □

Corollary 47 *Each endpoint of a given straight line can be the common endpoint of any number of adjacent straight lines not in straight line with the given straight line.*

▷ Let AB be any straight line [Ax. 3, Cr. 39].

There is a number greater than any given number of points not in straight line with AB [Cr. 46].

Join each of them with, for instance*, the endpoint A of AB [Cr. 39].

Each of these straight lines are adjacent at A to AB [Df. 24].

If any of them, for instance* AP, were in straight line with AB, they would be segments of the same straight line l [Df. 32],

P, A and B would be points of that straight line l [Df. 22],

P would be in straight line with A and B [Df. 32],

and then with AB [Df. 32],

which is not the case. □

Corollary 48 *If two adjacent straight lines are not in straight line, then no point of any of them, except their common endpoint, is in straight line with the other. And by producing any of them from their common endpoint, the production is also adjacent to the non-produced one.*

▷ Let AB and AC be two straight lines adjacent at A and not in straight line with each other [Cr. 47].

Let P be a point of, for instance*, AB [Cr. 25].

A, P and B belong to AB. So, if P were in straight line with AC, it would be in straight line with A and C [Cr. 44],

and it would also belong to AC, whether or not produced [Df. 32].

In such a case AB and AC would have two common points, A and P, they would be segments of the same straight line [Cr. 42],

and they would be in straight line with each other [Df. 32],

which is not the case. So, P is not in a straight line with AC.

On the other hand, if AQ is any production from A, for example* of AB, AQ is adjacent to AB and is in a straight line with AB [Cr. 40].

The common endpoint A is the only common point of AQ and AC, otherwise they would have at least two common points; and AQ and AC would be segments of the same straight line [Cr. 42],

and, consequently, AC and AB would also be in a straight line with each other [Cr. 43],

which is not the case. So, AQ and AC are also adjacent at A [Df. 24]. □

FUNDAMENTALS ON PLANES

Definition 33 *Plane: a surface that contains at least three points not in straight line and any straight line through any two of its points. A line is said in a plane iff all of its points are points of that plane. Lines in a plane are said plane lines. Points, or lines, or points and lines in the same plane are said coplanar. Two planes are said different if at least one of them has a point that is not in the other.*

Definition 34 *Sides of a given straight line in a plane: parts of the plane that contain all points of the plane, and only them, each part with* at least two common points and at least two non-common points, where a point is said common, or common to all parts, if it is in straight line with the given straight line; and non-common if it is not, being said non-common of a part iff it is in that part. Any other line is said to be in one of those parts iff all of its points between its endpoints are non-common points of that part.

Axiom 8 *Any three points lie in a plane, in which any straight line has two, and only two, sides. Any other line is in one of such sides iff its endpoints are in that side.*

Corollary 49 (A variant of Hilbert's Axiom I.5) *A plane has a number of points greater than any given number, any two of which can be joined by a unique straight line in that plane. And any given straight line is at least in a plane, in which it can be produced by any given length from any of its endpoints.*

▷ Let P, Q and R be any three points not in straight line [Cr. 46],

and Pl a plane in which they lie [Ax. 8].

Pl has at least the points P, Q and R and all points of any straight line [Cr. 25]

through any two of its points [Dfs. 33, Ax. 7, Cr. 41].

So, Pl has a number of points greater than any given number [Cr. 25].

Let, then, A and B be any two points of Pl. Join A and B [Cr. 39],

and produce AB from A and from B by any given length to the respective points A' and B' [Cr. 40].

Since $A'B'$ is a straight line [Cr. 40]

through two points A and B [Df. 22, Cr. 41]

of Pl, $A'B'$ is in Pl [Df. 33],

so that all points of $A'B'$ are in Pl [Df. 33],

and then all points of its segment AB are in Pl [Df. 22, Cr. 29].

Hence, Pl contains the unique straight line joining any two of its points A and B [Crs. 38, 39].

Let now AB be any straight line [Ax. 3, Cr. 39],

and P, Q and R any thee of its points between A and B [Cr. 25].

There is a plane Pl containing P, Q and R [Ax. 8],

and the straight line AB through P and Q is in Pl [Df. 33].

Produce AB from A and from B by any given length to the points A' and B' respectively [Cr. 40].

Since the produced straight line $A'B'$ is a straight line [Cr. 40]

through two points A and B [Cr. 41]

of Pl, it is a straight line of Pl [Df. 33]. □

Corollary 50 *A point of a plane can only be either common to both sides of a straight line in that plane, or non-common of one, and only of one, of such sides.*

▷ Let A, B and P be any three points of a plane Pl [Ax. 8].

Join A and B [Cr. 39].

AB is in Pl [Cr. 49].

Either P belongs to AB, whether or not produced [Cr. 40],

or it does not [Ps. C].

If P belongs to AB, whether or not produced [Cr. 40],

P is a point common to both sides of AB [Ax. 8, Df. 34].

If P does not belong to AB [Df. 34],

whether or not produced [Cr. 40],

P cannot be in both sides of AB [Df. 34],

and being a point of Pl, it can only be in one, and only in one, of the two sides of AB [Df. 34, Ax. 8].

So, it is a non-common point of that side, and only of it [Df. 34]. □

Corollary 51 *There is a plane containing any two adjacent straight lines not in straight line with each other, being each of them in the same side of the other. And there is a plane containing any two intersecting and non-adjacent straight lines.*

▷ Let AB and AC be two straight lines adjacent at A and not in straight line with each other [Cr. 47].

A, B and C are not in straight line [Cr. 48].

There is a plane in which lie A, B and C [Ax. 8]

and the adjacent straight lines AB and AC [Cr. 49].

The common endpoint A is a common point of both sides of AC [Df. 34],

B is not in straight line with AC [Cr. 48],

so it is a non-common point of one of the sides of AC [Df. 34].

Therefore AB is in that side of AC [Ax. 8].

For the same reasons AC is in one of the sides of AB. Let now l_1 and l_2 be any two non-adjacent straight lines that intersect at a unique point P [Cr. 45],

Q a point of l_1, and R a point of l_2 [Cr. 25].

There is a plane containing P, Q and R [Ax. 8],

the straight line l_1 through Q and P [Cr. 41, Df. 33],

and the straight line l_2 though R and P [Cr. 41, Df. 33]. □

Corollary 52 *All points between two points of a straight line in the same side of a given straight line lie in that side of the given straight line, and that side has a number of non-common points greater than any given number.*

▷ Let l be a straight line in a plane Pl [Cr. 49]

and P and Q be any two non-common points in the same side, for instance Pl_1, of l [Ax. 8, Df. 34].

Join P and Q [Cr. 39].

PQ is in Pl_1 [Ax. 8].

All points between P and Q are non-common points of Pl_1 [Df. 34].

So, Pl_1 has a number of non-common points greater than any given number [Cr. 25]. □

Corollary 53 *In a plane and in each side of a straight line in that plane, it is possible the existence of a number greater than any given number of straight lines, whether or not adjacent, none of which is in straight line with any of the others.*

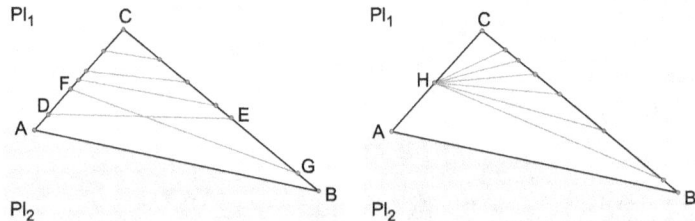

Figure C.6 – Corollary 53

▷ (Fig. C.6, left) Let A, B and C be any three points not in straight line [Cr. 46],

and Pl a plane in which they lie [Ax. 8].

Join A and B [Cr. 39]

and let Pl_1 and Pl_2 be the two sides of AB in Pl [Ax. 8].

C.3 Foundational basis of Euclidean geometry   395

C will be a non-common point [Df. 34]

of, for example*, Pl_1 [Cr. 50].

Join C with A and with B [Cr. 39].

CA and CB are not in straight line, otherwise A, C and B would be in straight line [Df. 32],

which is not the case. Join each of any number n of points of CA between C and A with a different point of CB between C and B [Crs. 29, 39],

and let DE and FG be any two of such straight lines, for example* D and F in CA, and E and G in CB. The straight lines DE and FG cannot be in straight line with each other, otherwise they would be segments of the same straight line [Df. 32],

and D, E, F and G would be in that straight line [Df. 22],

so that D would be in straight line with E and G, and then with CB [Df. 32],

which is impossible [Cr. 48].

The same argument applies to the n straight lines joining the same point H of CA between A and C (Fig. C.6, right) with n different points of CB between C and B [Crs. 29, 39],

being all of these straight lines adjacent at H [Df. 24].

Since CA and CB are in Pl_1 [Ax. 8],

all of these straight lines in Pl, whether or not adjacent, have their respective endpoints on Pl_1 [Df. 34],

so that all of them are in Pl_1 [Ax. 8]. □

Corollary 54 *The intersection point of two intersecting straight lines has its two sides in each of the intersecting straight lines in different sides of the other intersecting straight line in the plane that contains both straight lines.*

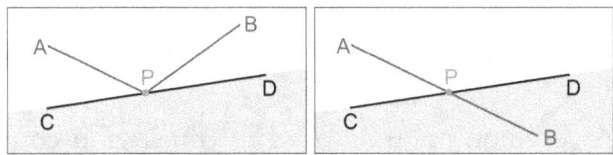

Figure C.7 – Corollary 54

▷ (Fig. C.7) Let P be the unique intersection point of two straight lines [Cr. 45]

AB and CD in a plane Pl [Cr. 51].

Since the only points of Pl common to both sides of CD in Pl are the points in straight line with CD [Df. 34],

and P is the only common point of AB and CD, even arbitrarily produced [Crs. 40, 45],

P is the only point of AB in straight line with CD [Df. 32],

and therefore the only point of AB that is a common point of both sides of CD in Pl [Df. 34].

Therefore, the endpoints A and B can only be non-common points of the sides of CD in Pl [Df. 34, Cr. 50].

So, if PA and PB were in the same side of CD in Pl, the endpoints A and B would be non-common points of that side [Ax. 8],

and being P between them [Cr. 28],

P would also be a non-common point of that side [Cr. 52],

which is impossible because it is a common point of both sides [Cr. 50].

So, A and B must be in different sides of CD in Pl [Cr. 50],

and the sides PA and PB of P are on different sides of CD in Pl [Ax. 8].

The same argument proves PC and PD can only be in different sides of AB in Pl. □

Corollary 55 *The straight line joining any two non-common points, each in a different side of another given coplanar straight line, intersects the given straight line, or a production of it, at a unique point.*

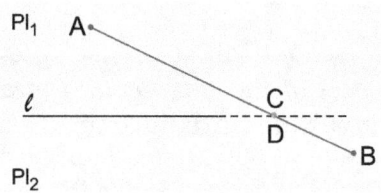

Figure C.8 – Corollary 55

▷ (Fig. C.8) Let Pl_1 and Pl_2 be the two sides of a line l in a plane Pl [Cr. 49, Ax. 8].

Let A be a non-common point of Pl_1, and B be a non-common point of Pl_2 [Cr. 52].

Join A and B [Cr. 39].

AB is in Pl [Cr. 49].

Except A and B, all points of AB are between A and B [Cr. 28].

If all points of AB between A and B were non-common points of Pl_1, AB, including B, would be in Pl_1 [Df. 32, Ax. 8],

which is not the case. Therefore, at least one point of AB between A and B is in Pl_2. So, AB contains points of Pl_2 other than B; and, for the same reason, points of Pl_1 other than A [Cr. 25].

So, AB has at least two points in each side of l. Since all points between two points of a straight line in the same side of another coplanar straight line are also in that side [Cr. 52],

AB has a segment AC whose points are all points of AB in Pl_1 [Cr. 36].

And for the same reasons it also has a segment BD whose points are all points of AB in Pl_2 [Cr. 36].

If C and D were different points, all points of AB between C and D [Cr. 29]

would be in no side of l in Pl, which is impossible because all points of AB are points of Pl [Df. 33],

and all points of Pl are points either of Pl_1, or of Pl_2, or of both of them [Ax. 8, Df. 34].

So, C and D are the same point. Since all points between A and C are in Pl_1, AC is in Pl_1 [Df. 34],

and C is also in Pl_1 [Ax. 8].

For the same reasons D is in Pl_2. Since C and D are the same point, and this point belongs to Pl_1 and to Pl_2, it is a point of l, whether or not produced [Cr. 40, Df. 34].

So, it is an intersection point of AB and l [Df. 23]

whether or not produced [Cr. 40].

And it is the unique intersection point of AB and l, otherwise the non-common point A of Pl_1 would be in straight line with at least two points of l and it would be a common point of Pl_1 and Pl_2 [Dfs. 34, 32],

which is impossible [Cr. 50]. □

Corollary 56 *A plane contains at least two non-intersecting straight lines, which can be intersected by any number of different coplanar straight lines.*

▷ Let l be a straight line in a plane Pl [Cr. 49],

Pl_1 and Pl_2 the two sides of l in Pl [Ax. 8],

A, B any two non-common points of Pl_1, and C, D any two non-common points of Pl_2 [Cr. 52].

Joint A with B; and C with D [Cr. 39].

AB is in Pl_1, and CD in Pl_2 [Ax. 8].

AB and CD cannot intersect with each other because the intersection point would be a common point of Pl_1 and Pl_2 [Df. 34],

while all points of AB and CD, even endpoints, are non-common points respectively of Pl_1 and of Pl_2 [Df. 34, Ax. 8].

On the other hand, AB and CD can be intersected by any number n of straight lines in Pl, each joining each of any n points of AB with a point of CD [Crs. 25, 39, 49]. □

Fundamentals on distances

Definition 35 *Distance between two points: length of the straight line joining both points.*

Definition 36 *Distance from a point not in a given line to the given line: the shortest distance between the point and a point of the given line, or of a production of the given line if the given line is a straight line and the point is not in straight line with it.*

Definition 37 *Distancing.-Two points of a straight line in the same side of another given straight line define, in the first straight line, a distancing direction with respect to the given straight line: from the point at the shortest distance to the given straight line to the point at the greatest distance to the given straight line. The difference between these distances is called the relative distancing with respect the given straight line of the segment defined by these two points of the first straight line.*

Definition 38 *Parallel straight lines.-A straight line is said parallel to another coplanar straight line, iff all of its points are at the same distance, said equidistance, from the other straight line.*

According to [Df. 35], the length of a straight line AB and the distance from A to B will be used as synonyms.

Axiom 9 *The distances from the points of a line to a fixed point or to another line vary in a continuous way. The distances from a point to itself and to a line to which it belongs are zero.*

Corollary 57 *The distance between any two given points is unique.*

▷ It is an immediate consequence of [Cr. 39, Df. 35, Ax. 9]. □

Fundamentals on circles

Definition 39 *Circle: a plane self-closed and non-self-intersecting line whose points are all points of the plane, and only them, at the same given* *finite distance, said radius, from a fixed point of that plane, said centre of the circle. A straight line joining any point of the circle with its centre is also said a radius of the circle. A segment of a circle is called arc, and the straight line joining its endpoints is a chord, or straight line subtending the arc. If the center of the circle is a point of a chord, the chord is said a diameter, and the corresponding arc a semicircle. Coplanar circles, and their corresponding segments, with the same centre are said concentric. The centre and any coplanar point at a distance from the centre less than its radius are said interior to the circle; if that distance is greater than the radius of the circle, the coplanar point is said exterior to the circle.*

Axiom 10 *Any point in a plane can be the center of a circle of any radius, and its complementary arcs are each on a different side of its chord.*

Corollary 58 *A circle has interior points, other than its centre, and exterior points. And any point coplanar with a circle is either in the circle, or it is interior or exterior to the circle.*

▷ Let O be the centre of a circle c in a plane Pl [Ax. 10],

and A any point of c [Df. 39].

Joint A with O [Cr. 39].

Produce OA from A by any given finite length to a point A' [Cr. 40].

OA' is in Pl [Cr. 49].

Let P be any point of OA between O and A [Cr. 29].

Since $OP < OA$ and $OA < OA'$ [Cr. 37],

P is interior and A' is exterior to c [Dfs. 35, 39].

Join now any point R of Pl with O [Crs. 39, 49].

It holds $RO \gtreqless OA$ [Ps. A],

and R will be either in c ($RO = OA$), or it will be interior ($RO < OA$) or exterior ($RO > OA$) to c [Dfs. 35, 39]. □

Corollary 59 *A plane line intersects a coplanar circle at a point between its endpoints iff it has points interior and exterior to the circle.*

▷ Let O be the centre and AO the finite radius of a circle c [Ax. 10]

in a plane Pl; BC a plane line in Pl [Df. 33, Ax. 8],

and P and Q two points of BC [Cr. 25]

such that P is interior and Q exterior to c [Cr. 58].

Being P interior to c, its distance to O is less than AO [Df. 39].

Being Q exterior to c, its distance to O is greater than AO [Df. 39].

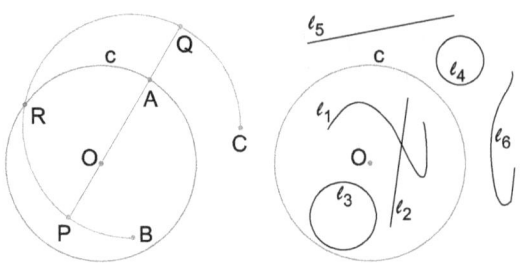

Figure C.9 – Corollary59

Therefore, there will be at least one point R in PQ, and then in BC [Cr. 25, 22],

whose distance to O is just AO [Ax. 9, Df. B].

And R will also be in c [Df. 39].

So, R is an intersection point of BC and c [Df. 23].

On the other hand, if all points of a line DE coplanar with c are interior (exterior) to c, none of its points is at a distance AO from O [Df. 39],

and then no point of DE is in c [Df. 39].

Therefore c and DE have no point in common, and they do not intersect with each other [Df. 23]. □

Corollary 60 *Any point of a circle defines a unique diameter and two unique complementary semicircles, each on a different side of that diameter.*

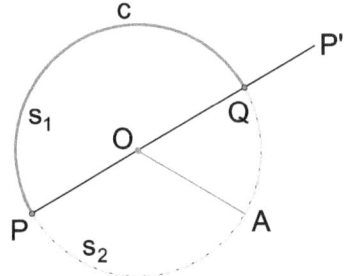

**Figure C.10 –]
Corollary 60**

▷ Let O be the centre, AO the finite radius, and P any point of a circle c [Ax. 10, Cr. 25].

Join P with the centre O of c [Cr. 39],

and produce PO from O by any given length greater than OA to a point P' [Cr. 40].

Since $OP' > OA$, PP' is a straight line with points interior, as any point of OP, and exterior, as P', to c [Cr. 40, Df. 39],

PP' intersects c at a point Q [Cr. 59].

P and Q are the common endpoints of two complementary semicircles s_1 y s_2 de c whose only common points are P and Q [Dfs. 27, 39, Ax. 10].

The center O of c is in the chord PQ of s_1 and s_2, which is the only straight line joining P and Q [Cr. 39].

Therefore, PQ is the only diameter defined by P [Df. 39].

And s_1 and s_2 are on different sides of PQ [Ax. 10]. □

FUNDAMENTALS ON ANGLES

Definition 40 *Rigid transformations of lines: metric and reversible displacements of lines that preserve the definition and the metric properties of the displaced lines, each of whose points moves from an initial to a final position along a line of finite length called trajectory, in any of the two opposite directions defined by the endpoints of the trajectory. If all points of the displaced line, except at most one, move around a fixed point and their trajectories are arcs of concentric and coplanar circles whose centre is the fixed point, the rigid transformation is called rotation.*

Definition 41 *Superpose two adjacent lines: to place them with at least two common points by means of rotations around their common endpoint. Lines with at least two common points are said superposed.*

Definition 42 *Angle.-Two straight lines are said to make an angle greater than zero iff they are adjacent, one of them can be superposed on the other by two opposite rotations around their common endpoint, and the other can be superposed on the one by the same two rotations, though in opposite directions. The least of the rotations, of both if they are equal, is said (convex) angle, the greater one is said concave angle. The angle is said to be in the side of one of the adjacent straight lines where the other adjacent straight line lies. The straight lines and their common endpoint are said respectively sides and vertex of the angle. A side is said to make an angle with the other side at their common vertex. A line joining a different point on each side of the angle is said to subtend the angle, its points are called interior to the angle. The non-interior points are called exterior to the angle.*

Definition 43 *Adjacent angles and union angle.-Two angles are said adjacent iff they have the same vertex, a common side, the first angle superposes its non-common side on the common side, and the second angle superposes the common side on its non-common side, both angles in the same direction of rotation. The angle that superposes the non-*

common sides of both angles in the same direction of rotation of both angles is their union angle, which can be concave. If two adjacent angles are equal to each other, they are said to bisect their union angle.

Definition 44 Straight angle.-Except endpoints, the angle that make the two sides of a point of a straight line at their common endpoint is said straight angle.

Definition 45 Acute, obtuse and right angles.-If a straight line cuts another given straight line and makes with it at the intersection point two adjacent angles that are equal to each other, both angles are said right angles, in which case, and only in it, the two sides of each angle are said perpendicular to each other, and the first straight line is also said perpendicular to the given one. Angles less (greater) than a right angle are said acute (obtuse).

Definition 46 Interior and exterior points and angles.-If two given coplanar straight lines are intersected by another coplanar straight line, said common transversal, a point of this transversal, different from the intersection points, is said interior to the given straight lines if it is between the intersection points of the transversal with both given straight lines; otherwise it is said exterior to them. Of the angles that a common transversal makes with the two given coplanar straight lines at their intersection points, those whose sides in the transversal have only exterior points are said exterior angles; and those whose sides in the transversal have interior points are said interior angles.

Definition 47 Alternate, corresponding and vertical angles.-Of the angles that a common transversal makes with two coplanar straight lines, the angles of a couple of non-adjacent angles are said alternate if they are both interior, or both exterior, and they are in different sides of the transversal; and corresponding if they are in the same side of the transversal, being the one interior and the other exterior. Of the angles that two intersecting straight lines make with each other at their intersection point, the couples of angles with no common side are said vertical angles.

Axiom 11 It is possible for two adjacent straight lines to make any angle at their common endpoint. The angle is zero iff both straight lines are superposed.

Corollary 61 Two straight lines make an angle greater than zero iff they are adjacent, being equal and unique the angle that each of the straight lines make with the other at their common endpoint, both rotations in opposite directions. And the adjacency point is their only common point, even arbitrarily produced from their non-common endpoints.

▷ Each of two coplanar adjacent straight lines [Cr. 51],

makes with the other the same angle greater than zero at their common endpoint, though in opposite directions [Df. 42, Ax. 11].

And being a metric transformation, that angle is unique [Dfs. 40, C].

The only common point of both sides, even arbitrarily produced from their non-common endpoints [Cr. 40],

is the vertex of the angle, otherwise both sides would be superposed [Df. 41],

and they would make an angle zero [Ax. 11].

which is not the case. On the other hand, if two straight lines make an angle zero they will be superposed [Ax. 11]

and they will not be adjacent [Dfs. 41, 24]. □

Corollary 62 *The superposition by rotation of two adjacent straight lines around their common endpoint is a unique straight line.*

▷ It is an immediate consequence of [Df. 41, Cr. 42]. □

Corollary 63 *An angle does not change by producing arbitrarily its two sides from their non-common endpoints. Nor if only one of the sides is produced from its non-common endpoint.*

▷ Let AB and AC be two adjacent straight lines [Cr. 51]

that make an angle $\alpha > 0$ at their common endpoint A [Cr. 61].

Apart from the common endpoint A, the angle α superposes at least one point P of AB with a point Q of AC [Dfs. 41, 42].

Produce AB from B and AC from C by any given length respectively to the points B' and C' [Cr. 40].

A is a common point of AB' and AC'; and P and Q are also points respectively of AB' and AC' [Cr. 40, Df. 22],

Therefore, the rotation α superposes AB' and AC' [Df. 41].

Suppose that a rotation α' smaller than α superposes two points R and S respectively of AB' and AC' but does not superposes AB and AC. Therefore, α' does not superpose P and Q [Df. 41].

The point R could not be between A and P; nor S between A and Q, otherwise α' would superpose AB and AC [Dfs. 41, 31],

which is not the considered case. Therefore P is between A and R; and Q is between A and S [Cr. 32].

We would then have two straight lines with non-common points, P and

Q, between two common points, the point A and the superposed R and S, which is impossible [Df. 31].

So, AB' and AC' also make at A an angle α. The same argument applies if only one of the sides, for instance* AB, is produced from B to B', now the points R and S being respectively in BB' and QC, in both cases between the corresponding endpoints. □

Corollary 64 *Three adjacent straight lines define three angles at their common endpoint. And two intersecting straight lines define with each other at most four angles at their intersection point.*

▷ Three coplanar straight lines AB, AC and AD adjacent at the same point A [Cr. 53]

define three couples of coplanar straight lines adjacent at that point: AB, AC; AB, AD; and AC, AD [Df. 24].

So, AB, AC and AD define three angles at that point A [Cr. 61].

For the same reason, two intersecting straight lines define at most four angles whose two sides are not in the same straight line. □

Corollary 65 *(Fig. C.11) Three straight lines adjacent at the same point define a couple of adjacent angles at that point.*

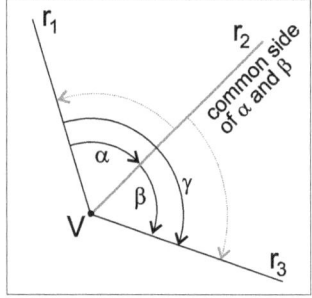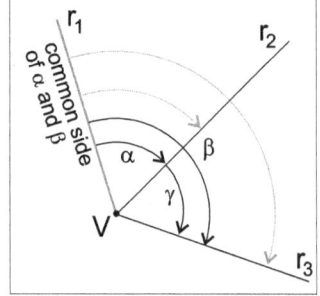

Figure C.11 – Corollary 65

▷ Three straight lines r_1, r_2, r_3 adjacent at V define three angles α, β and γ at V [Cr. 64],

and then three couples of angles: α and β; α and γ; and β and γ. Being only three sides, the two angles of each of such couples must have a common side [Df. 42].

The angles of such couples that superpose their common side on their respective non-common sides can only be rotations in the opposite direction, or in the same direction [Dfs. 21, 40].

In the first case (Fig. C.11, left), the angles of the couple, for instance α and β, are adjacent because either of them also superposes its non-common side on the common side in the same direction as the other superimposes the common side on its non-common side [Dfs. 42, 43].

In the second case (Fig. C.11, right), let r_1 be the common side of α and β. Assume α superposes r_1 on r_2; β can only superpose r_1 on r_3; and it will be different from α otherwise r_2 would be superposed on r_3 and they would not be adjacent [Dfs. 41, 24].

Since α and β are different, one of them, for instance α, will be less than the other [Ps. A],

in which case γ can only be the angle that, in the same direction of rotation as α, superposes r_2 on r_3. So, α and γ are adjacent [Df. 43].

So, in any case three straight lines adjacent at the same point define a couple of adjacent angles at that point. □

Corollary 66 *Two adjacent straight lines make a straight line iff they make a straight angle at their common endpoint.*

▷ If two adjacent straight lines l_1 and l_2 [Cr. 51]

make at their common endpoint P a straight angle, they are the two sides of the point P in a straight line l [Df. 44],

so that l_1 and l_2 make the straight line l [Cr. 26].

If two straight lines l_1 and l_2 adjacent at P make a straight line l, l_1 and l_2 are the sides in l of their common endpoint P [Df. 25]

so that they make a straight angle at P [Df. 44]. □

Corollary 67 *Except for the vertex, no point of either side of an angle is in straight line with the other side of the angle if the angle is not an straight angle and is greater than zero.*

▷ It is an immediate consequence of [Ax. 11, Crs. 66, 48] □

FUNDAMENTALS ON POLYGONS

Definition 48 *Polygon.-Three or more finite coplanar straight lines, called sides, each of which is adjacent at each of its two endpoints, called vertexes, to just one of the others, being not in straight line with each other, and being their common endpoints their only intersection points, are said to make a polygon. Two sides of the same or of different polygons are said equal iff they have the same length. Two polygons are said adjacent iff they have a common side; opposite iff they have two vertical angles at a common vertex; similar iff the angles of the one are equal to the angles of the other; and equal if they are similar and the sides of each angle of the one are equal to the sides of the corresponding equal angle of the other. Polygons with at least one concave angle are* *said concave. The angle each side makes with the production of another adjacent side is said exterior. A straight line joining two points each on a different side of a polygon is a divisor of the polygon; if the ends of a divisor are vertexes, the divisor is called diagonal. A divisor bisects a polygon if it is the common side of two adjacent polygons with the same area.*

Note.-The classical definition of diagonal is a particular case of the above general definition of divisor.

Definition 49 *Triangles and quadrilaterals. A polygon of three (four) sides is a triangle (quadrilateral). A triangle (quadrilateral) is said equilateral if its three (four) sides are equal to one another. A triangle is said isosceles if it has two equal sides; and scalene if the three of them are unequal. If one of its angles is a right angle, it is said a right-angled (or simply right) triangle. A rectangle is a quadrilateral all of whose angles are right angles. An equilateral rectangle is a square. And a parallelogram is a quadrilateral with two couples of equal and parallel sides. Polygons with more than four sides are named pentagons, hexagons, heptagons etc. A polygon is said to lie between two given lines iff its vertexes are in the given straight lines or in straight lines whose endpoints are points of the given straight lines.*

Axiom 12 *The area of a polygon is greater than zero, and is the sum of the areas of the two adjacent polygons defined by any of its divisors. Equal polygons have equal areas.*

Corollary 68 *Any two adjacent sides of a polygon make an angle greater than zero at their common endpoint, and the polygon has as many angles as sides. And twice as many exterior angles as angles.*

▷ Being coplanar all sides of a polygon [Df. 48],

each couple of its adjacent sides makes a unique angle greater than zero at their common endpoint [Cr. 61].

So, the polygon has as many angles as couples of adjacent sides. Since each couple of adjacent sides is defined by two adjacent sides, and each side defines two of such couples, one at each of its two endpoints [Df. 48],

the polygon has as many angles as sides. And since each side makes an exterior angle with the production of each of the other two adjacent sides at each of its two vertices [Df. 48],

the polygon has twice as many exterior angles as angles. □

The last element of this new foundational base of the Euclidean geometry is the following corollary, which is not strictly geometric because

C.3 Foundational basis of Euclidean geometry

the demonstration makes use of some basic results of set theory. Although the demonstration is simple, the corollary can be omitted and considered its statement as an additional hypothesis: the length of a line is finite whenever it has two well-defined endpoints. In any case, at the end of the demonstration, the corresponding used concepts are explained. A more complete and detailed proof is given in [212, Chap. 16].

Corollary 69 *In the Euclidean space R^3, the length of a line with two endpoints is always finite. And the distance between any two given points is always finite and unique.*

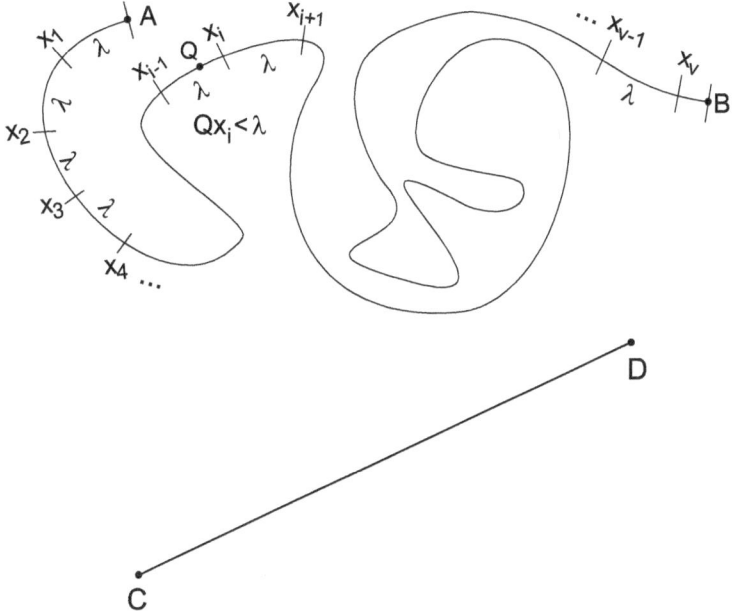

Figure C.12 – Corollary 69

▷ (Fig. C.12) Let AB be any line* in the Euclidean space $\mathbb{R}^3$, and $\lambda > 0$ any finite length [Ax. 3, 6].

Let $\mathbf{P} = AP_1, P_1P_2, P_2P_3 \ldots$ be a partition of AB all of whose parts have the same finite length $\lambda > 0$, except the last one, if any, that can be less than λ. A point X such that $XB < \lambda$ will belong to a part that can only be the last part or the penultimate part of $\mathbf{P}$. [Cr. 37].

So, $\mathbf{P}$ has a last part $P_\phi B$. Any point Y of the segment AP_i such that $YP_i < \lambda$, and any point Z of the segment P_iB such that $P_iZ < \lambda$ can only belong respectively to the parts $P_{i-1}P_i$ and P_iP_{i+1} of $\mathbf{P}$, for all $1 < i < \phi$ [Cr. 37].

Therefore, $\mathbf{P}$ has a first element AP_1, a last element $P_\phi B$, and each element has an immediate predecessor (except AP_1), and an immediate successor (except $P_\phi B$).

Let us suppose there exists an n-th element of **P** with a finite number of predecessors. The $(n+1)$th element of P, if any, will also have a finite number $n+1$ of predecessors, or n predecessors if it were the last element of the partition. Since P_1P_2 has a finite number of predecessors, just 1, we can inductively conclude that each element of **P**, including its last element $P_\phi B$, has a finite number of predecessors. So, **P** has a finite number of elements. And being finite the sum of any finite number of finite lengths, AB has finite length. Let us now join any two points C and D [Ax. 3, Cr. 39].

It has just been proved that CD has finite length. Therefore, the distance from C to D is also finite, and unique [Df. 35, Cr. 57]. □

List of figures

4.1	Square's side and diagonal.	36
5.1	Natural emergence of anisotropy	42
9.1	Crescas' Paradox	99
13.1	Three surfaces	142
13.2	Michelson-Morley interferometer.	144
13.3	MM interferometer: horizontal arm.	145
13.4	MM interferometer: vertical arm.	146
14.1	Self-rotating lines.	153
15.1	Velocity vector of light.	164
15.2	The elastic cord at rest.	166
16.1	Square's side and diagonal.	172
16.2	Pythagoras Discrete Theorem.	188
19.1	Galileo's mast.	222
19.2	Galileo's laser beam.	223
19.3	Two ways of observing Einstein's clock.	224
19.4	Inherited c_{ox}.	229
19.5	Absolute motion is undetectable.	232
19.6	Multiple superposed and simultaneous realities.	235
19.7	Absolute speed of light.	237
20.1	Z^*-points and Z-points	241
26.1	Twin robots.	315
26.2	Two pendulums in RF_o.	315
26.3	Two pendulums in RF_v.	317
30.1	The Principle of Inertia.	346
A.1	The power of the ellipsis	363
A.2	Hilbert machine	364
A.3	Finite Hilbert machine	368
B.1	The hollow cylinder	370
B.2	Last disk supertask	370
B.3	Last disk finite task	373

C.1 Corollary 24	379
C.2 Corollary 32	382
C.3 Corollary 34	383
C.4 Corollary 36	385
C.5 Corollary 40	387
C.6 Corollary 53	394
C.7 Corollary 54	395
C.8 Corollary 55	396
C.9 Corollary 59	400
C.10 Corollary 60	400
C.11 Corollary 65	404
C.12 Corollary 69	407

Bibliographic references

[1] BP Abbott, Richard Abbott, TDea Abbott, S Abraham, F Acernese, K Ackley, C Adams, RX Adhikari, VB Adya, Christoph Affeldt, et al. Gwtc-1: a gravitational-wave transient catalog of compact binary mergers observed by ligo and virgo during the first and second observing runs. *Physical Review X*, 9(3):031040, 2019.

[2] A. D. Aczel. *The Mystery of the Aleph: Mathematics, the Kabbalah and the Search for Infinity*. Pockets Books, New York, 2000.

[3] A. Alegre Gorri. *Estudios sobre los presocráticos*. Anthropos, Barcelona, 1985.

[4] R. Aloisio, A. Galante, Grillo A., Luzio E., and F. Méndez. A note on DSR-like approach to space-time. *Phys. Lett. B*, 610:101 – 106, 2005.

[5] R. Aloisio, A. Galante, A. F. Grillo, E. Luzio, and F. Méndez. Approaching space-time through velocity in doubly special relativity. *Phys. Rev. D*, 70:125012, Dec 2004.

[6] J. S. Alper and M. Bridger. Mathematics, Models and Zeno's Paradoxes. *Synthese*, 110:143 – 166, 1997.

[7] J. S. Alper and M. Bridger. On the Dynamics of Perez Laraudogotia's Supertask. *Synthese*, 119:325 – 337, 1999.

[8] J. S. Alper, M. Bridger, J. Earman, and J. D. Norton. What is a Newtonian System? The Failure of Energy Conservation and Determinism in Supertasks. *Synthese*, 124:281 – 293, 2000.

[9] J. Ambjorn, J. Jurkiewicz, and R. Loll. El universo cuántico autoorganizado. *Investigación y Ciencia*, 384:20–27, 2008.

[10] Amelino-Camelia, Giovanni, Lämmerzahl, Claus, Mercati, Flavio, Tino, and Guglielmo M. Publisher's Note: Constraining the Energy-Momentum Dispersion Relation with Planck-Scale Sensitivity Using Cold Atoms [phys. Rev. Lett. 103, 171302 (2009)]. *Phys. Rev. Lett.*, 104:039901, Jan 2010.

[11] G. Amelino-Camelia. A Phenomenological description of quan-

tum gravity induced space-time noise. *Nature*, 410:1065–1069, 2001.

[12] Aristotle. *Posterior Analitics*. Clarendon Press, Oxford, 1993 (2002).

[13] Aristotle. *Posterior Analytics*. Kessinger Publishing LLC, Whitefish, MT, 2004.

[14] Aristotle. *On the Heavens*. Forgotten Books, London, 2007.

[15] Aristotle. *Categories*. E-Bookarama Editions, 2023.

[16] Aristóteles. *Física*. Gredos (Kindle Edition), Madrid, 1995.

[17] Aristóteles. *Metafísica*. Espasa Calpe, Madrid, 1995.

[18] Aristóteles. *Acerca de la generación y la corrupción*. Gredos, Madrid, 1998.

[19] G. Arrigo and B. D'Amore. Lo veo pero no lo creo. Obstáculos epistemológicos y didácticos en el proceso de comprensión de un teorema de Cantor que involucra al infinito actual. *Educación matemática*, 11(1):5–24, 1999.

[20] J. Baez. The Quantum of Area? *Nature*, 421:702 – 703, February 2003.

[21] C. Barceló and G. Jannes. A Real Lorentz-FitzGerald Contraction. *Found. Phys.*, 38:1199–199, 2008.

[22] J. D. Barrow. *The Infinite Book*. Vintage Books (Random House), New York, 2006.

[23] M. Bartels. First Evidence of Giant Gravitational Waves Thrills Astronomers. *Scientific American*, June 2023.

[24] Jacob D. Bekenstein. La información en un universo holográfico. *Investigación y Ciencia*, 325:36–43, 2003.

[25] E. Beltrami. Note fisico-matematiche. *Rendiconti del Circolo Matematico di Palermo (1884-1940)*, 3(1):67–79, 1889.

[26] Eugenio Beltrami. Saggio di interpretazione della geometria non-euclidea. *Giornale di Mathematiche*, pages 285–315, 1868.

[27] Paul Benacerraf. Tasks, Super-Tasks, and Modern Eleatics. *J. Philos.*, LIX(24):765–784, 1962.

[28] Jean Paul Van Bendegem. In defense of discrete space and time. *Logique et Analyse*, 38:150 –152, 1997.

[29] Jean Paul Van Bendegem. Finitism in Geometry. In E. N. Zalta, editor, *Stanford Encyclopaedia of Philosophy*. Stanford University, URL = http://plato.stanford.edu, 2002.

[30] Henri Bergson. *Creative Evolution*. Dover Publications Inc., New York, 1998.

[31] Henri Bergson. The Cinematographic View of Becoming. In Wesley C. Salmon, editor, *Zeno's Paradoxes*, pages 59 – 66. Hackett Publishing Company, Inc, Indianapolis/Cambridge, 2001.

[32] George Berkeley. *A Treatise Concerning the Principles of Human Knowledge.* Renascence Editions, http://darkwing.uoregon.edu/ bear/berkeley, 2004.

[33] Alberto Bernabé. Introducción y notas. In Alberto Bernabé, editor, *Fragmentos presocráticos.* Alianza, Madrid, 1988.

[34] Károly Bezdek. *Classical Topics in Discrete Geometry.* Springer., New York, 2010.

[35] Joël Biard. Logique et physique de l'Infini au XIVe siècle. In Fran çoise Monnoyeur, editor, *Infini des mathématiciens, infinit des philosophes.* Belin, Paris, 1992.

[36] Erwin Biser. Discrete Real Space. *J. Philos.*, 38:518 – 524, 1941.

[37] Eftichios Bitsakis. Space and Time: Who was Right, Einstein or Kant? In Franco Selleri, editor, *Open Questions in Relativistic Physics*, pages 115–125. Apeiron, Montreal (Canada), 1998.

[38] M. Black. Achilles and the Tortoise. *Analysis*, XI:91 – 101, 1950 - 51.

[39] Simon Blackburn. *Oxford Dictionary of Philosophy.* Oxford University Press, New York, second edition edition, 2008.

[40] D. Blanco Laserna. *Las ondas gravitacionales.* RBA Editores, Mexico, 2017.

[41] David Blanco Ledesma. *El espacio es una cuestión de tiempo.* RBA Editores, 2012.

[42] D. Bohm. *La totalidad y el orden implicado.* Editorial Kairós, 1987.

[43] Max Born. *Einstein's theory of relativity.* Dover Publications Inc., New York, 1965.

[44] E. J. Borowski and J. M. Borwein. *The Harper Collins Dictionary of Mathematics.* Harper Collins Publisher, New York, 1991.

[45] E. J. Borowski and J. M. Borwein. *Web-linked Dictionary of Mathematics.* HarperCollins Publisher, New York, 2006.

[46] Rodney A. Brooks. *Fields of Colors.* Rodeney A. Brooks, 2016.

[47] John Burnet. *Early Greek philosophy.* Adam and Charles Black, London, 1920.

[48] Antonopoulos. C. A Bang into Nowhere. Comments on the Universe Expansion Theory. *Apeiron*, 10(1):40–68, January 2003.

[49] R. T. Cahill. Quantum Foam, Gravity and Gravitational Waves. *Relativity, Gravitation, Cosmology*, pages 168–226, 2003.

[50] Reginald T. Cahill. A new light-speed anisotropy experiment: absolute motion and gravitational waves detected. *Progress in Physics*, 4:73–92, 2006.

[51] Reginald T. Cahill and Kirsty Kitto. Re-Analysis of Michelson-Morley Experiments Reveals Agreement with COBE Cosmic

Background Radiation Preferred Frame so Impacting on Interpretation of General Relativity. *Apeiron*, 2002.

[52] Florian Cajori. The History of Zeno's Arguments on Motion. *American Mathematical Monthly*, XXII:1–6, 38–47, 77–82, 109–115, 143–149, 179,–186, 215–220, 253–258, 292–297, 1915. http://www.matedu.cinvestav.mx/librosydocelec/Cajori.pdf.

[53] Florian Cajori. The Purpose of Zeno's Arguments on Motion. *Isis*, III:7–20, 1920-1921.

[54] Georg Cantor. Grundlagen einer allgemeinen Mannichfaltigkeitslehre. *Mathematishen Annalen*, 21:545 – 591, 1883.

[55] Georg Cantor. *Gesammelte Abhandlungen*. Verlag von Julius Springer, Berlin, 1932.

[56] Georg Cantor. *Contributions to the founding of the theory of transfinite numbers*. Dover, New York, 1955.

[57] Georg Cantor. Foundations of a General Theory of Manifolds. *The Theoretical Journal of the National Caucus of Labor Committees*, 9(1-2):69 – 96, January - February 1976.

[58] Georg Cantor. On The Theory of the Transfinite. Correspondence of Georg Cantor and J. B: Cardinal Franzelin. *Fidelio*, III(3):97 – 110, Fall 1994.

[59] Georg Cantor. *Fundamentos para una teoría general de conjuntos*. Crítica, Barcelona, 2005.

[60] F. Capra. *EL TAO DE LA FÍSICA*. LUIS CARCAMO, editor, Madrid, 1984.

[61] Ana María Cetto. *La luz. En la naturaleza y en el laboratorio*. Fondo de Cultura Económica, 2012.

[62] A Chalumeau, S Babak, A Petiteau, S Chen, A Samajdar, R N Caballero, G Theureau, L Guillemot, G Desvignes, A Parthasarathy, K Liu, G Shaifullah, H Hu, E van der Wateren, J Antoniadis, A-S Bak Nielsen, C G Bassa, A Berthereau, M Burgay, D J Champion, I Cognard, M Falxa, R D Ferdman, P C C Freire, J R Gair, E Graikou, Y J Guo, J Jang, G H Janssen, R Karuppusamy, M J Keith, M Kramer, K J Lee, X J Liu, A G Lyne, R A Main, J W McKee, M B Mickaliger, B B P Perera, D Perrodin, N K Porayko, A Possenti, S A Sanidas, A Sesana, L Speri, B W Stappers, C Tiburzi, A Vecchio, J P W Verbiest, J Wang, L Wang, and H Xu. Noise analysis in the European Pulsar Timing Array data release 2 and its implications on the gravitational-wave background search. *Monthly Notices of the Royal Astronomical Society*, 509(4):5538–5558, 11 2021.

[63] Li M. Chen. *Digital and discrete geometry*. Springer, New York, 2014.

[64] Marios Christodoulou and Carlo Rovelli. On the Possibility of Experimental Detection of the Discreteness of Time. *Frontiers in Physics*, 8:207, 2020.

[65] I. Ciufolini, V. Gorini, U. Moschella, and P. Fré. *Gravitationsl Waves*. Institute of Physics Publishing, 2001.

[66] Brian Clegg. *A Brief History of Infinity. The Quest to Think the Unthinkable*. Constable and Robinson Ltd, London, 2003.

[67] W.K. Clifford. On the space theory of matter. *Proceedings of the Cambridge philosophical society*, 2:157–158, 1870.

[68] Frank Close. *Nothing*. Oxford University Press, New York, 2009.

[69] Jonas Cohn. *Histoire de l'infini. Le problème de l'infini dans la pensée occidentale jusqu''a Kant*. Leséditions du CERF, Paris, 1994.

[70] H. R. Coish. Elementary particles in a finite world geometry. *Phys. Rev.*, 114:383 – 388, 1959.

[71] Giorgio Colli. *Zenón de Elea*. Sexto Piso, Madrid, 2006.

[72] Nicolai Copernici Thorunensis. *De revolutionibus orbium caelestium libir sex*. Oldenburg, Munich, 1949.

[73] Lucas C. Céleri and Vasileios I. Kiosses. Unruh effect as a result of quantization of spacetime. *Physics Letters B*, 781:611 – 615, 2018.

[74] John Daintith, editor. *Dictionary of Physics*. Oxford University Press, New York, 2009.

[75] S. Das. Michelson-Morley experiment proves light speed is not constant. *Physics Essays*, 27(1):134–138, 2014.

[76] Josep W. Dauben. *Georg Cantor. His mathematics and Philosophy of the Infinite*. Princeton University Press, Princeton, N. J., 1990.

[77] Richard Dedekind. *Qué son y para qué sirven los números (Was sind Und was sollen die Zahlen (1888))*. Alianza, Madrid, 1998. Definición de conjunto infinito p. 115.

[78] Jean-Paul Delahaye. El carácter paradójico del Infinito. *Investigación y Ciencia (Scientifc American)*, Temas: Ideas del infinito(23):36 – 44, 2001.

[79] René Descartes. *Los Principios de la Filosofía*. RBA Coleccionables S.A., 2002.

[80] Satyan L. Devadoss and Joseph O'Rourke. *Discrete and computational geometry*. Princeton University Press, Princeton and Oxford, 2011.

[81] P. A. M. Dirac. *Quantum Mechanics*. Oxford University Press, London, 1958.

[82] William Dunham. *Journey Through Genius. The Great Theorems of Mathematics*. John John Wiley and Sons, New York, 1990.

[83] John Earman. Determinism: What We Have Learned and What We Still Don't Know. In Michael O'Rourke and David Shier, editors, *Freedom and Determinism*, pages 21–46. MIT Press, Cam-

bridge, 2004.

[84] John Earman and John D. Norton. Forever is a Day: Supertasks in Pitowsky and Malament-Hogarth Spacetimes. *Philosophy of Science*, 60(1):22–42, 1993.

[85] John Earman and John D. Norton. Infinite Pains: The Trouble with Supertasks. In S. Stich, editor, *Paul Benacerraf: The Philosopher and His Critics*. Blackwell, New York, 1996.

[86] John Earman and John D. Norton. Comments on Laraudogoitia's 'Classical Particle Dynamics, Indeterminism and a Supertask'. *The British Journal for the Phylosophy of Science*, 49(1):122 – 133, March 1998.

[87] A. Einstein. Zur Elektrodynamik bewegter Körper. *Ann. Phys.*, 17:891–921, 1905.

[88] A. Einstein. Über den Einfluss der Schwercraft auf die Ausbreitung des Lichtes. *Ann. Phys.*, 35:898–90, 1911.

[89] A. Einstein. Zum Relativitätsproblem. *Scientia*, 15:337–348, 1914.

[90] A. Einstein. Die Grundlage der allgemeinen Relativitätstheorie. *Ann. Phys.*, 354(7):769–822, 1916.

[91] A. Einstein. *The Meaning of Relativity. Four Lectures Deliverd at Princeton University, May 1921*. Princeton University Press, 1922.

[92] A. Einstein. PHYSIK UND REALITÄT. *Journal of the Franklin Institute*, 221(3):349–382, March 1936.

[93] A. Einstein. Cosmological considerations on the general theory of relativity. In *The Principle of Relativity*. Dover Publications Inc., 1952.

[94] A. Einstein. *The Principle of Relativity*, chapter III On the Electrodynamics of Moving Objects, pages 35–65. Dover Publications Inc., 1952.

[95] A. Einstein. The Problem of Space, Ether, and the Field In Physics. In Nick Huggett, editor, *Space from Zeno to Einstein*, pages 253 – 260. The MIT Press, Cambridge, 2002.

[96] A. Einstein. Sobre la electrodinámica de los cuerpos en movimiento. In A. Ruíz de Elvira, editor, *Cien años de relatividad. Los artículos clave de 1905 y 1906*, pages 88–139. nivola, Madrid, 2003.

[97] A. Einstein and L. Infeld. *The evolution of physics*. Cambridge University Press, 1938.

[98] A. Einstein and L. Infeld. *La evolución de la física*. Salvat Editores, Barcelona, 1986.

[99] A. Einstein, B. Podolsky, and N. Rosen. Can quantum mechanical description of physical reality be considered complete? *Physical Review*, 47:777– 780, May 1935.

[100] A. Einstein and N. Rosen. On Gravitational Waves. *J. Franklin Inst.*, 223:43–54, 1937.

[101] Ch. Eisele, A. Yu. Nevsky, and S. Schiller. Laboratory Test of the Isotropy of Light Propagation at the 10^{-17} Level. *Phys. Rev. Lett.*, 103:090401, Aug 2009.

[102] V. Ekroll, F. Faul, and J. Golz. Classification of apparent motion percepts basedon temporal factors. *Journal of Vision*, 8(4):1–22, 2008.

[103] Alberto Elena, Javier Ordóñez, and Mariano Colubi. *Después de Newton: ciencia y sociedad durante la Primera Revolución Industrial*. Anthropos, Barcelona, 1998.

[104] Berthold-Georg Englert, Marlan O. Scully, and Artur K. Ekert. La dualidad en la materia y en la luz. In *Misterios de la física cuántica*, pages 68–74. Prensa Científica. Investigación y Ciencia. Temas 10, 1997.

[105] Real Academia Española. *Diccionario de la lengua española*. Real Academia Española, 2014.

[106] Euclides. *Elementos*. Gredos, Madrid, 2000.

[107] M. Faraday. Thoughts on ray-vibrations. *The London, Edinburg and Dublin Philosophical Magazine and Journal of Science*, 28(188):345–350, 1846.

[108] Michael Faraday. On the Physical Character of the Lines of Magnetic Force. *The London, Edinburg and Doublin Philosophical Magazine and Journal of Science*, 3(20):407–437, 1852.

[109] Michael Faraday. *On the Physical Character of the Lines of Magnetic Firce*. HardPress, Miami, 2017.

[110] A. A. Faraj. The Ives Experiment. *The General Science Journal*, pages 1–15, 2016.

[111] Jesús Navarro Faus. *Heisenberg. El principio de incertidumbre*. RBA Editores, Barcelona, 2012.

[112] José L. Fernández Barbón. Geometría no conmutativa y espaciotiempo cuántico. *Investigación y Ciencia*, (342):60–69, Marzo 2005.

[113] Enrique Fernández Borja. *El vacío y la nada*. RBA Editores, Barcelona, 2015.

[114] José Ferreirós. Matemáticas y platonismo(s). *Gaceta de la Real Sociedad Matemática Española*, 2(3):446–473, 1999.

[115] Richard P. Feynman. *Electrodinámica cuántica: la extraña teoría de la luz y la materia*. Alianza Universidad, Madrid, 1992.

[116] Richard P. Feynman. *El carácter de la ley física*. Tusquets, Barcelona, 2000.

[117] Richard P. Feynman, Robert B. Leighton, and Matthew Sands. *Lectures on Physics*. Addison-Wesley Publishing Company,

Reading, Massachusetts, 1977.

[118] D. Finkelstein and E. Rodríguez. Quantum time-space and gravity. In R. Penrose and C. J. Isham, editors, *Quantum Concepts in Space and Time*, pages 247 – 254. Oxford University Press, Oxford, 1986.

[119] Fabrizio Fiore, Luciano Burderi, Tiziana Di Salvo, Marco Feroci, Claudio Labanti, Michelle R. Lavagna, and Simone Pirrotta. HERMES: a swarm of nano-satellites for high energy astrophysics and fundamental physics. In Jan-Willem A. den Herder, Shouleh Nikzad, and Kazuhiro Nakazawa, editors, *Space Telescopes and Instrumentation 2018: Ultraviolet to Gamma Ray*, volume 10699. International Society for Optics and Photonics, SPIE, 2018.

[120] E. Fischbach, H. Kloor, R. A. Langel, A. T. Y. Lui, and M. Peredo. New geomagnetic limits on the photon mass and on long-range forces coexisting with electromagnetism. *Phys. Rev. Lett.*, 73:514–517, Jul 1994.

[121] Robert Fogelin. Hume and Berkeley on the Proofs of Infinite Divisibility. *The Philosophical Review*, XCVII(1):47 – 69, January 1988.

[122] P. Forrest. Is Space-Time Discrete or Continuous? *Synthese*, 103:327 – 354, 1995.

[123] A. P. French. *Special relativity*. W. W. Norton and Company Inc., New York, 1968.

[124] G. Galilei. *Dialogue concerning the two chief world systems-Ptolemaic and Copernican*. University of California Press, Berkeley and Los Angeles, 1967.

[125] G. Galilei. *Diálogo sobre los dos máximos sistemas del mundo ptolemaico y copernicano*. Círculo de Lectores, Barcelona, 1997.

[126] G. Galilei. *The Collected works of Galileo Galilei*, chapter Dicourses and Mathematical Demonstrations Relating to Two New Sciences, pages 945–1321 (Kindle Edition). Delphi Publishing Ltd., 2017.

[127] C. F. Gauss and H. C. Schumacher. *C. F. Gauss Werke: Briefwechsel Mit Schumacher 3 vol.* Georg Olms, 1831 (1075).

[128] Henning Genz. *Nothingness. The science of empty space*. Perseus Publishing, Cambridge, MA, 1999.

[129] A. Gjurchinovskii. Reflection of light from a uniformly moving mirror. *American Journal of Physics*, 72(10):1316–1324, October 2004.

[130] James Gleick. *The information*. Fourth Estate, London, 2012.

[131] Pedro F. González Díaz. La energía fantasma y el futuro del universo. *Investigación y Ciencia*, (357):53–61, Junio 2006.

[132] A. Goswami. *Self-Aware Universe*. Putnam's Sons, New yoek,

1993.

[133] I. Grattan-Guiness. *Del cálculo a la teoría de conjuntos, 1630-1910*. Alianza Editorial, Madrid, 1984.

[134] Jeremy Gray. *Worlds Out of Nothing. A Course in the History of Geometry in the 19th Century*. Spriger Verlag, 2010.

[135] Marvi Jay Greenberg. *Euclidean and Non-Euclidean Geometries. Development and History*. W.H. Freeman, 2008.

[136] Marvin Jay Greenberg. *Euclidian and Non-Euclidian Geometries: History and Development*. W. H. Freeman, San Francisco, CA, 1994.

[137] B. Greene. *El tejido del cosmos*. Editorial Crítica. Drakontos Bolsillo., Barcelona, 2011.

[138] Brian Greene. *The Fabric of the Cosmos. Space, Time, and the Texture of Reality*. Alfred A. Knopf, New York, 2004.

[139] Adolf Grünbaum. Modern Science and Refutation of the Paradoxes of Zeno. *The Scientific Monthly*, LXXXI:234–239, 1955.

[140] Adolf Grünbaum. *Modern Science and Zeno's Paradoxes*. George Allen And Unwin Ltd, London, 1967.

[141] Adolf Grünbaum. Modern Science and Refutation of the Paradoxes of Zeno. In Wesley C. Salmon, editor, *Zeno's Paradoxes*, pages 164 – 175. Hackett Publishing Company, Inc, Indianapolis/Cambridge, 2001.

[142] Adolf Grünbaum. Modern Science and Zeno's Paradoxes of Motion. In Wesley C. Salmon, editor, *Zeno's Paradoxes*, pages 200 – 250. Hackett Publishing Company, Inc, Indianapolis/Cambridge, 2001.

[143] Adolf Grünbaum. Zeno's Metrical Paradox of Extension. In Wesley C. Salmon, editor, *Zeno's Paradoxes*, pages 176 – 199. Hackett Publishing Company, Inc, Indianapolis/Cambridge, 2001.

[144] Alan Guth. *The Inflationary Universe. The Quest for a New Theory of Cosmics Origins*. Basic Books, 1998.

[145] Alan H. Guth. Inflationary universe: A possible solution to the horizon and flatness problems. *Phys. Rev. D*, 23:347–356, Jan 1981.

[146] S. Hacyan. *Ondas gravitacionales*. Fondo de Cultura Económica, México, 2019.

[147] Michael Hallet. *Cantorian Set Theory and Limitation of Size*. Oxford University Press, 1984.

[148] Charles Hamblin. Starting and Stopping. *The Monist*, 53:410–425, 1969.

[149] Stephen W. Hawking. *El futuro del espaciotiempo*. Crítica, Barcelona, 2003.

[150] Thomas Heath. *The Thirteen Books of Euclid's Elements*, vol-

ume I. Dover Publications Inc, New York, second edition, 1956.

[151] Georg W.F. Hegel. *The Science of Logic*. Cambridge University Press, Cambridge, UK, 2010.

[152] Georg Wilhelm Frederich Hegel. *Lógica*. Folio, Barcelona, 2003.

[153] Michael Heller. *On Space and Time*, chapter Where physics meets metaphysics, pages 238–277. Cambridge University Press, 2008.

[154] G. Herrera Corral. *Agujeros negros y ondas gravitacionales*. Editorial Sexto Piso, Madrid, 2019.

[155] S. Herrmann, A. Senger, K. Möhle, M. Nagel, E.V. Kovalchuk, and A. Peters. Rotating optical cavity experiment testing Lorentz invariance at the 10(-17) level. *prd*, 80(10):105011, November 2009.

[156] D. Hilbert. *David Hilbert's Lectures on the Foundations of Aritmetic and Logic 1917-1993*. Springer-Verlag, Heidelberg, 2013.

[157] David Hilbert. Über das Unendliche. *Mathematische Annalen*, 95(1):161–190, 1926.

[158] David Hilbert. *The Foundations of Geometry*. The Open Court Publishing Company, La Salle, Illinois, 1950.

[159] Craig J. Hogan. Interferometers as Probes of Planckian Quantum Geometry. *Phys.Rev.*, D85:064007, 2012.

[160] M. L. Hogarth. Does General Relativity Allow an Observer to view an Eternity in a Finite Time? *Foundations of Physics Letters*, 5:173 – 181, 1992.

[161] John Horgan. Filosofía cuántica. In *Misterios de la física cuántica*, pages 36–45. Prensa Científica. Investigación y Ciencia. Temas 10, 1997.

[162] S. Hossenfelder. Interpretation of quantum field theories with a minimal length scale. *Phys. Rev. D*, 73:105013, May 2006.

[163] Pamela H. Huby. Kant or Cantor? That the universe, if real, must be finite In both space and time. In A. W. Moore, editor, *Infinity*, pages 121 –152. Dartmouth, Aldershot, 1993.

[164] Nick Huggett. *Space from Zeno to Einstein*. MIT Press, Cambridge, Massachusetts, 2002.

[165] Nick Huggett. Zeno's Paradoxes. In Edward N. Zalta (ed.), editor, *The Stanford Encyclopaedia of Philosophy (Summer 2004 Edition)*. Stanford University, 2004.

[166] David Hume. *A Treatise of Human Nature*. Outlook Verlag GmbH, Frankfurt am Main, Germany, 2020.

[167] H. E. Ives and G. R. Stilwell. An experimental study of the rate of a moving atomic clock. II. *Journal of the Optical Society of America*, 31(5):369, 1941.

[168] Herbert E. Ives. Light Signals Sent Around a Closed Path. *Jour-

nal of the Optical Society of America, 28(8):296–299, Aug 1938.

[169] Herbert E. Ives. Xlviii. derivation of the lorentz transformations. *The London, Edinburgh, and Dublin Philosophical Magazine and Journal of Science*, 36(257):392–403, 1945.

[170] Herbert E. Ives. Historical Note on the Rate of a Moving Atomic Clock. *Journal of the Optical Society of America*, 37(10):810–813, Oct 1947.

[171] Herbert E. Ives. The Measurement of the Velocity of Light by Signals Sent in One Direction. *Journal of the Optical Society of America*, 38(10):879–884, Oct 1948.

[172] Herbert E. Ives. Lorentz-Type Transformations as Derived from Performable Rod and Clock Operations. *Journal of the Optical Society of America*, 39(9):757–761, 1949.

[173] Herbert E. Ives. Revisions of the Lorentz transformation. *Proceedings of the American Philosophical Society*, 95(2):125–131, 1951.

[174] Herbert E. Ives and G. R. Stilwell. An Experimental Study of the Rate of a Moving Atomic Clock. *Journal of the Optical Society of America*, 28(7):215–226, Jul 1938.

[175] N. Jafari and A. Shariati. Projective interpretation of some doubly special relativity theories. *Phys. Rev. D*, 84:065038, Sep 2011.

[176] Max Jammer. *Concepts of Space: The History of Theories of Space in Physics*. Dover Publlcations, Inc., New York, 1993.

[177] A. Jannussis. Einstein and the Development of Physics. In Franco Selleri, editor, *Open Questions on Relativistic Physics*, pages 127–130. Apeiron, Montreal (Canada), 1998.

[178] Immanuel Kant. *Gesammelte Schriften*, chapter Neuer Lehrbegriff der Bewegung und der Ruhe, pages 13–25. De Gruyter, 1905 (1758).

[179] Immanuel Kant. *Crítica de la razón pura*. Alfaguara, Madrid, 1989.

[180] Immanuel Kant. *Principios metafísicos de la ciencia de la naturaleza (Metaphysische Anfangsgründe der Naturwissenschaft)*. Editorial Tecnos S.A., 1991 (1786).

[181] Immanuel Kant. *Critique of pure reason*. Cambridge University Press, 1998.

[182] Hugh Kearney. *Orígenes de la ciencia moderna: 1500 - 1700*. Guadarrama, Madrid, 1970.

[183] Al Kelly. *Challenging Modern Physiscs. Questionin Einstein's Relativity Theories*. BrownWalker Press, 2005.

[184] Daniel Kennefick. Einstein Versus the Physical Review. *Physics Today*, 58(9):43, 2005.

[185] Ludwik Kostro. The Physical and Philosophical Reasons for A. Einstein Denial of the Ether in 1905 and its Reintroduction in 1916. In Franco Selleri, editor, *Open Questions in Relativistic Physics*, pages 131–139. Apeiron, Montreal (Canada), 1998.

[186] Ludwik Kostro. *Einstein and the ether*. Apeiron, Montreal, 2000.

[187] H. Kragh and B. Carazza. From Time Atoms to Space-Time Quantization: the Idea of Discrete Time, ca 1925-1936. *Studies in History and Philosphy of Science*, 25:437 – 462, 1994.

[188] Lawrence M. Krauss. *Un universo de la nada*. Pasado y Presente, S.L, Barcelona, 2013.

[189] Robert Laughlin. *Un universo diferente. La reinvención de la física en la edad de la emergencia*. Katz, Buenos Aires, 2007.

[190] Shaughan Lavine. *Understanding the Infinite*. Harvard University Press, Cambridge MA, 1998.

[191] Marc Lchièze-Rey. *L' infini. De la philosophie à l'astrophysique*. Hatier, Paris, 1999.

[192] A. León Sánchez. Coevolution: New Thermodynamic Theorems. *J. Theor. Biol.*, 147(2):205 – 212, 1990.

[193] A. León Sánchez. Living beings as informed systems: towards a physical theory of information. *Journal of Biological Systems*, 4(4):565 – 584, 1996.

[194] A. León Sánchez. Sobre las nociones físicas de orden y organización. Curso de doctorado, 2000.

[195] A. León Sánchez. The aleph-zero or zero dichotomy. *Cogprints*, pages 1–7, September 2006. https://arxiv.org/abs/0804.2934.

[196] A. León Sánchez. Proving unproved Euclidean propositions on a new foundational basis. *International Journal of Scientific Research in Mathematics and Statistical Sciences*, 7(3):61–68, 2020.

[197] A. León Sánchez. Physics and the problem of change. *Preprint version*, 2020. PDF.

[198] A. León Sánchez. Infinity one by one. *Preprint*, 2021.

[199] A. León Sánchez. A critique of selfreference: what Gödel theorem really proves. *The General Science Journal*, pages 1–9, 2021. PDF.

[200] A. León Sánchez. *New Elements of Euclidean Geometry*. Amazon's Kindle Direct Publishing, 2021. PDF.

[201] A. León Sánchez. *Paradoxes and theorems*. Amazon's Kindle Direct Publishing, 2021. PDF.

[202] A. León Sánchez. *The Physical Meaning of Entropy*. Amazon's KDP, 2021. PDF.

[203] A. León Sánchez. Towards a discrete cosmology: Paper 14: Relativity and discreteness. *The General Science Journal*, 2022.

[204] A. León Sánchez. Towards a discrete cosmology. Paper 3: Finite versus infinite. *The General Science Journal*, 2022.

[205] A. León Sánchez. Towards a discrete cosmology: Paper 5: A consistent and discrete universe. *The General Science Journal*, 2022.

[206] A. León Sánchez. Towards a discrete cosmology: Paper 8: Infinite regress. *The General Science Journal*, 2022.

[207] A. León Sánchez. Towards a discrete cosmology: Paper 9: Discrete space. *The General Science Journal*, 2022.

[208] A. León Sánchez. *Apparent relativity*. Amazon's KDP, 2022. PDF.

[209] A. León Sánchez. Towards a discrete cosmology: Paper 7: Preinertia. *The General Science Journal*, 2022. PDF.

[210] A. León Sánchez. Infinity, Language and non-Euclidean Geometries. *The General Science Journal*, 2023.

[211] A. León Sánchez. Special relativity as a proof of space-time discreteness. *The General Science Journal*, 2023.

[212] A. León Sánchez. *Infinity put to the test*. Amazon's KDP, 2023 (2021). PDF.

[213] A. León Sánchez. *Towards a discrete cosmology*. Amazon's KDP, 2023. PDF.

[214] A. León Sánchez. Gravitational waves as empirical proof of space reality. *The General Science Journal*, 2023. PDF.

[215] A. León Sánchez. The shame of physics. *The General Science Journal*, 2023. PDF.

[216] Keren Li, Youning Li, Muxin Han, Sirui Lu, Jie Zhou, Dong Ruan, Guilu Long, Yidun Wan, Dawei Lu, Bei Zeng, and Raymond Laflamme. Quantum spacetime on a quantum simulator. *Communications Physics*, 2(1):122, 2019.

[217] John Locke. *Essay concerning human understanding*. Clarendon Press, Oxford, 1950.

[218] O. J. Lodge. *The Ether of Space*. Alpha Editions, 2021.

[219] H. A. Lorentz. Electromagnetic phenomena in a system moving with any velocity smaller than that of light. *Proceedings of the Royal Netherlands Academy of Sciences and Arts*, 6:809–831, 1904.

[220] Seth Loyd and Y. Jack Ng. Computación en agujeros negros. *Investigación y Ciencia (Scientifc American)*, (340):59 – 67, Enero 2005.

[221] Lucrecio. *De rerum natura*. Cátedra, Madrid, 1994.

[222] Juan de Lugo. *Cómo se puede explicar la composición del continuo por solo indivisibles finitos, según la opinión de los filósofos actuaes*, chapter Las categorías de tiempo y espacio en el pensamiento de la Escolástica tardía, pages 84–106. Ediciones

Universidad de Salamanca, Salamanca, España, 2004.

[223] Jean Pierre Lumient and Marc Lachièze-Rey. *La physique et l'infini.* Flammarion, Paris, 1994.

[224] Jean-Pierre Luminet, Glenn D. Starkman, and Jeffrey R. Weeks. ¿Es finito el espacio? *Investigación y Ciencia*, (273):6–13, Junio 1999.

[225] Charles Lyell. *Principles of Geology.* Penguin Books, 1997.

[226] Peter Lynds. Time and Classical and Quantum Mechanics: Indeterminacy vs. Discontinuity. *Foundations of Physics Letters*, 16:343 – 355, 2003.

[227] Peter Lynds. Zeno's Paradoxes: A Timely Solution. *philsci-archives*, pages 1 – 9, 3003. http://philsci-archives.pitt.edu/archive/00001197.

[228] Shahn Majid. Quantum space time and physical reality. In Shahn Majid, editor, *On Space and Time*, pages 56–140. Cambridge University Press, New York, 2008.

[229] Eli Maor. *To Infinity and Beyond. A Cultural History of the Infinite.* Pinceton University Press, Princeton, New Jersey, 1991.

[230] A. Marini, N. Barcellona, and M. Tinelli. *Diccionario de Ciencia y Tecnología.* Gruppo Editoriale Jackson, 1987.

[231] Mariano Martínez. *Espacio y tiempo en la Escuela de Salamanca*, chapter Prólogo, pages 11 – 16. Ediciones Universidad de Salamanca, Salamanca, 2004.

[232] Joel Gabàs Masip. *Maxwell. La naturaleza de la luz.* Nivola, 2012.

[233] Tim Maudlin. *Filosofía de la física I. El espacio y el tiempo.* Fondo de Cultura Económica, México, 2014.

[234] Tim Maudlin. *Philosophy of Pysics. Space and Time.* Princeton University Press, New Jersey, 2015.

[235] James Clerk Maxwell. *Materia y movimiento.* Crítica, Barcelona, 2006.

[236] Joseph Mazur. *The Motion Paradox.* Dutton, 2007.

[237] William I. McLaughlin. Una resolución de las paradojas de Zenón. *Investigación y Ciencia (Scientifc American)*, (220):62 – 68, Enero 1995.

[238] William I. McLaughlin and Silvia L. Miller. An Epistemological Use of non-Standard Analysis to Answer Zeno's Objections Against Motion. *Synthese*, 92(3):371 – 384, September 1992.

[239] J. E. McTaggart. The unreality of time. *Mind*, 17:457 – 474, 1908.

[240] Brian Medlin. The Origin of Motion. *Mind*, 72:155 – 175, 1963.

[241] A. Meessen. Is it logically possible to generalize physics thro-

ugh space-time quantization? In P. Weingartner and G. Schurz, editors, *Philosophie der Naturwissenschaften. Akten des 13 Internationalen Wittgenstein Symposium*, pages 19 – 47. Hölder-Pichler-Tempsky, Vienna, 1989.

[242] Constantin Meis. *Light and Vacuum. The Wave-Particle Nature of the Light and the Quantum Vacuum.* World Scientific Publishing, 2017.

[243] N. David Mermin. *It's about time. Understanding Einstein's relativty.* Princeton University Press, Princetona and Oxford, 2009.

[244] H. Meschkowski. *Georg Cantor. Leben, Werk und Wirkung.* Bibliographisches Institut, Mannheim, 1983.

[245] C. W. Misner, K. S. Thorne, and J. A. Wheeler. *Gravitation.* W. H. Freeman and Company, San Francisco, 1973.

[246] Leónard Mlodinow. *Euclid's Windows. The Story of Geometry from Parallel Lines to Hyperspace.* Penguin Books, 2002.

[247] M. Moliner. *Diccinario de uso del español. Edición abreviada.* Editorial Gredos, Madrid, 2008.

[248] Andreas W. Moore. Breve historia del infinito. *Investigación y Ciencia (Scientifc American)*, (225):54 – 59, 1995.

[249] Andreas W. Moore. *The Infinite.* Routledge, New York, 2001.

[250] Christopher J. Moore and Alberto Vecchio. Ultra-low-frequency gravitational waves from cosmological and astrophysical processes. *Nature Astronomy*, 5(12):1268–1274, 2021.

[251] Richard Morris. *Achilles in the Quantum Universe.* Henry Holt and Company, New York, 1997.

[252] Richard Morris. *La historia definitiva del infinito.* Ediciones B S.A., 2000.

[253] Chris Mortensen. Change. In E. N. Zalta, editor, *Stanford Encyclopaedia of Philosophy*. Stanford University, URL = http://plato.stanford.edu, 2020.

[254] Michael Moyer. Is space digital? *Sci. Amer.*, 306(2):30–37, 2012.

[255] Robert Munafo. Notable Properties of Specific Numbers, 2013.

[256] George Musser. Filosofía del tiempo. *Investigación y Ciencia (Scientifc American)*, (314):14 – 15, Noviembre 2002.

[257] José Luis Muñoz. *Riemann. Una nueva visión de la geometría.* Nivola, 2009.

[258] Moritz Nagel, Stephen R. Parker, Evgeny V. Kovalchuk, Paul L. Stanwix, John G. Hartnett, Eugene N. Ivanov, Achim Peters, and Michael E. Tobar. Direct terrestrial test of Lorentz symmetry in electrodynamics to 10(-18). *Nature Communications*, 6(1):1–6, 2015.

[259] Isaac Newton. *Mathematical Principles of Natural Philosophy.* Daniel Adee Publishing, New York, 1846.

[260] Isaac Newton. *Principios matemáticos de la filosofía natural.* Alianza, Madrid, 1987.

[261] Carlos Garcia Nu nez, Gavin Wallace, Lewis Fleming, Kieran Craig, Shigeng Song, Sam Ahmadzadeh, Caspar Clark, Simon Tait, Iain Martin, Stuart Reid, Sheila Rowan, and Des Gibson. Amorphous dielectric optical coatings deposited by plasma ion-assisted electron beam evaporation for gravitational wave detectors. *Appl. Opt.*, 62(7):B209–B221, Mar 2023.

[262] J. D. Norton. Mach's principle before einstein. In *Mach's Principle*. The Center for Einstein Studies, 1995.

[263] John D. Norton. A Quantum Mechanical Supertask. *Found. Phys.*, 29(8):1265 – 1302, 1999.

[264] J.J. O'Connor and E.F. Robertson. Non-Euclidean Geometry History, 1996. McTutor History of Mathematics. Accesed: 2016-02-16.

[265] Pappus of Alexandria. *Mathematicae Collectiones.* Apud Francifcum de Francifcis Senenfem, 1589. Translated to Latin by Federico Commandino. Digitalized by Google.

[266] Javier Ordoñez, Victor Navarro, and José Manuel Sánchez Ron. *Historia de la Ciencia.* Espasa Calpe, Madrid, 2004.

[267] A. Pais. *Subted is the Lord. The Science and Life of Albert Einstein.* Oxford University Press, New York, 1982.

[268] S. Palazzo. *Presocráticos. Los albores de la filosofía.* Prisanoticias Colecciones y Empse Edapp S.L., 2020.

[269] Alba Papa-Grimaldi. Why mathematical solutions of Zeno's paradoxes miss the point: Zeno's one and many relation and Parmenides prohibition. *The Revew of Metaphysics*, 50:299–314, December 1996.

[270] Derek Parfit. *Reasons and Persons.* The Clarendon Press, Oxford, 1984.

[271] Parménides. Acerca de la naturaleza. In Alberto Bernabé, editor, *De Tales a Demócrito. Fragmentos presocráticos*, pages 159 – 167. Alianza, Madrid, 1988.

[272] Giuseppe Peano. *Arithmetices Principia. Nova Methodo Exposita.* Libreria Bocca, Roma, 1889.

[273] Igor Pikovski, Michael R. Vanner, Markus Aspelmeyer, M. S. Kim, and Caslav Brukner. Probing Planck-scale physics with quantum optics. *Nat. Phys.*, 8:393–397, 2012.

[274] I. Pitowsky. The Physical Church Thesis and Physical Computational Complexity. *Iyyun: The Jerusalem Philosophical Quarterly*, 39:81–99, 1990.

[275] Josep Pla i Carrera. *Las matemáticas presumen de figura.* Ediciones RBA, 2012.

[276] Max Planck. Zur Theorie des Gesetzes der Energieverteilung im

Normalspektrum. *Verh. Dtsch. Phys. Ges.*, 2:237–243, 1900.

[277] Max Planck. Interview. *The Observer*, page 17, January 1931.

[278] Platón. *Timeo*. Greenbooks Editore. Edición de Kindle., 2020.

[279] J. Playfair. *Elements of Geometry*. W.E. Dean Printer and Publisher, New York, 1846.

[280] John Playfair. *Elements of Geometry*. Number London. Forgotten Books, 2015.

[281] H. Poincaré. Sur la dynamique de l'electron. *Académie des Sciences*, pages 1504–1508, June 1905.

[282] Mary Potter, Carl Hagmann, and Emily McCourt. Banana or fruit? Detection and recognition across categorical levels in RSVP. *Psychonomic Bulletin and Review*, 22(2):578–585, 2015.

[283] Proclus and Gottfried Friedlein. *Procli Diadochi in primum Euclidis Elementorum librum commentarii*. in aedibus B. G . Teubneri, 1873. Digitalized by Google.

[284] Jon Pérez Laraudogoitia. A Beautiful Supertask. *Mind*, 105:49–54, 1996.

[285] Jon Pérez Laraudogoitia. Classical Particle Dynamics, Indeterminism and a Supertask. *British Journal for the Philosophy of Science*, 48(1):49 – 54, 1997.

[286] Jon Pérez Laraudogoitia. Infinity Machines and Creation Ex Nihilo. *Synthese*, 115:259 – 265, 1998.

[287] Jon Pérez Laraudogoitia. Why Dynamical Self-Excitation is Possible. *Synthese*, 119(3):313 – 323, 1999.

[288] Jon Pérez Laraudogoitia. Supertasks. In E. N. Zaltax, editor, *The Stanford Encyclopaedia of Philosophy*. Standford University, URL = http://plato.stanford.edu, 2001.

[289] Jon Pérez Laraudogoitia, Mark Bridger, and Joseph S. Alper. Two Ways of Looking at a Newtonian Supertask. *Synthese*, 131(2):157 – 171, 2002.

[290] Martin Rees. *Just Six Numbers. The deep forces that shape the universe*. Phoenix. Orion Books Ltd., London, 2000.

[291] Hans Reichenbach. *The Philosophy of Space and Time*. Dover Publications Inc, New York, 1957.

[292] Thomas Reid. *Inquiry Into the Human Mind: on the Principle of Common Sense*. Edinburgh University Press, 1997 (1764).

[293] Nicholas Rescher. Process Philosophy. In Edward N.. Zalta, editor, *Stanford Encyclopedia of Philosophy*. Stanford University, URL = http://plato.stanford.edu, 2002.

[294] Pietro Riccardi. *Saggio di una bibliografia euclidea*. Tipografia Gamberini e Parmeggiani, 1887. Digitalized by Google.

[295] F.J. Romero Mora and García García J.A. *La física de la luz*.

RBA Editores, Barceloma, 2017.

[296] Michele Ronco. *Quantum Space-Time: theory and phenomenology*. PhD thesis, Rome U., 2018.

[297] Boris Abramovich Rosenfeld. *A History of Non-Euclidian Geometry. Evolution of the Concept of a Geometric Space*. Spriger Verlag, New York, 1988.

[298] Francesc Rossell i Pujols. *El infinito*. EMSE EDAPP, Barcelona, 2019.

[299] Tony Rothman. The Secret History of Gravitational Waves. *American Scientist*, 106(2):95, 2018.

[300] Brian Rotman. *The Ghost in Turing Machine*. Stanford University Press, Stanford, 1993.

[301] Carlo Rovelli. On quantum mechanics. *arXiv preprint hep-th/9403015*, 1994.

[302] Carlo Rovelli. Quantum spacetime: What do we know? In Craig Callender and Nick Huggett, editors, *Physics meets Philosophy at the Plank scale*, pages 101 – 122. Cambridge University Press, Cambridge, 2001.

[303] Carlo Rovelli. *La realidad no es lo que parece. La estructura elemental de las cosas*. Tusquets, 2015.

[304] Chandrasekhar Roychoudhuri, A.F. KrackLauer, and Hatherine Creath, editors. *The nature of light. What is a photon?* CRC Press, 2019.

[305] Rudy Rucker. *Infinity and the Mind*. Princeton University Press, Princeton, 1995.

[306] Bertrand Russell. *Misticismo y lógica y otros ensayos*. Aguilar, Madrid, 1973.

[307] Bertrand Russell. Sobre la inducción. In Richard Swinburne, editor, *La justificación del razonamiento inductivo*. Alianza, Madrid, 1976.

[308] Bertrand Russell. *Historia de la Filosofía Occidental*. Espasa Calpe, Madrid, 1997.

[309] Bertrand Russell. *Mysticism and logic*. Spokesman Books, 2007.

[310] Andrew M. Ryan. *The Substance of Spacetime: Infinity, Nothingness and the Nature of Matter*. Gadfly LLC, Virginia, 2 edition, 2016.

[311] Bardi. J. S. *The Fifth Postulate: How Unraveling a Two Thousand Year Old Mystery Unraveled the Universe*. John Wiley & Sons, Hoboken, New Jersey, 2009.

[312] Wesley C. Salmon. Introduction. In Wesley C. Salmon, editor, *Zeno's Paradoxes*, pages 5 – 44. Hackett Publishing Company, Inc, Indianapolis, Cambridge, 2001.

[313] Gurcham S. Sandhu. Fundamental Invalidity of all Michelson-

Morley Type Experiments. *Applied Physics Research*, 8(3):45–57, 2016.

[314] Barbara Sattler. *Space. A History*, chapter Space in Ancient Time: From the Beginning to Aristotle, pages 11–51. Oxford University Press, 2020.

[315] Steven Savitt. Being and Becoming in Modern Physics. In Edward N. Zalta, editor, *The Stanford Encyclopedia of Philosophy*. The Stanford Encyclopedia of Philosophy, 2008.

[316] Mark J. Schiefsky. New technologies for the study of Euclid's Elements. http://www.archimedes.fas.harvard.edu, February 2007. Accessed: 2016-02-13.

[317] Erwin Schrödinger. *La naturaleza y los griegos*. Tusquets, Barcelona, 1996.

[318] Jan Sebestik. La paradoxe de la réflexivitédes ensembles infinis: Leibniz, Goldbach, Bolzano. In Françoise Monnoyeur, editor, *Infini des mathématiciens, infini des philosophes*, pages 175–191. Belin, Paris, 1992.

[319] J. Selbie. *La Física de Dios*. Editorial Sirio S.A., Málaga (Spain), 2017.

[320] Manuel Sellés and Carlos Solís. *La Revolución Científica*. Síntesis, Madrid, 1994.

[321] Raymond A. Serway and Clement J. Moses aand Curt A. Moyer. *Modern Physics*. Thomson Learning, Inc., 2005.

[322] Burra G. Sidharth. Comments on the mass of the photon, 2006.

[323] Z. K. Silagadze. Zeno meets modern science. *Philsci-archieve*, pages 1–40, June 2005.

[324] Hourya Sinaceur. ¿Existen los números infinitos? *Mundo Científico (La Recherche)*, Extra: El Universo de los números:24 – 31, 2001.

[325] Lee Smolin. *The Life of the Cosmos*. Phoenix, London, 1998.

[326] Lee Smolin. *Three roads to quantum gravity. A new understanding of space, time and the universe*. Phoenix, London, 2003.

[327] Lee Smolin. Átomos del espacio y del tiempo. *Investigación y Ciencia*, (330):58 – 67, Marzo 2004.

[328] M. C. Solaeche. La Controversia entre L. Kroneckery G. Cantor acerca del Infinito. *Divulgaciones Matemáticas*, 3(1/2):115–120, 1995.

[329] Carlos Solís and Luis Sellés. *Historia de la ciencia*. Espasa Calpe, Madrid, 2005.

[330] R. A. Sorensen. Lorentz contraction, a real change of shape. *Am. J. Phys*, 63:413–415, 1995.

[331] Glenn Stark. Light. In *Encyclopedia Britannica*. Encyclopedia Britannica, 2021.

[332] Paul J. Steinhard. The inflation debate: Is the theory at the heart of modern cosmology deeply flawed? *Scientific American*, 304(4):18–25, 2011.

[333] Wolfgang Steinicke. Einstein and the Gravitational Waves. *Astronomische Nachrichten*, 326(7):640–641, 2005.

[334] Victor J. Stenger. *The comprehensible cosmos*. Prometheus Books, New York, 2006.

[335] Angus Stevenson. *Oxford Dictionary of English. Kindle Edition*. Oxford University Press, 2021.

[336] Robert R. Stoll. *Set Theory and Logic*. Dover, New York, 1979.

[337] Steven Strogatz. *SYNC. The Emerging Science of Spontaneous Order*. Penguin Books, London, 2004.

[338] Patrick Suppes. *Axiomatic Set Theory*. Dover, New York, 1972.

[339] Leónard Susskind. Los agujeros negros y la paradoja de la información. *Investigación y Ciencia (Scientifc American)*, (249):12 – 18, Junio 1997.

[340] José Manuel Sánchez-Ron. *El origen y desarrollo de la relatividad*. Alianza Universidad, Madrid, 1985.

[341] Eusebio Sánchez Álvaro. *LOS INGREDIENTES SECRETOS. Materia Oscura, Energía Oscura y las Nuevas Ideas sobre el Universo*. Cultiva Libros, Madrid, 2015.

[342] N. A. Tambakis. On the Question of Physical Geometry. In Franco Selleri, editor, *Open Questions in Relativistic Physics*, pages 141–147. Apeiron, Montreal (Canada), 1998.

[343] Edwin F. Taylor and John Archibald Wheeler. *Spacetime physics. Introduction to special relativity*. W. H. Freeman and Company, New York, 1997.

[344] Richard Taylor. Mr. Black on Temporal Paradoxes. *Analysis*, 12:38 – 44, 1951 - 52.

[345] James F. Thomson. Tasks and SuperTasks. *Analysis*, 15:1–13, 1954.

[346] Paul A. Tipler and Ralph A. Llewellyn. *Modern Physics*. W.H. Freeman and Company, New York, 2000.

[347] Paul A. Tipler and Gene Mosca. *Physics for sicentists and engineers*. W.H. Freeman and Company, New York, 2008.

[348] Roberto Torretti. Nineteenth Century Geometry. In Edward N. Zalta, editor, *The Stanford Encyclopedia of Philosophy*. Stanford University, 2014.

[349] Liang-Chen Tu, Jun Luo, and George T. Gillies. The mass of the photon. *Reports on Progress in Physics*, 68:77–130, 2005.

[350] D. Turner. *The Einstein Myth and the Ives Papers*, chapter Part 1. Absolute motion. Hope Publishing House, Pasadena (California), 1979.

[351] D. Turner and R. Hazelett. *The Einstein Myth and the Ives Papers.* Hope Publishing House, Pasadena (California), 1979.

[352] Agustín Udías Vallina. *Historia de la Física, de Arquímedes a Einstein.* Editorial Síntesis, 2004.

[353] varios autores. *Gran Enciclopedia Larousse.* Editorial Planeta, 1988.

[354] Gabriele Veneziano. El universo antes de la Gran Explosión. *Investigación y Ciencia (Scientifc American)*, (334):58 – 67, Julio 2004.

[355] J. P. Vigier. New non-zero photon mass interpretation of the Sagnac effect as direct experimental justification of the Langevin paradox. *Physics Letters A*, 234:75–85, 1997.

[356] John Carl Villanueva. How Many Atoms Are There in the Universe? Universe Today, 2009.

[357] Eugenio Villar García. *Breve Historia de la Física: sus artífices.* Ediciones Universidad de Cantabria, 2012.

[358] Gregory Vlastos. Zeno's Race Course. *Journal of the History of Philosophy*, IV:95–108, 1966.

[359] Gregory Vlastos. Zeno of Elea. In Paul Edwards, editor, *The Encyclopaedia of Philosophy.* McMillan and Free Press, New York, 1967.

[360] Woldemar Voigt. Über das Doppler'sche Princip. *Göttinger Nachr.*, (8):41–51, 1887.

[361] G. H. Von Wright. *Time, Change and Contradiction.* Cambridge University Press, Cambridge, 1968.

[362] Jearl Walker. *Fundamentals of Physics.* John Wiley and Sons, Inc., 2008.

[363] David Foster Wallace. *Everything and more. Acompact history of infinity.* Orion Books Ltd., London, 2005.

[364] John Watling. The sum of an infinite series. *Analysis*, 13:39 – 46, 1952 - 53.

[365] J. Weber. Detection and generation of gravitational waves. *Physical Review*, 117:306–313, January 1960.

[366] J. Weber. Evidence for discovery of gravitational radiation. *Physical Review Letters*, 22:1320–1324, Jun 1969.

[367] Jeffrey R. Weeks. *The shape of space.* Marcel Derkker INC., 2002.

[368] Eric W. Weisstein. Continuum. From MathWorld. a Wolfram Web Resource, 2022.

[369] H. Weyl. *Philosophy of Mathematics and Natural Sciences.* Princeton University Press, Princeton, 1949.

[370] Frank Wilczek. *The lightness of being. Big questions, real an-*

swers. Allen Lane. Penguin Books. (Kindle ebook edition), 2009.

[371] Frank Wilczek. *Las diez claves de la realidad*. Editorial Planeta. (Kindle ebook edition), Barcelona, 2022.

[372] Heng Xu, Siyuan Chen, Yanjun Guo, Jinchen Jiang, Bojun Wang, Jiangwei Xu, Zihan Xue, R. Nicolas Caballero, Jianping Yuan, Yonghua Xu, Jingbo Wang, Longfei Hao, Jingtao Luo, Kejia Lee, Jinlin Han, Peng Jiang, Zhiqiang Shen, Min Wang, Na Wang, Renxin Xu, Xiangping Wu, Richard Manchester, Lei Qian, Xin Guan, Menglin Huang, Chun Sun, and Yan Zhu. Searching for the nano-hertz stochastic gravitational wave background with the chinese pulsar timing array data release i. *Research in Astronomy and Astrophysics*, 23(7):075024, jun 2023.

[373] Gideon Yaffe. Reconsidering Reid's geometry of visibles. *The Philosophical Quarterly*, 52(209):602–620, 2009.

[374] Mark Zangari. Zeno, Zero and Indeterminate Forms: Instants in the Logic of Motion. *Australasian Journal of Philosophy*, 72:187–204, 1994.

[375] Eberhard Zeidler, W. (with Hackbush, and H.R.) Schwarz. *Oxford Users' Guide to Mathematics*. Oxford University Press, 2004.

[376] Paolo Zellini. *Breve storia dell'infinito*. Adelphi Edicioni, Milano, 1980.

Alphabetical index

AT-points, 248
AT-points (Section), 249
ω-Asymmetry (Subsection), 262

α, β, γ and δ movements, 37
A discreet model to start with (Section), 354
A discrete model: cellular automata (Section), 216
A discrete solution to Zeno Contradiction (Section), 251
A preinertial argument on the nature of motion (Section), 327
A real Newton's experiment (Section), 125
A relativistic conflict on the reality of space (Section), 198
A revolution in three words (Section), 358
A short proof of inconsistency (Section), 18
A Special Relativity Inconsistency (Chapter), 333
A thought experiment: Newton's rotating globes (Section), 130
Absolute and relative motions, 235
Absolute motion, 143, 233, 325, 338
Absolute motion (Section), 279
Absolute motion cannot be detected, 360
Absolute motion is undetectable, 233
Absolute reference frame, 324
Absolute rotations, 335
Absolute space, 123, 290
Achilles and the Tortoise, 248
Achilles, the tortoise and the speed of light (Chapter), 247
Actual infinity, 11–17, 25, 28, 39, 40, 47, 174, 179, 192, 200, 201, 236, 310
Additional reasons for the paradigm shift (Section), 294
Adjacency, 194, 196, 240, 349
Agrippan Trilemma, 45
Al Farisi, 96
al-Tutsi-Legendre version of Euclid's 5th Postulate, 86
Alexandria, 78
Alfonso VI of Castile, 96
Alhazen, 96
Ammonius Sachas, 89

Amyclas of Heraclea, 79
An elementary preamble on rotations (Section), 326
Anaxagoras, 59
Anaximander, 59
Anaximenes, 60
Anomaly of Mercury's orbit, 339
Antonopoulos, C., 298
Anything but discrete (Section), 264
Apeiron, 59, 60
Apollonius of Perga, 78
Arab and Judeo-Christian ideas about space (Section), 92
Arab arithmetic and trigonometry, 95
Arab scientific splendor, 95
Arab world, 58
Arabic numerals, 94
Archimedes, 77
Archytas, 65, 79
Archytas space, 65
Arché, 59
Argument on the discrete space, 275
Aristaeus, 79
Aristotelian infinite regress, 157
Aristotle, 10, 13, 27, 40, 47, 57, 62, 68, 70, 96, 101, 121, 242, 265
    and one to one correspondences, 242
    Categries, 70
    Characteristic of place, 71
    Concept of place, 70
    Contiguity, 70
    Continuity, 70
    Definition of place, 72
    Forced motions, 71
    Natural motions, 71
    On Heaven, 58
    Physics, 58, 70
    Quintaessence, 70, 135
    Successiveness, 70
    Vacuum, 281
Arithmetic operations, 94
As firm as a rock (Section), 257
Astral geometry, 141
Atomists, 57
Atoms, 27
Authors that tried to solve the problem of parallels, 81
Averroes, 96
Axiom of Infinity, 1, 12, 15, 17, 21, 24, 25, 29, 34, 44, 47, 57, 62, 149, 158, 176, 177, 180, 192, 200, 204, 240, 246, 255, 259, 260, 292, 301, 357, 358, 367, 369, 374
Axiom of the Whole and the Part, 15
Axioms, 80

Beltrami, E., 85, 142, 321

Berkeley, G., 116, 183
Big-Bang theory, 305
Bohm's implicit order, 219
Bohm, D., 219
Bolyai, J., 141
Bolzano's proof, 15
Bolzano, B., 10, 15
Borel, E., 196, 281, 286, 350
Born, M., 165
Brouwer, L.E.J., 29
Brouwer,L. E. J. , 17

C. MacLaurin's absolute space, 137
Cabala, 90
CALMs, 176, 191, 200, 201, 203, 206, 319, 320
Campanella, Tomasso, 99
Canonical change, 213, 218
Canonical changes (Section), 213
Cantor's proof, 16
Cantor, G., 10, 15, 27, 174, 239, 256, 260, 279, 310, 357
    Theorem §15 A, 239
Cantorian Infinitism, 48
Celestial objects uniformly moving, COUMs, 336, 337
Cellular automata, 293
Cellular Automata Like Models, 176
Cellular Automata Like Models (Chapter), 207
Cellular Automata Like Models (Section), 292
Cellular Automata Like Models, CALMs, 57, 59, 61, 69, 176, 216, 280, 293, 354, 361
Cellular Automata Like Models, CALMs objects, 217
Cellular Automata-Like Models, CALMs, 294
Centrifugal force, 128
Change problem, the, 217
Chasm, 59
Cicero, 92, 102
Circular arguments, 80
Clarke, S., 101, 115, 119, 127
Clarke-Leibniz discussion on Newton's bucket, 125
Clarke-Leibniz epistolar debate, 326, 327
Clifford, W. K., 269
Closed universe, 86
Commensurable lengths, 35
Complete totality, 2, 13
Conclusion (Section), 25, 246, 267
Confirmation of special relativity
    Acausal, 167, 331, 340
    Symmetry, 167, 330, 340
    Universality, 167, 330, 340
Consequences of the relativistic time deformations (Section), 318
Consequences on the theory of special relativity (Section), 330
Consistency of non-Euclidean geometries, 142

Constantinus Lascaris, 78
Contemporary Euclidean definition of straight line, 152
Contiguity, 240
Continuous magnitudes, 181
Continuous or discrete? (Section), 351
Continuous variables, 27
Conventions (Section), 3
Conventions and general fundamentals (Section), 376
Conventions and symbols (Chapter), 3
Copernicus, N., 102
Corollary
    of Discrete Threshold, 209, 359
    of Discrete Values, 182, 359
    of Finite Space and Time, 210, 359
    of Lines Joining Points, 186
    of Primitive Concepts, 48
    of the Closed Lines, 186
    of the Finite Distances, 186
    of the Finite Mass-Energy, 359
    of the First Cause, 49, 62, 121, 122, 266, 359
    of the Infinite Lengths, 186
    of the Physical Laws, 35, 43
Cosmic Microwave Background, CMB, 144
Cosmogony, 59
Countable, 4
Crescas Paradox, 99
Crescas, H., 99
Critias, 68
Critical energy density of the universe, 86
Criticism of Newton's bucket experiment (Section), 127
Critique of Aristotle's theory of motion, 96
Critique of Newton's absolute space (Section), 114
Cusa, N., 99
Cyzicenus of Athens, 79

Damascius' space, 90
De rerum natura, 92
Decimal numbering system, 94
Dedekind's proof, 16
Dedekind, R., 10, 15, 256, 357
Definition
    of Complete Totality, 4
    of consistent law, 42
    of discrete magnitude, 30
    of physical space, 281, 352
    of preinertia, 323
    of straight line:
      E. Beltrami, 83
      Heron of Alexandria, 83
      J. Playfair, 83
      Proclus, 83

Productive definition, 84
Definition of Preinertia (Section), 221
Democritus, 27, 66
Democritus' argument, 27
Democritus' argument (Section), 27
Denumerable, 4
Descartes' laws of motion, 100
Descartes, R., 100
Dimension Problem, 174, 260
Dingle, H., 312
Diogenes Laertius, 9
Dirac, P., 15, 301
Discontinuity of Z-points and Z*-points, 241
Discrete and continuous magnitudes (Section), 181
Discrete arithmetic, 34, 37
Discrete arithmetic (Section), 35
Discrete conclusions (Chapter), 345
Discrete electric charge, 29
Discrete energy, 29
Discrete geometries, 188
Discrete magnitudes, 30, 181
Discrete Magnitudes and Functions (Chapter), 27
Discrete magnitudes and functions (Section), 29
Discrete mass, 29
Discrete matter, 29
Discrete Pythagoras theorem, 36
Discrete space, 27
Discrete space and time, 187, 214, 217, 218, 229
Discrete time, 27, 290
Discrete versions of Pythagoras Theorem, 353
Discrete versus continuous (Chapter), 171
Discrete versus continuous (Section), 215
Discussion (Section), 343, 367, 371
Down quark, 29

e-Theorem of the Discrete Space, 276
e-Theorem of the Discrete Time, 278
Ecphantus, 102
Eficetas of Syracuse, 102
Einstein's clock, 224
Einstein's theory of special relativity, 135, 162–164
Einstein's theory of special relativity (Section), 162
Einstein, A., 142, 147, 169, 197, 198, 211, 270, 282, 302
Electric permittivity, 159, 204, 207
Electromagnetic spectrum, 208
Electron, 29
Elliptic and spherical geometries, 142
Elliptic Axiom, 85, 142
Enumerable, 4
Eratosthenes, 77
Euclid, 58, 77

Books, 92
Definition of straight line, 151
Definitions of point, line and straight line, 83
Elements, 77
Enigma, 81
Euclid's Elements
   Book I, 80
   Influences, 79
   Network of formal relatons, 79
   Number of foundational elements, 79
   Number of proposirions, 79
   Relevance, 79
   Success, 79
Euclid's geometric postulates, 80
   Postulate 1, 80
   Postulate 2, 80
   Postulate 3, 80
   Postulate 4, 80
   Postulate 5, 80, 84, 85
No royal road to geometry, 78
Other works, 78
Scientific works, 78
Unsatisfactory definitions, 151
Euclid (Section), 77
Euclid of Megara, 78
Euclid's Elements (Section), 78
Euclid's fifth postulate, 321
Euclidean enigma of parallels, 135
Euclidean geometry, 46, 77, 142
Euclidean space, 74
Euclidean space (Chapter), 77
Euclidean time and distance, 5
Eudoxus, 79
Euler's absolute space, 137
Euler's concept of place, 136
Euler, L., 136
European medieval scholastic, 58
Expanding and contracting the space continuum (Section), 300
Expanding geometrical space and physical space (Section), 204
Experimental confirmations of special relativity (Section), 167
Experiments and theories (Section), 299
Expo-factorial numbers, 178

ϕ-phenomenon, 37
Facticio, 1
Faraday, M., 158, 205, 269
Faraday-Maxwell electromagnetism, 158
Fascinating questions, 346
Feynmann, R., 231, 347
Fields and CALMs (Section), 205
Fifth Postulate

Enigma, 81
Euclid's statement, 80, 81
Other statements:
  Clairaut, 82
  Farkas Bolyai, 82
  Gauss, 82
  Legendre, 82
  Posidonius and Geminus, 82
  Proclus, 81, 82
  Proclus and Playfair, 81
  Saccheri, 82
  Thabit ibn Qurra, 82
  Wallis, Carnot and Laplace, 82
  Worpitzky, 82
Finite but non-computable natural numbers (Section), 176
Finite Hilbert machine, 367
Finite lengths and distances (Section), 185
Finite versus infinite (Chapter), 9
First Law of Mechanics, 345
First Principle of Relativity, 162
FitzGerald, G. F., 160, 302
FitzGerald-Lorentz contraction, 165, 166, 264, 287, 303, 339
  the elastic cord, 165
Formal consequences of inconsistent actual infinity, 358–359
Formal elements for a new theory of space and time (Section), 51
Foundational basis of Euclidean geometry (Section), 378
Fraunhofer diffraction, 175
Frege, G., 10, 256
French, A. P., 165
Fundamental questions about motion, 133

Gaia, 59
Gaius Plinius Secundus, 93
Galileo Galilei, 128, 158, 221, 287, 329, 338, 345, 360
Galileo's mast, 221, 222
Galileo's ship, 222
Galileo's transformation, 112
Galileo-Newton theory of relativity, 158
Gassendi, P., 99
al-Gauhary, 81
Gauss fictitious experiment on Euclidean space, 84
Gauss, C. F., 11, 84, 141, 260, 321
General relativity, 86, 142, 270
GEO 600 laser system, 270
Geodesics, 152
Geometrical points and physical points (Section), 193
Gorgias, 67
Gorgias' proof of sace finitude, 67
Gravitational mass, 132, 325
Gravitational waves, 123, 270, 360

Gravitational Waves as Empirical Proofs
    of Space Reality (Chapter), 269
Gravity from the CALM perspective (Section), 201
Great circles, 152
Grosseteste, R., 98
Gödel's famous incompleteness theorems, 39
Gödel's incompleteness theorems, 150

Heath, T. L., 79
Hegel, H. W. F., 280
Hegemony of the actual infinity, 10
Hegemony of the potential infinity, 10
Heisenberg's Uncertainty Principle, 210, 261
Heisenberg, W., 175, 187
Heraclides, 102
Heraclitus, 59, 61
Hermite, C., 17
Hesiod, 59
Hilbert Hotel, 192, 363
    A way of getting rich, 363
    A way of violating physical laws, 363
Hilbert Hotel (Section), 363
Hilbert machine
    Argument, 365–368
    Contradiction, 366, 367
    Definition, 364
    Restriction, 365
Hilbert machine (Section), 364
Hilbert machine contradiction (Section), 366
Hilbert, D., 10, 47
Hippocrates of Chios, 79
Hippocrates of Kos, 79
Hume, D., 183, 185, 196, 281, 286, 350
Huygens, C., 196, 281, 286, 350
Hyperbolic Axiom, 85, 142
Hyperbolic geometry, 142
Hyperfactorial numbers, 263
Hypothesis of the Actual Infinity, 13–15, 18, 24, 25, 29, 158, 204, 212,
    239, 240, 243, 245–247, 253, 292, 358, 365
Hypothesis of the Potential Infinity, 14

Ilves-Stilwell experiment, 312
Imaginary geometry, 141
Immediate predecessor, 240
Immediate successiveness, 63, 194, 196, 349
Immediate successor, 240
Implicit order and cellular automata, 219
Important warning (Chapter), 1
Incommensurability, 63
Incompatibility between quantum mechanics and general relativity, 29
Inconsistency of the actual infinite divisions (Section), 182

Inconsistency of the non-Causal Relativism (Section), 337
Inconsistent language of physics, 252
Indian decimal numbering system, 94
Indivisible units of space and time (Section), 207
Inductive arguments
    Hilbert machine, 366
    The last disk, 370
Inertial length contraction, 165
Inertial local simultaneity, 165
Inertial mass, 132, 325
Inertial reference frames, 160
Inertial time dilation, 165
Inextensive points, 27
Infinite regress, 39, 150
Infinite regress (Chapter), 39
Infinite regress of arguments, 80, 121
Infinite regress of causes (Section), 49
Infinite regress of definitions, 82
Infinite regress of definitions (Section), 48
Infinite regress of proofs (Section), 46
Infinity and ordinary language (Section), 256
Infinity, language, and non-Euclidean geometries (Chapter), 149
Infinity, physics and language (Chapter), 255
Instants have neither duration nor contiguity (Section), 310
Integer division, 64
Interacting mode, 218, 293
Interconnectivity between different sciences., 291
International System of Units, 29
Internet, 30
Interpretation of light as particle rays, 95
Interpretation of the rainbow, 95
Introduction (Section), 9, 27, 39, 51, 57, 77, 89, 103, 135, 149, 171, 207, 221, 239, 247, 255, 285, 297, 309, 321, 345, 369, 376
Introduction: Gravitational waves (Section), 269
Introductory definitions (Section), 239
Irrational numbers, 57
It is impossible to exaggerate the importance of ... (Section), 357
Ives, H. E., 299, 311, 339

Jammer, M., 73
John the Grammarian, 91

KAGRA interferometer, 270
Kaku, M., 197, 282
Kalam discrete matter, 97
Kalam discrete motion, 98
Kalam discrete space and time, 97
Kalam discrete world, 251
Kant
    absolute space, 138

axiom of movement, 139
definition of rest, 139
definition of space, 139
force of attraction, 139
relational space, 137
theorem of matter divisibility, 139
theorems on matter, 139
Kant's
definition of matter, 139
Kant, I., 137, 169, 198
Keill, J., 136
Kleene, S., 17
Kronecker, L., 17, 180
König, J., 17

Lagrange, J. L., 81, 139
Lambert, J. H., 81
Language abuses in geometry, 153
Language abuses in non-Euclidean geometries (Section), 152
Laplace, P. S., 139
Law of the reflection of light, 95
Legendre, A. M., 81
Leibniz's argument against absolute motion, 334
Leibniz's first cause, 122
Leibniz's kinematic shift argument, 115
Leibniz's Principle of Identity of Indiscernibles, 127
Leibniz's Principle of Sufficient Reason, 115, 127
Leibniz's relational space, 114
Leibniz's static shift argument, 115
Leibniz, G. W., 99, 114, 119, 127, 196, 205, 281, 286, 333, 350
Leon, 79
Leucipus, 66
LIGO interferometer, 270
Lobachevsky, N., 141
Local Cluster, 144
Local Group, 144
Local Supercluster, 144
Locke, J., 101
Lorentz factor, 302
Lorentz theory of inertial relativity (Section), 160
Lorentz Transformation, 164, 230, 264, 305, 354
Lorentz's factor, 168, 339
Lorentz's theory of inertial relativity, 160–162
Lorentz's theory of special relativity, 135
Lorentz, H. A., 160, 169, 198, 302
Lucius Annaeus Seneca, 93
Lucretius, 66
Lynds, P., 239

Mach's critique of Newton bucket experiment, 140
Mach's Principle, 116, 132, 141

Mach's Principle (Section), 139
Mach, E., 27, 127, 131, 135, 140, 169, 196, 198, 281, 286, 326, 333, 350
MacLaurin, C., 137
Magnetic constant, 159
Magnetic permeability, 159, 207
Majid, S., 29
Marcus Terentius Varro, 93
Marcus Tullius Cicero, 93
Marcus Vitruvius Pollio, 93
Martianus Minneus Felix Capella, 94
Mass and Mach's Principle (Section), 132
Maudlin, T., 273, 357
Maxwell equations, 159
Maxwell's absolute space, 140
Maxwell, J. C., 139, 158, 159, 211
McTaggart, J. M. E., 280
Medieval scholastics, 10
Melissus, 66
Michelson-Gales experiment, 312
Michelson-Morley experiment, 135, 143, 145–147, 159, 160, 234, 302
Michelson-Morley experiment (Section), 143
Michelson-Morley experiment and preinertial objects, 144
Michelson-Morley interferometer, 143
Modus Tollens arguments
    Hilbert machine, 365
More's theory of space, 100
More, H., 100, 113
Morgan Manuscript, 169, 198
Motions of the Earth, 144
Münchhausen Theorem (Section), 45
Münchhausen Trilemma, 40, 45

n-Expo-factorial numbers, 178
n-Hyperfactorial numbers, 263
Nasiraddin at-Tusi, 81
Negative results of Michelson Morley experiment, 147
Neo-Platonism, 89, 98
Neoplatonic space, 90
New axioms for Euclidean geometry, 154
New Elements of Euclidean Geometry, 83, 149, 154
New foundation of Euclidean Geometry, 49
Newton
    absolute space, 104, 114, 117, 122, 135, 136
    bucket experiment, 110, 125, 128, 326, 333
    Corollary V, 113
    divine sensorium, 98
    Euclidean space, 103
    First Law of Mechanics, 112
    First Rule for Philosophizing, 114
    Principia, 103, 119, 125, 136

Books, 104
  Formal elements, 104
 rational mechanics, 103
 rotating globes, 111, 127
Newton absolute space (Chapter), 103
Newton's bucket and absolute rotations (Chapter), 125
Newton, I., 58, 70, 89, 99, 101, 103, 125, 127, 233, 345
Newton-Maxwell conflict, 159
Newtonian space, 123
Nicomachean Arithmetic, 92
Non-commensurable lengths, 35
Non-countable, 5
Non-denumerable, 5
Non-Euclidean geometries, 135
Non-preinertial objects, 144
Numbers with infinitely many decimals (Section), 179
Numerable, 4

ω-Dichotomy, 244
ω-Ordered objects, 370
ω^*-Dichotomy, 242
ω, the first transfinite ordinal, 239, 365, 369
ω-Asymmetry, 262
ω-Division, 183
ω-Order, 4, 239, 240, 365, 367, 369, 374
ω-Ordered objects, 4, 7, 240, 244, 246, 364–367, 369–371, 373
ω-Ordered sequences, 196
ω-Ordered sets, 182
ω^*-Ordered, objects, 4
ω^*-Discontinuity, 241
ω^*-Order, 239
ω^*-Ordered objects, 240–242, 246
ω^*-Ordered sequence, 240
Ockham's razor, 85, 329
Ockham, W., 85, 101
Omar Khayyam, 81
On space deformations (Chapter), 297
On spooky actions and double slits (Section), 341
On the empirical confirmation of special relativity (Section), 339
On the substantiality of physical space (Section), 288
On time deformations (Chapter), 309
One to one correspondence, 10
Open questions on the physical nature of space, 73
Open universe, 86
Ouranus, 59

Pappus of Alexandria, 77
Paradise of the actual infinity, 10
Parallel enigma, 81
Parmenides, 9, 40, 61, 280, 357
Parmenides and Zeno of Elea (Section), 61

Parmenides' Principle of Identity, 61
Parmenides' Principle of Non-Contradiction, 61
Peano's Axiom of the Successor, 12, 14, 310
Peano's Axioms, 177
Permanence mode, 218, 293
Perpetuum mobile, 363
Philippus of Mende, 79
Philo, 89
Philoponus' place and space, 91
Philoponus, J., 91
Photon mass, 132, 287, 326
Photons are preinertial (Section), 224
Physical space (Section), 196
Physical space as a fiction, 123
Physical space is a real physical object (Section), 273
Physical space is discrete (Section), 274
Physical space is Euclidean (Section), 84
Physical space is real, 274
Physical space is real and discrete (Section), 286
Physical versus geometrical space (Chapter), 191
Physicists and physical space (Appendix), 281
Physics and mathematics (Section), 157
Physics contradictory ordinary language sentences, 261
Physics inconsistent ordinary language, 255
Physics wrong ordinary language sentences, 256, 259
Pied Piper of Hamelin, 310
Planck constant, 29
Planck constants, 181
Planck scale
    Planck length, 175, 187, 207, 210
    Planck time, 132, 175, 187, 207, 210, 326, 360
Planck volume, 290
Planck, M., 28, 29, 211
Plane universe, 86
Plato, 9, 57, 68, 77
Plato's receptacle, 69
Plato's Timaeus, 68, 92
Playfair's Axiom, 85, 143
Playfair's definition of straight line, 151
Playfair's Elements of Geometry, 151
Playfair, J., 47, 151
Plotinus, 89
Poincaré, H., 17, 29, 162, 196, 281, 286, 350
Points and instants of the spacetime continuum (Section), 258
Points have neither extension nor shape (Section), 299
Points of non-null extension, 35
Posidonius and the terrestrial tides, 74
Potential infinity, 11, 13–16, 179
Power of the continuum, 173
Pre-Socratic atoms, 66

Pre-Socratic extensive points, 251
Precession and nutation, 129
Preinertia, 30, 86, 132, 193, 204, 229, 234, 287, 291, 325, 360
    and absolute motion, 231
    and the nature of light, 235
    concept, 222
    photons are preinertial, 224
    Principle of Inertia, 230
    Santiago del Collado experiment, 229
    the reasons for, 234
    Theorem of Preinertia, 230
    universal preperty, 223
Preinertia and absolute motion, 324–325
Preinertia and absolute motion (Section), 231
Preinertia and the nature of light (Section), 235
Preinertia: the vectorial inheritance of motion (Section), 323
Preinertial objects
    Photons are preinertial, 229
    Michelson-Morley experiment, 144
    velocity, 226
Prince of Mathematics, 260
Principle
    of Actualism-Uniformism, 41, 291
    of anti-Identicality, 61
    of Autonomy, 367
    of Conservation of Matter, 96
    of Directional Evolution, 39–41, 119–121, 150, 279, 291
    of Execution, 366, 367, 371
    of Identity of Indiscernibles, 119, 333
    of Inertia, 323, 345, 346
    of Relativity, 113, 264, 337, 353
    of Sufficient Reason, 119, 333
    of the Constancy of the Speed of Light, 264
    ofthe Constancy of the Speed of Light, 163
Principles of relativity, 162, 163
Principles of relativity, compact forms, 163
Problem of Change, 40
Problems posed by the infinity, 57
Problems with space deformations, 272
Proclus, D., 77, 81
Productive definition of straight line, 152, 321
Productive definitions, 83
Properties of physical space, 199
Proton, 29
Ptolemy I Soler, 77
Ptolemy, C., 81
Pythagoras Discrete Theorem, 188
Pythagoras Discrete Theorem (Section), 187
Pythagoras Theorem, 64, 168, 300, 339
Pythagorean discrete points, 171

Pythagorean metric, 64
Pythagorean points, 35
Pythagoreans, 57, 79
Pythagprean's discrete space, 63

Quantum electric charge, 29
Quantum mass, 132, 287, 326
Quantum mechanics, 28, 29
Questioning Leibniz's Principle of Sufficient Reason (Chapter), 119
Questions on inertia, 230
Qusit, 63, 188, 203, 289, 324, 354, 360
Qutit, 203, 219, 291, 311, 354, 360

Real and apparent velocity changes (Section), 335
Real or fictitious? (Section), 349
Reid, T., 141
Relativistic consequences on space and time. (Section), 165
Relativistic dilation of time (Section), 314
Relativistic local simultaneity (Section), 315
Rest mass of an electron, 204
Riemann geometry, 142
Riemann, B., 141, 142
Riemannian straight lines, 152
Ritz, W., 311
River Mesopotamian cultures, 57
Robinson, A., 17
Roman Empire, 92
Rotations are always absolute motions (Section), 334
Russell's metaphor of the chicks, 47
Russell, B., 192, 266
Rydberg constant, 132, 204, 287, 326
Rydberg mass, 326

Saccheri, G., 81, 141
Sagnac effect, 312
Santiago del Collado experiment, 234, 238, 291, 361
Schlick, M., 169, 198
Schweikart, F., 141
Scientific Revolution, 89
Second Law of Thermodynamics, 121
Second Principle of Relativity, 229
Sensory perception of the physical world, 251
Sextus Empiricus, 75
Sextus Iulius Frontinus, 93
Simplicius, 9
Smolin, F., 168
Snell Law, 300, 304
Socrates, 68
Solutions to Zeno paradoxes, 239
Some classical questions to start with (Section), 345
Some discrete conclusions (Section), 34

Sources of gravitational waves, 271
Space according to Plato (Section), 68
Space deformations (Section), 271
Space discrete units, 275
Space in ancient Greece (Chapter), 57
Space in post-Aristotelian Classical Greece (Section), 74
Space in the 20th century (Section), 168
Space in the XVIII and XIX centuries (Chapter), 135
Space isotropy, 290
Space light and Gold (Chapter), 89
Space substance, 1
Space, time and motion in the Principia (Section), 105
Spacetime continuum, 1, 23, 27, 29, 35, 47, 171, 173–176, 187, 193, 195, 196, 201, 208, 210, 212–216, 218, 219, 236, 293, 299, 310, 311, 352
Special relativity, 12, 169, 174, 193, 198, 210, 212, 230, 231, 234, 235, 300, 305, 318, 337
Special relativity inconsistency, 337
Special relativity is not compatible with discreteness (Section), 353
Speed of light, 207, 237
Speed of light in vacuum, 159
Speed of the electromagnetic waves, 159
Spherical (Riemannian) geometry, 152
Standard Model, 204
Stilwell, G. R., 311, 339
Stoics' infinite void, 74
Stoics' space, 74
Straight lines and parallelism (Section), 151
Strato of Lampsacus, 74
Study of the refraction of light, 95
Supertask
    Definition, 12
Supertasks, 5
    the last disk, 370–371
Symbols (Section), 6

TAMA 300 interferometer, 270
Tambakis, N. A., 168, 197, 282
Tartarus, 59
Teleological motion, 71
Tentament, 141
Thabit ibn Kurra, 95
Thales of Miletus, 59
The actual and the potential infinity (Section), 10
The Aristotelian infinite regress (Section), 39, 265
The Aristotelian space (Section), 70
The atomists (Section), 66
The Axiom of Infinity and Zeno Contradiction (Section), 250
The axiom of infinity is inconsistent (Section), 21
The birth of non-Euclidean geometries, 141–143
The birth of non-Euclidean geometries (Section), 141

The definition of straight line (Section), 82
The enigma of the parallel straight lines (Section), 80
The expansion of intergalactic space (Section), 305
The first cosmologists (Section), 58
The first universities, 97
The formal language of Newton's Principia (Section), 103
The Formal Scenario (Chapter), 51
The formal setting of the discussion (Section), 120
The geometry of visibles, 141
The gravitational deformation of space (Section), 307
The inevitable incompleteness of human knowledge (Section), 150
The infinity of the Axiom of Infinity (Section), 16
The initial success of Newtonian absolute space (Section), 136
The Ives-Stiwell experiment (Section), 311
The language of physics, 158
The last disk (Section), 370
The mathematical language of physics (Section), 191
The memory of a historic debate (Section), 333
The model R^+ of time and the problem of change (Section), 311
The nature of space according to Kant (Section), 137
The Neoplatonics (Section), 89
The Newton-Maxwell relativistic conflict (Section), 158
The pending revolution in physics (Chapter), 357
The Principle of Sufficient Reason (Section), 121
The problem of change, 211–213, 216, 219, 309
The problem of change (Section), 211, 264
The problem of the continuous (Section), 171
The Pythagoreans (Section), 63
The reality of physical space, 287
The relativistic contraction of space (Section), 302
The scandal of elementary geometry, 141
The shame of geometry, 221
The shame of physics, 30, 221
The Shame of Physics (Chapter), 321
The shameful part of mathematics, 141
The spacetime continuum (Section), 173
The speed of light and absolute motion (Section), 237
The substance of physical space (Chapter), 285
The ultraviolet catastrophe (Section), 28
The Umayyads, 95
The universe as a cellular automata, 276
The universe is consistent (Section), 41
The use of zero, 94
The Vedas, 94
Theaetetus, 79
Theophrastus, 74
Theorem
    of Adjacency, 35, 210
    of Change, 214, 311, 359
    of Finite Distances and Durations, 187

of Formal Dependence, 40, 43, 45, 49, 120, 121, 150, 202, 205, 306
of Identicality, 43, 61, 120
of Inconsistent Infinity, 119
of Indivisible Units, 209, 359
of non-extensive points, 208
of the Axiom of Infinity, 21
of the Canonical Changes, 214
of the Consistent Universe, 42, 44, 120, 191, 218, 353, 359
of the Discrete magnitudes, 31
of the Discrete Space and Time, 34
of the Discrete Threshold, 34, 201
of the Finite Distances and Durations, 359
of the Finite Divisions, 182
of the Finite Lengths, 185, 208
of the Finite Number of Universes, 359
of the Finite Universe, 359
of the First Element, 40, 45, 46, 121, 266
of the Incompletable Regress, 45
of the Inconsistent Continuum, 299, 310, 359
of the Inconsistent Dense Order, 358
of the Inconsistent Divisions, 185
of the Inconsistent Infinity, 22, 120, 351, 358
of the Indexed Sets, 240
of the Physical Laws, 208
of the Physical Space, 274
of the Reference Frames, 44, 359
of the Strictly Ordered Sets, 359
Theorem of the First Element (Section), 46
Theories of inertial relativity (Chapter), 157
Theories of relativity, 27
Theudius of Magnesia, 79
Timaeus of Lycritus, 68
Time in CALM (Section), 319
Time is a discrete magnitude (Section), 277
Titus Lucretius Carus, 92
Translations into Latin of Greek works, 96
Twin robot paradox, the, 315
Two fallacies in modern physics (Chapter), 341
Two false assertions in modern physics (Section), 342
Two key questions (Section), 234
Two Leibniz's Principles (Section), 119

Ultraviolet catastrophe, 29
Uncomputable numbers, 178
Universal preinertia (Chapter), 221
Up quark, 29

Virgo Cluster, 144
Virgo interferometer, 270
Virgo Supercluster, 144

Wallis, J., 81
Weber, C. H., 270
Weyl, H., 17
Wilczek, F., 168, 282
Witelo, 98
Wittgenstein, L., 17, 29, 196, 281, 286, 350
Works by Euclids, 78

Z-points and Z*-points, 240
Zeno Contradiction, 248
Zeno Contradiction (Section), 250
Zeno Dichotomies (Chapter), 239
Zeno Dichotomy, 185, 211, 239, 248
Zeno Dichotomy (Section), 249
Zeno Dichotomy I , 244–245
Zeno Dichotomy I (Section), 244
Zeno Dichotomy II, 242–244
Zeno Dichotomy II (Section), 242
Zeno of Elea, 9, 40, 62, 196, 239, 242, 248, 357
Zeno paradoxes, 10, 247
Zeno, Aristotle and Cantor (Section), 9

www.ingramcontent.com/pod-product-compliance
Lightning Source LLC
Chambersburg PA
CBHW082201220526
45470CB00010B/3002